Plumbing 301

Plumbing 301

Third Edition

**PHCC EDUCATIONAL FOUNDATION
PLUMBING APPRENTICE & JOURNEYMAN TRAINING COMMITTEE**

THOMSON

DELMAR LEARNING

Australia Canada Mexico Singapore Spain United Kingdom United States

THOMSON

DELMAR LEARNING

Plumbing 301. Third Edition

PHCC EDUCATIONAL FOUNDATION
PLUMBING APPRENTICE & JOURNEYMAN TRAINING COMMITTEE

Vice President, Technology Professional Business Unit:
Gregory L. Clayton

Product Development Manager:
Ed Francis

Product Manager:
Stephanie Kelly

Editorial Assistant:
Jaclyn Ippolito

Marketing Director:
Beth A. Lutz

Executive Marketing Manager:
Taryn Zlatin

Marketing Specialist:
Marissa Maiella

Director of Production:
Patty Stephan

Production Manager:
Andrew Crouth

Content Project Manager:
Kara A. DiCaterino

Art Director:
Benj Gleeksman

Director of Technology:
Paul R. Morris

Technology Project Manager:
Jim Ormsbee

Library of Congress Cataloging-in-Publication Data:

ISBN-10: 1-4180-6534-X
ISBN-13: 978-1-4180-6534-8

NOTICE TO THE READER

Publisher does not warrant or guarantee any of the products described herein or perform any independent analysis in connection with any of the product information contained herein. Publisher does not assume, and expressly disclaims, any obligation to obtain and include information other than that provided to it by the manufacturer.

The reader is expressly warned to consider and adopt all safety precautions that might be indicated by the activities herein and to avoid all potential hazards. By following the instructions contained herein, the reader willingly assumes all risks in connection with such instructions.

The publisher makes no representation or warranties of any kind, including but not limited to, the warranties of fitness for particular purpose or merchantability, nor are any such representations implied with respect to the material set forth herein, and the publisher takes no responsibility with respect to such material. The publisher shall not be liable for any special, consequential, or exemplary damages resulting, in whole or part, from the readers' use of, or reliance upon, this material.

Table of Contents

Acknowledgements

I would like to extend thanks and give credit to the following subject matter experts and reviewers for their support and guidance; only through them could so much material have been presented:

Kirk Alter, Purdue University, West Lafayette, IN

Charlie Chalk, EBL Engineering, Baltimore, MD

Charles E. Greene, Greene & Associates, Inc, Macon, GA

Larry W. Howe, Howe Heating & Plumbing, Inc, Sioux Falls, SD

Eric L. Johnson, Quality Plumbing & Mechanical, Kodak, TN

Michael J. Kastner, Jr., Kastner Plumbing & Heating, West Friendship, MD

Richard R. Kerzetski, Universal Plumbing & Heating Co., Las Vegas, NV

Robert Kordulak, The Arkord Company, Belmar, NJ

Frank R. Maddalon, F. R. Maddalon Plumbing & Heating, Hamilton, NJ

Robert Muller, J. Muller Plumbing & Heating, Matawan, NJ

John Rattenbury, RMS Engineering LLC, Cohasset, MA

James S. Steinle, Atomic Plumbing, Virginia Beach, VA

Don Wolf, Rheem Manufacturing, Montgomery, AL

Larry Rothman, Roto-Rooter Services Co., Cincinnati, OH

I also wish to thank Ed Francis, Stephanie Kelly and the rest of the team at Delmar Learning as well as Merry Beth Hall and the rest of the PHCC organization.

Credit for authorship of the original series of PHCC Plumbing Apprentice Manuals is extended to:

Ruth H. Boutelle

Patrick J. Higgins

Richard E. White

About the Author

Edward T. Moore

Edward Moore is the author of the Residential Construction Academy Plumbing Video Set. He holds a Bachelor of Science degree in Mechanical Engineering. Ed is currently the Program Coordinator and an Instructor for the Building Construction Trades and Air Conditioning Program at York Technical College in Rock Hill, SC. He has also served as an instructor for the Industrial Maintenance and Welding programs as well. A licensed Master Plumber for South Carolina, Ed is also the owner of Moore Plumbing and Cabinetry. He has earned the NATE certification in Air Conditioning and Heat Pumps and is currently working on a Masters degree in Manufacturing Engineering. Ed currently resides in Clover, SC with his wife and 3 children.

CHAPTER 1

Pre-Planning, Productivity, and Profitability

The student will:

- Explain the importance of pre-planning.
- Describe the concept of prefabrication.
- List opportunities that save the company money.

TAKING OWNERSHIP AND RESPONSIBILITY

In order for companies to survive in today's market, they must be able to produce a quality product at a reasonable profit. For that, they need employees who take pride in their work and promote the company in a positive manner. In many cases, you, the worker, will represent the face of the company and your actions will be the basis of the customer's opinion of the company. If you, the worker, show up when expected and present yourself as a well-organized professional, the customer will be more trusting and view your professionalism as an added value. If you cannot be on time, call ahead and inform the customer of the delay. This shows respect for the customer's time.

A company with a good reputation does not have to have the lowest prices. People are willing to pay a little extra for a reputable company with a history of fair dealings. In fact, lowering bid prices to obtain work or the promise of future work is probably the worst course of action. Satisfied customers will tell a few of their friends and family about the good service they received, but unhappy customers will tell everyone about a bad experience.

PRE-PLANNING

As a valued employee, you must perform your work as quickly and professionally as possible. The best way to do this is to pre-plan your work. If you don't attend the pre-construction meetings, then discuss what was covered with your supervisor. Discuss the work before the job begins and "brainstorm" with co-workers on how to approach the job. Review plans to see if there are any conflicts with the other trades and identify any changes that might be beneficial to your company.

For example, the engineer may not have laid out the piping in the most efficient manner. Remember to keep a record of any changes so they can be added to the drawings when the job is completed. When you understand the full scope of a project, you can devise a cost-saving plan that will reduce or eliminate delays.

Think about the job and list the steps necessary to complete it. Include materials needed, safety and power equipment, and the coordination with other trades. Determine how much of your work will depend on the work of other trades already being in place. On some large jobs, or jobs that are not well engineered, there may be overlaps in responsibilities. For example, if the roofing contractor is responsible for all roof penetrations, approach him early on with the locations you would prefer.

Material Handling

By pre-planning a job, you know how much materials are needed and when. Make sure you have in writing where and when you can store your materials on the job-site. This will prevent a memory-challenged general contractor from forcing you to move materials before you're ready to use them. Remember, there is a cost every time a worker has to pick up a piece of material. If the general contractor has a designated storage area, it must be large enough and secure. If it isn't suitable, try to have your suppliers deliver only certain items and wait on the others. Whatever you save by ordering in bulk is easily consumed by having to spend time moving material to secure locations. If possible, make sure that the material is labeled clearly and organized on pallets; this will allow you to move it in bulk with motorized equipment and you will save on labor costs by not having to rummage through the whole delivery. Also, invest in wheeled and/or motorized carts to move tools and materials around on the jobsite. If you pre-plan, you can figure these carts into the cost of doing business and spend less time carrying items by hand. Figure 1-1 shows a motorized wheelbarrow that can haul heavy loads with minimum effort.

Figure 1-1 Motorized wheelbarrows (sometimes called wheelbarrels) can be a great time-saving tool because they can carry heavy loads with ease.

PREFABRICATION (PREFAB)

As mentioned earlier, time on the jobsite should be spent performing the install, not looking for materials or waiting for others. While pre-planning, you may determine that some of the tasks for the job can be done beforehand. This is referred to as *prefab*. Prefab is when any part of the job can be assembled before its final installation.

Look at the job and identify any tasks that appear to be repetitive, such as installing faucets on lavatories or banking together carriers for urinals or water closets. Small tasks like this can performed in the shop ahead of time. The shop usually provides better working conditions and allows newer employees to gain hands-on experience. Typically, the work can be performed at a more productive pace because the workspace and all the necessary tools are available. This strategy also allows for the man-hours to be spread out over a longer time frame and reduces the amount of overtime. Figure 1-2 shows a worker installing garbage disposals ahead of time.

Figure 1-2 Here a worker installs garbage disposals in an assembly line fashion. This will be a great time-saver when compared to doing it on the jobsite. **Courtesy of Marlin Mechanical Corporation, Phoenix, AZ.**

PROFITABILITY

As a third-year apprentice, it is unlikely that you will be selecting who your company performs work with, but you should watch out for companies that want to play only by their rules. The general contractor who changes the rules as he goes along erodes profits by allowing changes in the sequence of construction and the rerouting of systems. In cases like this, you should notify your supervisor as soon as possible. It is likely that your supervisor may ask for a *change order* or some other type of consecution. A change order is when part of the job requirement has been changed from what was originally bid on. Your company may ask for additional payment or some other type of compensation. If the general contractor is unwilling to issue a change order, he may allow a change in another part of the job that could be in your favor. For example, he may be willing to loan you a worker for clean up or provide the equipment for digging a trench. When you ask for compensation, always be prepared with several options that might benefit you. You can't be accused of not being a "team player" if you offer someone choices.

As you progress through the rest of this book, try to use the material learned when out on the job. A good worker should always be looking for ways to improve the quality and efficiency of his work. By staying focused on the task at hand, a worker can complete the job in a professional manner and thereby help assure that the company stays profitable and in business.

REVIEW QUESTIONS

1 What would be a good example of an item that could be prefab?

2 What is a change order?

3 Why notify the customer of a change in a scheduled appointment?

CHAPTER 2

Residential Fixtures and Appliances

LEARNING OBJECTIVES

The student will

- Describe the operation of common residential fixtures and appliances.
- Identify the various sizes and materials available for many fixture types.

This chapter is concerned with the operation and available choices of plumbing fixtures and appliances most often found in residences. While many vary in size and form, they are similar in terms of intended use. To avoid confusion concerning how items are classified or regulated, it may be wise to review some common terms and definitions used in the plumbing industry. Most definitions can be found in either Chapter 1 of the 2006 *National Standard Plumbing Code,* or Chapter 2 of the 2006 *International Plumbing Code.*

DEFINITIONS

Fitting

A *fitting* is a part of a pipeline other than nipples, valves, or straight lengths of pipe. An example of a fitting would be a trap.

Fixture Fitting

A *fixture fitting* is a manual device that controls the flow of water to a fixture.

Plumbing Appliance

A *plumbing appliance* is a plumbing fixture designed for a specific function that may depend on energized components for proper operation. Appliances may operate automatically, manually, or by other mechanisms such as timers, pressure switches, or temperature controls. A water cooler is an example of an appliance.

Plumbing Appurtenance

A *plumbing appurtenance* is a device (factory or field assembled) placed in the basic plumbing and piping system to perform a useful function in the system. It requires no additional water supply nor does it add to the fixture-unit load of the drainage system. A grease trap is an example.

Plumbing Fixture

A *plumbing fixture* is a device connected to both the water system and the drainage system and requires water for its operation. Fixtures discharge waste to the drainage system. A water closet is an example.

Table 2-1 shows a list of some fixtures and appliances commonly found in homes.

Table 2-1   Residential Fixtures and Appliances

Fixtures	Appliances
Water closet	Food waste grinder
Lavatory	Dishwasher
Bathtub	Automatic clothes washer
Shower	Ice maker
Sink	Hot water dispenser
Bidet	Whirlpool bathtub
Laundry tray	Spa
	Bidet assemblies

WATER CLOSET

The water closet, more commonly referred to as a toilet, receives human waste and discharges it into the drainage system. The water closet must operate to maintain a sanitary condition and not permit sewer gas to enter the building. Modern water closets must meet the following requirements as dictated by standards:

1. A water closet must contain a water seal to prevent sewer gas from entering the room.
2. Each flush must clear the contents of the bowl and must demonstrate compliance with hydraulic performance tests.
3. The flushing action must scour the walls of the bowl.
4. The material used must provide smooth, nonabsorbent surfaces.
5. The bowl must withstand load testing without cracking or breaking.
6. The assemblies must contain components or designs that will prevent backflow.

Vitreous china, stainless steel, plastic, and cultured marbles are some of the materials used to make water closets. Over the years, residential bowls have been classified as either siphon jet, reverse trap, siphon vortex, siphon wash, or wash down. These terms are gradually fading from use because the standards for water closets are becoming more oriented toward performance. Performance is typically based on clearing the contents of the bowl with the minimum amount of water. In the past, a 3.5-gallon per flush toilet was common, but in current construction and renovations, the 1.6-gallon per flush (6L) low-consumption water closet is the standard.

In residential applications, the water used to accomplish the flush is introduced into the bowl by either a gravity flush tank or a flushometer tank. Traditional design calls for the tank to be located on the back of the water closet and exposed. This is typically referred to as *close-coupled,* or simply put, the tank is bolted directly to the bowl. Because the tank is attached to the bowl, it will cause the bowl to extend farther away from the wall. This is important to remember because code requires at least 21 inches of clearance in the front of the water closet. To avoid this problem, some manufacturers make a concealed (in the wall) tank that allows the bowl to be bolted directly to the wall. There are two concerns about this design. First, it is usually special order, and therefore, typically more expensive than a standard design toilet. Second, it requires a 6" wall cavity for the tank. If you run across one, it is critical to pre-plan so you have the needed wall cavity free of studs and other piping.

One-Piece Unit

Another common tank/bowl arrangement is the one-piece unit. This style has the tank and bowl molded and fired as one unit. Because the unit is more expensive than a close-coupled model, it is typically considered an upgrade. The rough-in of the water and waste must be performed with precision and the available pressure must be at least 20 psi (flowing).

The typical flush tank (Figure 2-1) stores sufficient water for the flush in a wall-mounted or fixture-mounted tank. The water is under atmospheric pressure and flows out of the tank due to gravity.

When the flush lever is tripped, water is released through a large passageway, flows under the force of gravity and fills the flushing rim and any waterways into the trap with water. The energy contained in this water flow starts the water in the bowl moving through the trap and continues until the flush tank is empty. When the

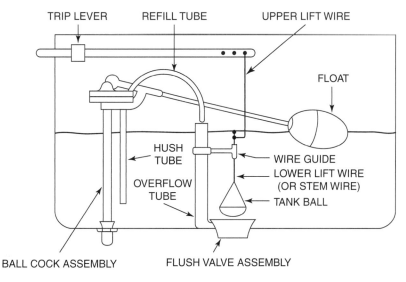

Figure 2-1 Operating components of a flush tank.

Figure 2-2 Flushometer tank, a pressurized container within a standard flush tank. Photography courtesy of American Standard Bath & Kitchen.

water level falls to a certain height, the flapper will close and the tank will begin to refill. As the tank is refilled, a small amount of water is sent through the refill tube so there is enough water in the bowl to seal the trap.

The flushometer tank is a device that stores water under pressure, so a smaller amount of water delivers enough energy to develop the flush (see Figure 2-2). The main difference in operation is that now the water is being supplied to the bowl at a pressure above atmospheric pressure. This delivers the water more quickly and aids the flushing process.

Corner Tank

The corner tank is triangular in shape and used primarily for special layout problems. This special tank assembly is very expensive and may have limited availability.

Closet Bowls

Historically, water closet bowl design has been one of five types as described below.

NOTE: Current and future designs may have variations on these features. As designs swing to more water-saving features, trapways and water surfaces will shrink.

The siphon jet (Figure 2-3) is one of the most efficient bowl types ever designed. It includes a deep-water seal, a large water surface, and a large trapway. Although perhaps more expensive than other types, the siphonic action of the trapway jet develops a rapid, complete flush with 1.6 gallons of water.

The siphon vortex (Figure 2-4) design is used in gravity-type, one-piece tank-bowl combinations. The moving water develops a swirling vortex that clears the bowl. The siphon vortex requires a minimum of 20 psig flowing pressure to fully develop the flush. A larger supply pipe to the tank can compensate for somewhat lower pressure, but usually a 1/2" pipe with a 3/8" ID supply tube is the minimum size required for this design.

Figure 2-3 Cross-sectional view of a siphon jet water closet bowl.

Figure 2-4 Cross-sectional view of a siphon vortex water closet bowl.

The reverse trap (Figure 2-5) is similar to the siphon jet except that the water seal volume and water surface area are each less than the siphon jet model. The term *reverse trap* originates from the fact that this design, developed later than the washdown style, uses a trap reversed in orientation to the washdown type. Many 1.6 gpf (6 lpf) gravity-tank assisted bowls use a reverse trap design.

Washdown bowls (Figure 2-6) are seldom installed now, but many are still in use. It was an inexpensive design with a large surface area located toward the front of the bowl. Unfortunately this surface was located above the water line, which consequently made it difficult to remove any solid wastes located there. The water level in the bowl typically has to be higher, compared to other designs, in order to develop the head pressure necessary to begin the flush.

The term *water spot* is commonly used by manufacturers to indicate the surface area of the water remaining in the bowl once the flush is completed. The shape of the bowl and the height of the internal trapway dam determine the maximum size.

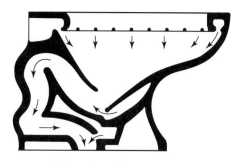

Figure 2-5 Cross-sectional view of a reverse trap water closet bowl.

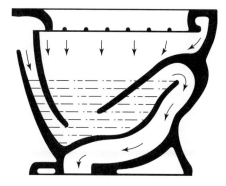

Figure 2-6 Cross-sectional view of a washdown water closet bowl.

Bowl Shape

Both the *National Standard and International Plumbing Codes* require that all closet bowls for commercial use be elongated fronts with open-front seats. The most common exception to this code is that water closets for residences, hotels (since toilets are cleaned between customers), and private offices may be round front with a closed front seat. The elongated bowl extends 1" to 2" farther outward from the wall, which increases the water seal area and makes the toilet more sanitary. Figure 2-7 shows a wall-mounted water closet with an elongated bowl.

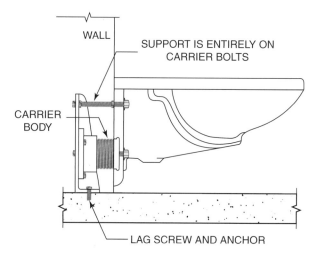

Figure 2-7 Wall-mounted water closet with an elongated bowl. Courtesy of Plumbing-Heating-Cooling Contractors–National Association.

Optional Features

Air-assisted product—flushometer tank and other similar air-assisted devices are available. These products use the water supply system to hold a volume of water under pressure so when the flush cycle starts, the contents of the sealed tank are directed into the bowl.

Grinder type—a pump/grinder assembly is attached to the bowl and is capable of grinding human waste and paper products into a slurry that can be pumped out. Many are being used in remodel jobs that require the new water closet or entire bathroom group to be located below the building drain. Figure 2-8 shows a typical macerator device.

Water volume—all codes call for 1.6 gallons per flush, but the plumbing industry is working on low consumption products for conservation reasons. One manufacturer has developed a water closet with a dual flush feature. When flushing only liquids,

Figure 2-8 Macerating toilets are used when the fixture is located below the building drain. Very popular for remodel work. Courtesy of Saniflo.

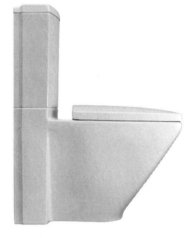

Figure 2-9 Dual flush toilets are gaining in popularity because of the amount of water that they save. Photo courtesy of Kohler.

Figure 2-10 Pedestal sink–shelf-back or slab lavatory that mounts using wall brackets and a vitreous china pedestal. The drain rough-in should be perfect as the pedestal does not provide much room for mistakes.

the user presses the "number 1 button," and a half flush (0.8 gpf) occurs. When flushing solid wastes, the user presses the "number 2 button," and a full flush (1.6 gpf) occurs (see Figure 2-9).

LAVATORY

Lavatories are sinks that are intended for personal hygiene such as washing the hands and face. The three most common forms are the pedestal-mounted (see Figure 2-10), the wall-mounted, and the cabinet-mounted fixtures. Common sizes include 22" × 19", 21" × 19", 19" × 16", 21" × 13" (rectangular); 20" × 17", 19" × 15" (oval); and 18" or 19" diameter (round) (see Figure 2-11).

Lavatories must comply with the following standards:

1. Ceramic, non-vitreous; ASME A112.19.9M
2. Enameled cast-iron; ASME A112.19.1M
3. Enameled steel; ASME A112.19.4M
4. Plastic; ANSI z124.3
5. Stainless steel; ASME A112.19.3M
6. Vitreous china; ASME A112.19.2M

All of the above standards specify the materials and testing standards for plumbing fixtures with the following requirements for lavatories:

1. One-piece construction.
2. Equipped with a properly sized drain opening.
3. Made of smooth, non-absorbent materials.
4. Load tests done on the front edge (for some materials).

Lavatories are usually made of vitreous china, stainless steel, enameled cast iron, enameled steel, cultured marble, or plastic. Hot and cold water are supplied to the fixture and controlled by faucets. The water is discharged into the lavatory basin and then down the drain or the drain may be stoppered to retain the water until the washing task is completed. Figure 2-12 shows a typical lavatory stopper arrangement.

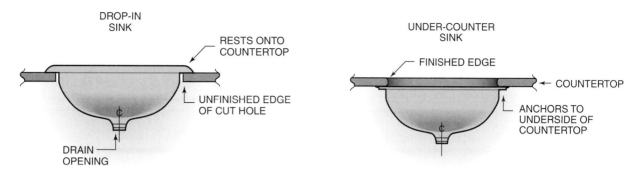

Figure 2-11 Drop-in sinks can be installed on top of the countertop or below the countertop. Make certain the person responsible for cutting the hole in the countertop is aware of how the sink will be installed.

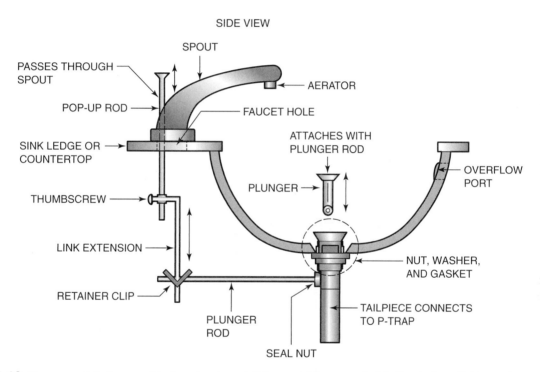

Figure 2-12 Stopper and drain assembly for a lavatory. A P-trap must be connected to the drain outlet to seal out sewer gases.

Even though lavatory basin designs have changed to become more artistic and customized to reflect the owner's taste, they still must be able to drain completely when the drain is opened. Figure 2-13 shows a more modern design.

Many lavatories are provided with an overflow passage from near the top of the basin to a place below the stopper position at the fixture outlet. The overflow reduces the chance of running water over the rim of a stoppered lavatory with the faucets on. The overflow was once a mandatory part of a lavatory, but present standards make it optional to the manufacturer. When provided, an overflow must be sized not less than 1-1/8th square inches or discharge not less than 3.0 gpm.

An overflow provides an avenue for quick venting the tailpiece below the lavatory when the stopper is removed from a full lavatory, but this is generally considered a small benefit in the operation of the fixture. The main objection to an overflow is that it may easily become fouled and germ laden.

Figure 2-13 Modern lavatories can be customized to the customer's taste.

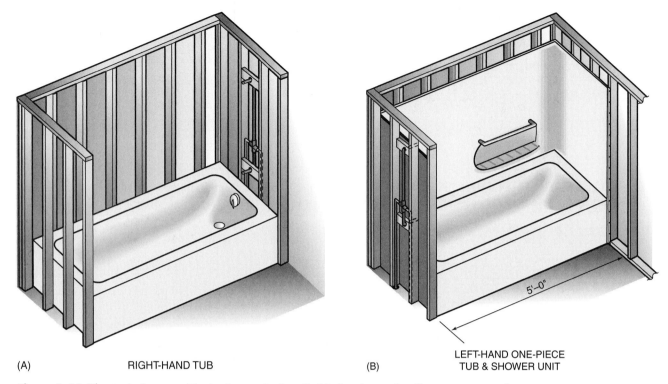

(A) RIGHT-HAND TUB

(B) LEFT-HAND ONE-PIECE TUB & SHOWER UNIT

5'-0"

Figure 2-14 The typical recessed bathtub must be installed before internal walls are constructed.

BATHTUB

The modern bathtub has become a very important fixture for personal hygiene. Its purpose is to aid in bathing the entire body. Bathtubs are available in many styles. The principal models are five-foot recess, one-piece modular units, bathing pools, and whirlpool bathtubs. When dealing with the molded one-piece modular units, the key to success is support. Any failure to support this device could cause cracking at the fixture. Units are also available in four-piece kits for remodeling purposes.

These units have standard dimensions of 48", 54", 60", 66", or 72". Bathtubs are usually identified as either right hand or left hand, which simply describes the location of the drain and fixture. If a person were preparing to step into the bathtub shown in Figure 2-14(B), the tub spout, fixture, and drain would be located on the left hand, so this tub would be referred to as a left-hand tub.

Bathtub construction must meet the following requirements:

1. Bathtubs must be made of smooth, non-absorbent surfaces (abrasive bottom surfaces for safety are allowed).
2. They should have proper-sized drain (1-1/2" minimum) and overflow openings.
3. The bottom must be sloped to the drain when the fixture rim is level.

The fixture is designed to hold bathwater in the sump. An overflow arrangement minimizes the risk of having water flood over the rim of the fixture if the faucets are left on. The drain opening may be stoppered with a lever, a cable-operated mechanical stopper arrangement, or a toe-tap arrangement. When the stopper is opened, the water should drain away completely. While the toe-tap is a simple and more trouble-free design, some people object to reaching into the water to release the stopper. Figure 2-15 shows the lever-type and the toe-tap arrangement. Figure 2-16 shows the cable-operated stopper.

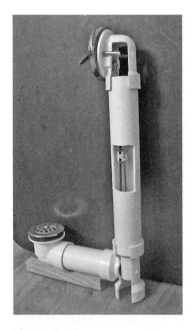

Figure 2-15 Lever-operated bath waste and overflow. It is usually installed before the unit is put in place.

Whirlpool bath units are becoming more commonplace in residential construction. They contain a circulating pump and related piping. These units provide for increased comfort and relaxation. Since they are drained after each use, the internal piping must be arranged to drain as completely as possible to minimize the amount of retained water mixing with the next filling of the fixture. Figure 2-17 shows a whirlpool tub before the internal piping has been covered.

Accessories for added safety include the abrasive (i.e., slightly roughened) bottom surfaces referred to above, grab bars, and thermostatically controlled or pressure-balanced mixing faucets to control the temperature of the water delivered to the tub. Other special accessories may include ventilated enclosures (to reduce moisture in the bathroom) and steam features.

SHOWER

Showers are designed to make cleaning the entire body quicker and easier than using a bathtub. Figure 2-18 shows a modern one-piece, fiberglass unit. Due to its one-piece construction, the unit must be brought into the building before the walls are sheathed. Common sizes are 32" × 32", 36" × 36", and 36" × 48".

Showers can be manufactured in one-piece and knocked-down versions for quick, easy installations or field-fabricated by providing waterproof walls around a bathtub or shower. They are sometimes constructed by building up a stall and base as a separate fixture, such as with a tiled enclosure. The walls and base must be made of smooth, non-absorbent material with waterproof joints. Walls are made of ceramic tile, plastic, marble, cultured marble, or other approved non-absorbent material. The base can be pre-cast or field-fabricated of ceramic tile, cement, terrazzo, or plastic. Field-fabricated shower floors require a waterproof membrane under the final base material. The shower floor must be sloped and provided with an adequate drain to carry away the shower water as fast as it reaches the base. More information concerning the installation of shower safety pans can be found in Chapter 4, "Installation Methods."

The complete installation requires a trap below the shower drain, properly sized water pipes, shower faucet, and showerhead. The bather stands under the water discharged from the showerhead and the water that reaches the floor flows out through the drain. According to the latest industry standards, showerhead flow rates are required to deliver a maximum flow of 2.5 gallons per minute at 80 psi. As with bathtubs, recommended accessories include abrasive floor surfaces, grab bars, and thermostatic or pressure-balanced mixing valves to control the temperature of the water delivered to the showerhead.

Figure 2-16 Cable-operated bath waste and overflow. It is usually installed before the unit is put in place.

Figure 2-18 Modern one-piece fiberglass unit. Make sure to install before walls are constructed.

Figure 2-17 Whirlpool rough-in. Remember to allow access to the pump.

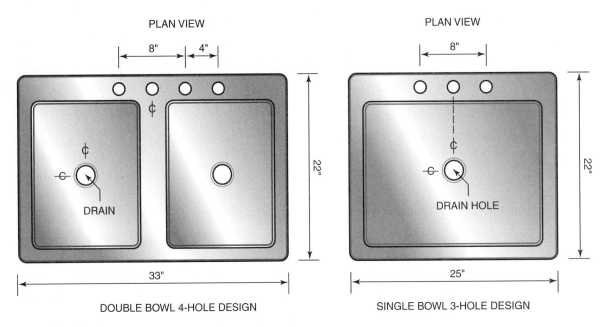

PLAN VIEW

PLAN VIEW

DRAIN

DRAIN HOLE

DOUBLE BOWL 4-HOLE DESIGN

SINGLE BOWL 3-HOLE DESIGN

Figure 2-19 Single- and double-compartment kitchen sinks.

SINK

Many sink designs and options are available for residential use. The most noticeable design consideration is the number of compartments. Traditional residential kitchen sinks come with either one or two compartments and in a variety of depths as shown in Figure 2-19.

Another noticeable design consideration would be the mounting style, either under mount or top mount. Top mount sinks are a very common style in which the sink is placed in a cutout slightly smaller than the sink lip. A bead of waterproof sealant is added to prevent moisture from leaking between the sink and the countertop. The sealant also helps hold the sink in place. Clips are sometimes used to help secure the sink to the countertop (see Figure 2-20).

Under mounted sinks are more common with higher end construction such as granite, Corian, and other solid surface countertops. The sink is typically glued to the underside of the countertop prior to the installation of the countertop.

Remember to make sure that any sealant or plumber's putty is compatible with the sink and countertop materials; some putties may stain marble and granite.

The construction of a sink should include these features:

1. Sinks must be made of smooth, non-absorbent materials.
2. A 3-1/2" drain outlet.
3. Bottom slope to the drain.
4. Faucet openings must be located on the ledge (if ledge-type).
5. The sump or sumps must be at least 6" deep.

The sink operates with a faucet supplying water and the drain opening(s) conveying the water away. The water can flow out the drain or the drain openings can be stoppered and water retained in the sink. The faucet has a spout that can be positioned (left-right directions) to deliver water to different compartments or to permit access to the sink bowl for large pots, pans, or dishes. The usual sump depth is 6-1/2" to 8" and the spout is 2" to 4" above the sink rim, so fairly large items can be placed in the sink and filled with water. Overflow passages are rarely built into kitchen sink designs as these seldom-washed passages are unsuitable in a fixture used for food preparation. In many cases, a traditional basket strainer will be

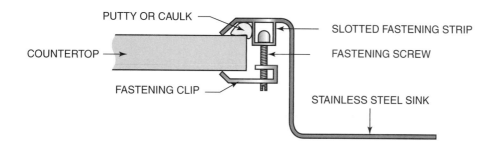

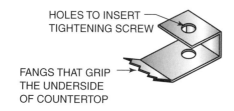

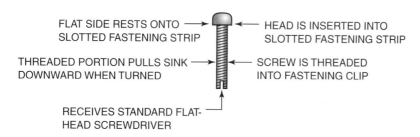

Figure 2-20 Mounting arrangements for stainless steel sinks.

replaced with a garbage disposal or food grinder. It is also common to find the discharge of the dishwasher attached to either the tailpiece of the sink or fed directly into the food grinder.

BIDET

The *bidet* is a fixture that is similar in appearance to a water closet bowl. The user typically straddles the fixture facing the controls and uses the water spray to clean external genital and anal areas. The bidet is supplied with both hot and cold water and controlled with a thermostat-mixing valve that the user can adjust to a comfortable setting. The water can be delivered from either a spray head located at the bottom of the bowl or from the flushing rim in the fixture. A vacuum breaker must be installed in the water line ahead of the spray fitting if the spray head is submerged in the bowl. Over-the-rim faucets do not require the vacuum breaker.

The bowl may be equipped with a pop-up assembly that can be used to retain water in the fixture for either soaking or foot-washing purposes. An overflow is provided to reduce spillage during soaking operations. Figure 2-21 shows a modern bidet.

Figure 2-21 A bidet used for personal hygiene. Image courtesy of American Standard Bath & Kitchen.

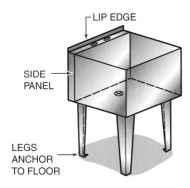

Figure 2-22 Free-standing plastic laundry tray.

LAUNDRY TRAY

Laundry trays are large capacity sinks typically found in a laundry room or garage. Depending on your local code requirements, they may be used to soak clothing, as a receiver of washing machine discharge, or as a general cleanup sink. The fixture is made with a much deeper well than a kitchen sink and of smooth, non-absorbent materials. Many older laundry trays were made of soapstone or concrete, but these fixtures were extremely heavy and the surfaces were generally not non-absorbent! Present fixtures are made of enameled cast iron, vitreous china, fiberglass, or plastic. Figure 2-22 shows a single compartment, floor-mounted sink.

The fixture can be used with the water flowing out the drain or with the drain stoppered and water retained in the fixture sump. When used as a receptor for an automatic clothes washing machine, the drain is frequently washed and the water slowly drains away as it is discharged from the washing machine.

The faucet is normally a 4" spread with a swivel spout. Some spouts have a threaded end that allows the user to attach a garden hose. Because this arrangement would allow the outlet to be located below the flood rim of the sink, a vacuum breaker is required (see Figure 2-23).

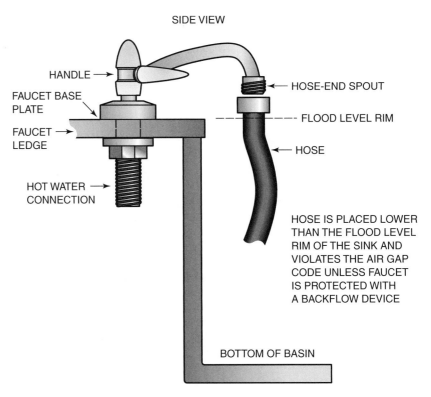

Figure 2-23 On a laundry sink faucet that is capable of having a hose attachment, a vacuum breaker is required.

FOOD WASTE GRINDER

Food waste grinders or *garbage disposals* are motorized grinders that connect directly to the kitchen sink through the drain outlet. This device is used to reduce food scraps to small particle size so they can be conveyed away by the drainage system. Most residential models have stainless steel internal components, a 1/2 to 1 horsepower motor, and a plastic housing. Some models have auto reverse that

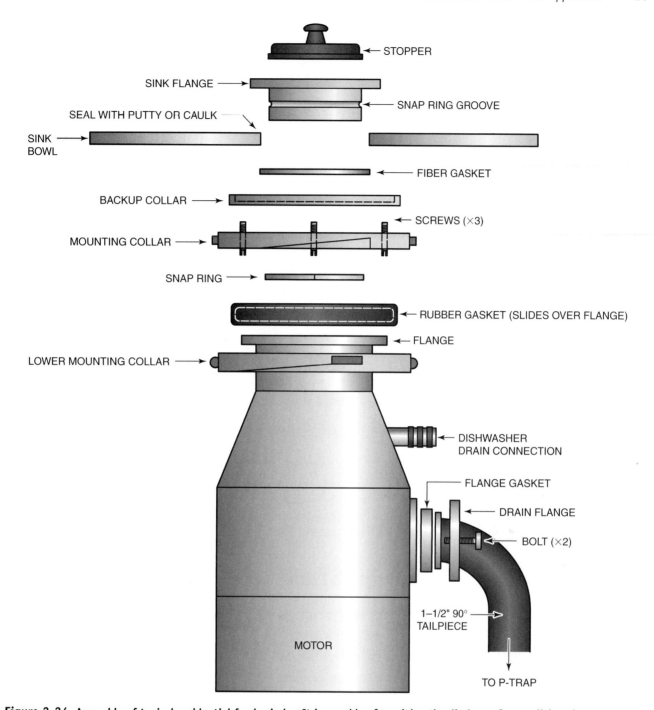

Figure 2-24 Assembly of typical residential food grinder. It is capable of receiving the discharge from a dishwasher.

causes the internal cutter to rotate in the opposite direction if the motor becomes overloaded (see Figure 2-24).

Other models are considered septic tank friendly because they periodically inject liquid containing microorganisms that aid in the breakdown of food waste.

DISHWASHER

The *dishwasher* is an appliance used to clean and sanitize dishes. It is fed by a 3/8" OD minimum hot water line. Check manufacturer specifications for minimum line size. The wastewater is discharged by a pump to either a special tapping in the food

Figure 2-25 Modern dishwasher with stainless steel front.

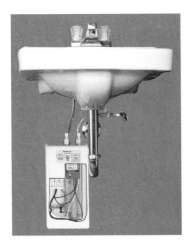

Figure 2-26 Under counter mounted water heaters can reduce the amount of water consumed because the water is instantly hot.

waste grinder or through a wye tailpiece into the sanitary drain above the trap. Also, check local codes for specific installation requirements. Modern dishwashers have plastic housings and a grinder pump for complete waste emulsification. Other added features such as delayed start and quiet operation are quite popular as are hidden controls and custom front panels. Figure 2-25 shows a popular trend with an all stainless steel front.

ICEMAKER

Most refrigerators are equipped with an *icemaker* or fitted to accept an icemaker. The line that serves such devices is usually a 1/4" OD tubing installed in a loop arrangement to facilitate movement of the refrigerator.

HOT WATER DISPENSER

With the rising cost of water and energy, single point-of-use *hot water dispensers* have become more of an energy saver than a luxury appliance. Figure 2-26 shows an example of a point-of-use water heater that mounts under the sink and provides hot water in the 180° ± 10° F range for drink preparation purposes. This eliminates the need to run the water until it gets hot, which can take a long time if the water heater is on the opposite side of the building.

These devices are usually fed by a 1/4" OD tubing and are usually 110 V.

SPA

These large soaking vessels circulate hot water for recreational bathing. They differ from whirlpool bathtub appliances because they retain water between uses. Water quality is maintained by filtration and chemical treatment. In order to regulate comfort, heaters are also installed in line.

REVIEW QUESTIONS

❶ What is a challenge of a concealed tank for a water closet?

❷ What is the normal spread for a lavatory faucet?

❸ What is the typical flow rate for a showerhead?

❹ What is a macerating toilet?

❺ What does left-hand bathtub refer to?

CHAPTER

3

Commercial, Industrial, and Institutional Fixtures and Appliances

LEARNING OBJECTIVES

The student will:

- Describe the intended use of common commercial, industrial, and institutional fixtures and appliances.
- Describe the basic design of commercial, industrial, and institutional fixtures and appliances.
- Explain the theory of operation for commercial, industrial, and institutional fixtures and appliances.
- Apply knowledge of safety and accessibility concerns when working in the three different environments.

This chapter is concerned with the operation and choices available for plumbing fixtures and appliances most often found in commercial, industrial, and institutional environments. While many vary in size and form, they are similar in terms of intended use. In fact, many of the same fixtures and appliances can be found in any of the three settings. To avoid redundancy, most items will be discussed in the commercial section. Only differences or specialized options will be discussed in the two remaining settings.

COMMERCIAL FIXTURES AND APPLIANCES

Table 3-1 shows some of the most common appliances and fixtures used in commercial buildings. If needed for reference, the definitions of plumbing appliances and fixtures can be found in the beginning of Chapter 2.

Table 3-1   Commercial Fixtures and Appliances

Fixtures	Appliances
Water closet	Dishwasher
Urinal	Food waste grinder
Lavatory	Ice-making machine
Shower	Air-conditioning equipment
Sink	
Drinking fountain	
Floor drain	

WATER CLOSET

As mentioned in Chapter 2, this fixture receives human waste and discharges it into the drainage system. Any of the following three flushing types are used for commercial water closets:

- Flush tank
- Flushometer tank
- Flushometer valve

In turn, the three flushing types can have any of these three discharge options:

- Floor mounted, floor discharge
- Wall mounted, wall discharge
- Floor mounted, wall discharge

Bowl types, as described in Chapter 2, are selected on the basis of the flushing type available. Table 3-2 shows the bowl type commonly used with a particular flushing type. Table 3-3 gives a brief description of how the mounting arrangements are configured.

Table 3-2   Bowl Design per Flushing Type

Flushometer Valve	Flushometer Tank	Flush Tank
Siphon jet, floor or wall mounted	Siphon jet, floor or wall mounted	Siphon jet, floor or wall mounted
Reverse trap, floor mounted	Reverse trap, floor mounted	Reverse trap, floor mounted
Blowout, wall mounted		Siphon vortex, floor mounted

Figure 3-1 shows a cross-sectional view of a blowout bowl.

Since flush tanks and flushometer tanks were described in Chapter 2, only the flushometer valve will be discussed here. Bowls with flush tanks are sometimes used in commercial applications where the restroom traffic is relatively light.

Flushometer Valve

Most commercial installations use *flushometer valve* fixtures. This arrangement is more resistant to vandalism and has a rapid flush operation (requires about 9 seconds) to accommodate high-frequency use. The flushometer valve seldom needs maintenance

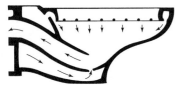

Figure 3-1 Cross-sectional view of a blowout bowl.

Table 3-3 **Mounting Arrangements**

FLOOR MOUNTED, FLOOR OUTLET	Bowl is floor mounted to closet flange and operated by a flush tank or flushometer valve. Standard rough-in is 12".
FLOOR MOUNTED, BACK OUTLET	Bowl outlet discharges from the rear of the bowl. Bowl is securely mounted to the wall and the floor. Tank and bowl are a close-coupled arrangement.
WALL MOUNTED, TOP SPUD	Bowl is wall mounted off the floor, elongated style for better sanitation. Operated by a flushometer valve mounted on its top surface.
WALL MOUNTED, BACK SPUD	Bowl is same as above, except that the inlet spud is mounted through the back.
WALL MOUNTED, CLOSE-COUPLED TANK	Mounting of the bowl is similar to that of all wall-hung units used with carriers. However, the tank mounts like a close-coupled unit.

and the vigorous flush minimizes bowl stoppages and promotes bowl cleanliness. Figure 3-2 shows a sectional view of the manually operated diaphragm-type flushometer valve.

The flushometer valve is connected through a vacuum breaker to either a top or back spud on the bowl. To minimize vandalism, the flushometer valve can be installed in a plumbing chase behind the wall and connected to the closet bowl with a rear spud. The use of electronic presence sensors has simplified installation because only a small box containing the sensor must be mounted on the wall above the fixture. There are a wide variety of presence-sensing devices available; usually, they are either battery powered or have a low voltage power supply wired to them. Figure 3-3 shows a diagram of how the piping and sensor are placed to install an electronic-activated flushometer valve in a plumbing chase.

Because of their high efficiency, the siphon jet and blowout bowl are the two most popular designs. The blowout bowl, as shown in Figure 3-1, gets its name because of the strong blowout action of its flush. Because the flushing action is so quick and forceful, the blowout bowl can only be used with flushometer valve installations as an immediate high-volume flow is required. This makes it ideal for high-traffic areas. This style is only available in wall hung, which can be recognized by the pattern of three mounting bolts. Siphon jet wall-hung bowls require four mounting bolts.

Section 7.4.3 of the 2006 *National Standard Plumbing Code* requires that water closet bowls designed for public installations be elongated; this design provides a large water surface and fixture opening. The code also requires that the seat type

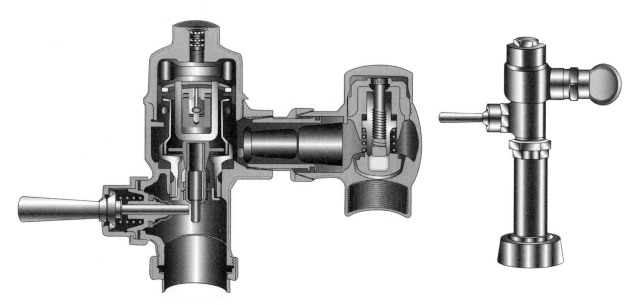

Figure 3-2 **Sectional view of manually operated diaphragm-type flush valve.**
Courtesy of Sloan Valve Company. All rights reserved.

Figure 3-3 Electronic flush valve is concealed behind the finished wall while the presence sensor is flush mounted to the wall. Courtesy of Sloan Valve Company. All rights reserved.

consist of an open front design without a cover. The combination of this seat and bowl shape provides the maximum sanitary condition at the fixture. Bowl variations available include child, juvenile, and handicap designs as well as other special variations. Bowls for carrier wall mounting are used in the highest quality installations for easy toilet room cleaning. Chapter 4 will give a more detailed discussion of the carrier fittings needed to install these bowls.

Operation of Diaphragm-Type Flushometer Valves

The *diaphragm-type* flushometer valve has a large segmented diaphragm in the upper portion of the valve, separating the passage for the water supply from the passage leading into the bowl. Normally the diaphragm is subjected to equal pressure on both the top and bottom sides, but because the topside has a larger surface area exposed to the incoming water pressure, a resultant downward force keeps the valve closed.

When the user depresses the handle on the flushometer valve in any direction, an internal connecting plunger tilts the relief valve and relieves pressure from the top portion of the diaphragm. Since the pressure on the lower portion of the diaphragm is now greater, the diaphragm, relief valve, disc, and guide are raised as one, and water is allowed to flow down through the valve outlet.

While the valve is allowing water to flow, a small amount of water flows through the bypass port in the diaphragm, which gradually refills the upper chamber of the valve. Once the pressures on the top and bottom of the diaphragm equalize, the diaphragm and other internal parts return to their original positions and the water flow stops.

When the flushometer valve is opened, a large initial flow of water is introduced into the bowl to start the flushing action and produce bowl scouring. After the initial surge of water, a final reduced flow is provided to reseal the bowl trap. The flushometer valve has somewhat higher noise levels when operating and it is associated with higher installation cost because the water pipe sizing must be larger than that required for flush tanks. Proper operation of these valves requires a minimum of 15-psi flowing pressure. The flushometer valve may be fitted with water-saving accessories that can limit total flow to 1.6 gallons per flush or less.

Manufacturers are making considerable effort to obtain satisfactory flushing with very low consumption (1.3 gpf) flushometer water closets.

URINAL

Urinals are vitreous china fixtures designed to receive human urine and convey it to the waste piping system. Fixture designs vary in details, but are generally one-piece in construction. All interior surfaces are washed during each flush and the trapway is resealed after each flush. Trapway designs for all urinals include siphon jet, blowout, and washout. Urinals operate in a manner similar to a water closet in that they receive waste, discharge waste to the drainage piping with a water flush, and maintain a trap seal to prevent sewer gases from entering the room. Water conserving designs require 1.0 gallon per flush or less. A number of manufacturers are currently making waterless urinals that use a replaceable or refillable cartridge instead of a water seal.

Figure 3-4 shows how the cartridge has a layer of sealant fluid that allows liquid waste to pass through. As more waste is added, it will start to fill the cartridge until it spills over the vertical dam and into the drain outlet. The liquid sealant will eventually be depleted and the cartridge must be replaced.

Through the years, there have been many different types of urinal designs; some are no longer permitted by current plumbing codes, but are still in use today. These include *stall, trough, floor trough,* or any type that is not completely washed at each flush. Designs that include integral strainers over the trap seal are also banned. Listed below are the names and a brief description of urinals that may still be found.

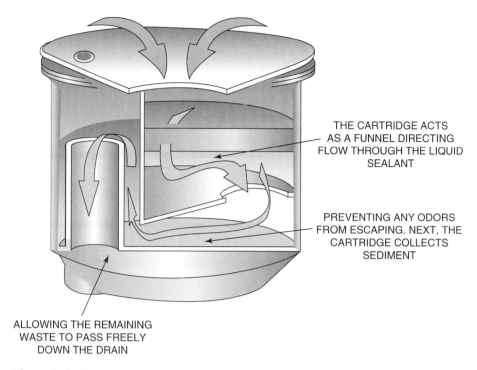

THE CARTRIDGE ACTS AS A FUNNEL DIRECTING FLOW THROUGH THE LIQUID SEALANT

PREVENTING ANY ODORS FROM ESCAPING. NEXT, THE CARTRIDGE COLLECTS SEDIMENT

ALLOWING THE REMAINING WASTE TO PASS FREELY DOWN THE DRAIN

Figure 3-4 Sloan waterless urinal cartridge.

Stall

Often installed in battery (several fixtures side by side) arrangements, these urinals are the washout type. This fixture has either a straight or sloped front with an integral flush spreader. The top spud opening will accommodate a flushometer valve or flush tank. The standard size is 18" × 42", of which approximately 4" is extended below the floor. Installed with the front lip level with (or slightly lower than) the floor, they will drain any water spilled on the floor. A separate trap below the fixture is required. Stall urinals are now prohibited by many codes because they have no visible water seal and are considered unsanitary.

Trough

Mounted on wall brackets, this washout urinal can be used by more than one person at a time. Even though chrome flushing spreader assembly is provided, the fixtures are generally poorly washed. That characteristic, the lack of privacy, and the fact that the trap is not integral with the fixture, make this urinal prohibited by the *National Standard Plumbing Code*.

Wall-Hung

This urinal can use any integral trap design. Since it is above floor level, restroom cleaning is made easy. Extended-shield models are available for battery installations, eliminating the need for privacy shields between fixtures (see Figure 3-5).

Wall-Hung, Blowout

This urinal is made of vitreous china, with an integral trap and a flushing rim. It mounts on a carrier and connects to the discharge flange. A flushometer valve is normally used to operate this fixture. The discharge outlet is usually 2" or 3" in diameter.

Figure 3-5 Wall-hung urinal with presence sensor and automatic flush.

Wall-Hung, Siphon Jet

The wall-hung, siphon jet is like the wall-hung blowout fixture except that the trap-way is a siphon jet design. The fixture may also have extended shields and many have an integral cast strainer, removable strainer, or open trapway.

Wall-Hung Trough, Washdown

Wall-hung trough washdown urinals are usually made of enameled cast iron. The 4', 5', or 6' trough-type design urinal has a base that slopes toward its center outlet and a trap is installed below the fixture. A special rim-washing sprayer is installed horizontally across the top of the fixture (prohibited by some codes except for temporary use).

Wall-Hung Washout

This type of urinal uses an exposed trap and a small volume of water to flush (prohibited by some codes).

Pedestal

This fixture is not as popular as it once was, so it is only available by special order. It is installed on a rough-in identical with that required for a floor water closet. This style was available for males or females. In most cases, however, females consider them unsuitable. New styles of female urinals are available—their success is yet to be determined.

Pedestal, Siphon Jet

This type of urinal is floor-mounted to a closet flange. Generally, the trap design is siphon jet. This unit is normally used in conjunction with a flushometer valve. Replacements may be hard to find as this fixture is only made on special order.

URINAL TRAPWAY DESIGNS

Siphon Jet

A high tank or flushometer valve may flush the *siphon jet* trapway. This fixture is used where high efficiency and minimum noise are desired. The urinal includes an integral flush spreader, a siphon jet trapway, and a top or back inlet. This type is now available only in wall-hung models.

Blowout

Blowout designs are installed for frequent-use applications, especially where vandalism may be prevalent. An integral flushing rim and clear trapway are used in wall-hung style, with or without extended shields.

Washout

For light-use applications, the head of water from a flushometer valve or a high flush tank accomplishes the flush. These urinals use a separate trap and incorporate a flushing rim, top or back inlet, and removable strainer. Some codes do not permit the use of washout urinals, usually on the basis that the fixture does not incorporate an integral trap.

For maximum protection against vandalism, integral-trap urinals may be installed on carriers with back supply connection and concealed flushing device and piping. Urinals are also made of stainless steel if additional vandalism protection is desired.

LAVATORY

A *lavatory* is used for hand and face cleaning. It receives water, facilitates the washing, and discharges the water to the drain. The designs suitable for commercial installations emphasize the aspects of vandalism resistance, minimum water use, and heavy-duty trim for minimum maintenance and long wear. Drain assemblies are most often grid strainer-type, rather than pop-up or plug styles. Residential style lavatories are often used in light-traffic commercial installations.

For added strength, commercial wall-hung installations include carrier mounting or countertop versions. Most wall-hung lavatories have a raised back, a ledge, or a shelf back (see Figure 3-6). Wheelchair lavatories (see Figure 3-7) that have a longer tapered front are also available.

The faucet most often used is the 4" *centerset* model. Spring return or self-metering types are recommended for minimum water consumption. The faucet for wheelchair lavatories typically has a longer neck and handles with long blade-type handles, both of which extend toward the front of the fixture and user. It is also worth mentioning that the elderly or persons with arthritis can grab the cross or lever-type handles easier than the knob-type. New presence-sensing faucets are gaining in popularity because they do not require the user to physically touch the faucet. This promotes a cleaner fixture. Figure 3-8 shows a lavatory with built-in faucet and presence-sensor controls.

Figure 3-6 Wall-hung lavatory with a ledge or shelf back.

Figure 3-7 Wall-hung lavatories that have extended fronts allow a person in a wheelchair easier access.

SINKS

Service

Service sinks are used to fill wash buckets, receive their contents, and discharge those contents into the drainage system. They are one-piece construction with the faucet connection either through the back or above the back. Some fixtures are cast with faucet holes while others are not. The sink is usually set on a cast-iron P-trap that provides the principal support for the fixture. Most service sinks are either enameled cast iron or vitreous china. A stainless steel rim guard is usually placed on the front and side rims for protection of these surfaces. Spouts for service sink faucets should include a hose-thread, bucket hook, and support rod. The faucet should have an integral vacuum breaker.

Mop

Mop sinks are usually floor-mounted receptors that are similar in construction to shower receptors except that the sides are higher, usually 8"-10" (see Figure 3-9). A mop sink is basically a floor-mounted service sink that allows for easier filling and emptying of mop buckets. Mop sinks are made of terrazzo, plastic, or some other composite material; these receptors are installed in lieu of or in conjunction with service sinks. They are also available for corner installation.

A faucet with an integral vacuum breaker is mounted on the wall for bucket filling and fixture rinsing. The fixture has a deep well for sufficient capacity to hold the contents of a bucket and minimize splashing. The drain opening is a minimum of 2" in diameter and is equipped with strainer plates to keep significant solids from entering the trap and drain piping.

Figure 3-8 Faucets with presence sensors allow the user to use the fixture without having to touch it.

Wall-Mounted Service Sink Faucet

This faucet may be used in conjunction with either the mop or service sink. The support arm and bracket are optional. However, when the bucket hook is provided, the support arm and bracket are mandatory. The support arm is usually located on the top of the faucet. Some faucet models that do not come with a support come with a 5-foot hose this eliminates the need for lifting the bucket when filling.

Figure 3-9 Mop sink. This 8" center-to-center faucet should be equipped with a 3/8" hose thread and integral vacuum breaker.

Culinary

Culinary sinks are used for food soaking, food processing, and dishwashing. They are usually multiple compartments and made of stainless steel with welded joints. The basins are generally deeper than residential-type sinks in order to accommodate the greater amount of food or dishes in a commercial establishment. Many sinks are available with drain boards, which are horizontal surfaces slightly sloped toward the sink basin. This surface can also be used for food preparation and clean up. Wall-mounting may be employed, but the sinks are usually set in a frame that is floor-mounted.

To ensure sanitary conditions in these commercial sinks, the waste connections are made indirectly to the drainage system. Figure 3-10 shows two ways to make the indirect connection: either by an air gap fitting on each drain outlet or by an air break arrangement with the drain from each compartment discharging into a receptor in or above the floor.

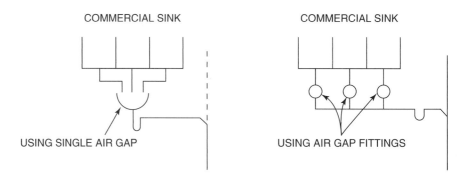

Figure 3-10 Indirect sink waste connection helps keep a more sanitary condition.

Bar

Bar sinks are used to fill drinking glasses and wash ware in bars and restaurants. As with other sinks, water can be discharged into the sink and drained into the waste system. Bar sinks are single or multiple compartment sinks, usually made of stainless steel or enameled cast iron, and are mounted in a counter top. Faucet mounting holes may be provided. The faucet is usually equipped with a high gooseneck spout for convenience in filling large containers. The drain opening is equipped with a standard 3-1/2" basket strainer or with a 2" strainer. Some codes require indirect waste connection from a bar sink.

P-Trap with Cleanout and Support Base

P-traps with cleanouts and support bases are installed in conjunction with the wall-mounted service sink. This fitting is normally made of cast iron, with or without an enameled interior. Its inlet is made of brass and chrome parts, with its outlet tapped to accommodate threaded pipe. Standard trap sizes are 2" or 3" in diameter. Figure 3-11 shows a P-trap with a built-in cleanout.

Drinking Fountain and Water Cooler

A *drinking fountain* or *water cooler* is used to provide drinking water in a sanitary way without the need for a cup or glass. They are usually made of vitreous china or stainless steel. The supply nozzle is elevated above the flood rim and has an integral strainer, a mouth guard, a stream regulator, and an integral automatic stop. Drinking fountains and water coolers are similar in function, but water coolers have a small refrigeration unit that delivers refrigerated water at approximately 50° F. Refrigerated drinking fountains and water coolers must comply with ARI 1010 and UL 399 standards. Figure 3-12 shows a wall-hung water cooler that is wheelchair accessible.

Figure 3-11 P-traps with a built-in cleanout allow for easier maintenance.

FOOD WASTE GRINDER

Food waste grinders or *garbage disposals* are used to reduce food scraps into particles small enough to be washed down the drainage piping. Commercial units are larger than residential types and are made of stainless steel with heavy-duty construction. Motor horsepower ratings are also greater than residential models. The units are installed in a special round-bottom sink and are the continuous feed-type so whenever the grinder is in use, water must be flowing into the sink (see Figure 3-13).

If no water is flowing, a flow switch in the water line will prevent operation. If a hose is connected to the water outlet or if the water inlet is below the flood level rim, a vacuum breaker should be installed in the water line above the flushing rim of the grinder sink.

IMPORTANT: The food waste grinder should not discharge into a grease interceptor because the high-velocity flow from the food waste grinder would defeat the purpose of the interceptor.

The *National Standard Plumbing Code* also dictates that the food waste grinder be separately trapped from any sink compartment or other fixture. Unlike residential models, which can be on a 1-1/2" drain line, a commercial food grinder requires a 2" or larger drain line.

Figure 3-12 Wheelchair accessible water fountains have a clearance space underneath for leg room.

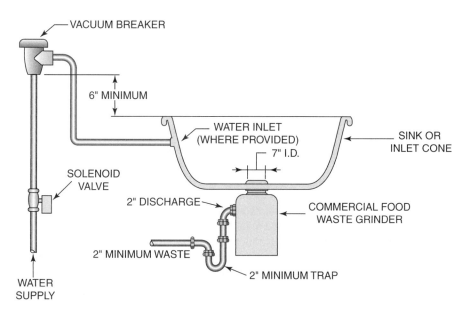

Figure 3-13 Industrial food grinder with constant flow of water.
Courtesy of Plumbing-Heating-Cooling Contractors–National Association.

DISHWASHER

Commercial *dishwashers* are used to wash dishes in restaurants and other establishments. The appliance automatically feeds racks of dishes through the washing cycle for high-volume output of clean ware. The water supply to the product is protected by an air gap and the waste discharge is usually indirectly connected to the drainage system. The dishes are loaded into a rack, pre-rinsed, and the rack placed on an automatic conveyor. A wash cycle and one or more rinse cycles complete the operation. The final rinse uses 180° F water to sanitize the dishes and ensure that all cleaning material is removed.

According to the 2006 *National Standard Plumbing Code,* commercial dishwashers must be indirectly connected to the drainage system through either an air gap or an air break. When the machine is within 5 feet of developed length from a trapped and vented floor drain, it may be indirectly connected to the drain inlet (see Figure 3-14).

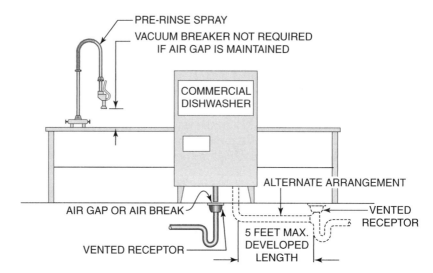

Figure 3-14 Drain arrangement for a commercial dishwasher.
Courtesy of Plumbing-Heating-Cooling Contractors–National Association.

GREASE INTERCEPTORS AND GREASE RECOVERY UNITS

A *grease interceptor* or *recovery device* may be required in the drain line from culinary sinks or any other fixture that could be expected to discharge fats, oils, or grease. They are either located inside the building, usually under the sink, or somewhere outside the building, usually buried. They are required to have a removable airtight lid that allows for periodic removal of trapped greases. While water is allowed to pass through, greasy substances in the sink effluent will be collected in the interceptor. The grease trap is designed so that oily and greasy solids collect in its baffle system, with water being able to discharge freely. Since grease and oil are lighter than water, a series of baffles is installed in the upper portion of the tank (see Figure 3-15).

A flow control valve in the inlet to the grease interceptor is required to limit flow rates to values that will not dislodge the collected grease. A grease interceptor installed under the sink must have a vent installed between the fixture it is being fed from, and the interceptor itself. Figure 3-16 shows a normal installation where the fixture has a trap and a vent. Figure 3-17 shows a common arrangement that allows the grease interceptor to serve as the fixture trap. Notice the 4' maximum separation distance.

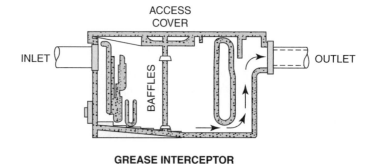

GREASE INTERCEPTOR

Figure 3-15 Grease interceptor.
Courtesy of Plumbing-Heating-Cooling Contractors–National Association.

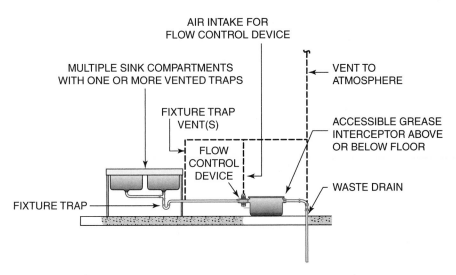

Figure 3-16 Normal grease trap installation where the interceptor has a trap and a vent.
Courtesy of Plumbing-Heating-Cooling Contractors–National Association.

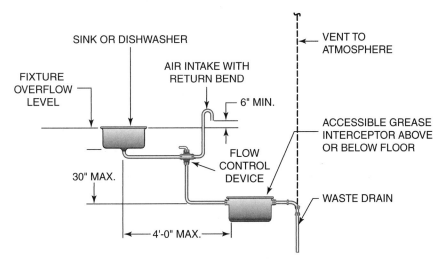

Figure 3-17 The grease interceptor serves as the fixture trap.
Courtesy of Plumbing-Heating-Cooling Contractors–National Association.

Sizing Grease Interceptors

Grease interceptors are typically designed in accordance with the expected amount of flow and the expected flow rate through the interceptor unit. The flow can be estimated by examining the size of the fixture or fixtures feeding into it or by the type and size of the building. It is important to note that in some locations, health departments have some control over the installation and operation of these devices. Always check with the local code before beginning this type of work.

The first method, known as the *fixture drainage method*, addresses the size of the fixtures that will be feeding into the interceptor. It examines the maximum amount of water that they can hold and how fast they can drain into the interceptor. The second method, known as the *seating method*, assumes that a restaurant will generate a certain amount of grease based on the number of people it can seat, the number of hours a day it is open, and how close it is to a high-traffic area.

Example One: Fixture Drainage Method

Assume that a facility has a three-compartment kitchen sink where each compartment is 24" wide $\times$ 24" long $\times$ 18" deep. The total cubic gallons could be determined as follows:

$$\text{Total volume} = (3 \text{ compartments}) \times (24" \times 24" \times 18")$$

$$\text{Total volume} = 3 \times (10368 \text{ cubic inches})$$

$$\text{Total volume} = 31104 \text{ cubic inches}$$

$$\text{Total gallons} = \frac{31104 \text{ cubic inches}}{1} \times \left(\frac{1 \text{ gallon}}{231 \text{ cubic inches}} \right)$$

$$\text{Total gallons} = 135 \text{ gallons}$$

Due to the dishes and the compartments not filling to the maximum, it is assumed that only 75% of the total volume of the sink will be used. So . . .

$$\text{Total gallons} = 135 \text{ gallons} \times .75$$

$$\text{Total gallons} = 101 \text{ gallons}$$

It can be assumed that this fixture will discharge in either 1 minute or 2. The 2-minute assumption is usually considered a reasonable choice. This would indicate that the flow rate from the sink would be 50.5 gallons per minute (101 gallons/2 minutes). Based on this method, the installer should choose an interceptor rated for 101 gallons at a flow rate of 51 gallons per minute.

Example Two: Seating Method

A new restaurant is being constructed in a large mall. The facility has a seating capacity for 50 people and is in operation from 11:00 a.m. to 8:00 p.m. Assume the flow per diner is 5 gallons per seat available. Using the following equation, a capacity in gallons can be determined:

$$Q = \frac{(D)(HR)(GL)(LF)}{2}$$

Where:
Q is the liquid capacity of the grease interceptor in gallons
D is the number of seats in the dining area
HR is the number of hours open per day
GL is the gallons of wastewater per meal
LF is a loading factor depending on restaurant location
　　1.25 for near interstate freeways and major city streets
　　1.0 for recreation areas
　　0.8 for a state numbered road
　　0.5 for other roads and highways

$$Q = \frac{(50)(9 \text{ hours})(5 \text{ gallons per seat})(1.25)}{2}$$

$$Q = \frac{2813}{2}$$

$$Q = 1406$$

Therefore, a grease interceptor rated for 1406 gallons or more will be required. It should be pointed out that some codes require a 1000-gallon minimum for any external grease interceptor.

A grease recovery device differs from the conventional grease interceptor in that it has a heater, skimmer assembly, motor, and a collection device for grease removal from the device. They are typically located near the point-of-use and are smaller than a normal grease interceptor. Like grease interceptors, they rely on a controlled flow of fluids to enter a holding chamber. The heavier solids settle to the bottom while the lighter greases and oils float on top. A skimmer dips into the upper layer of grease and lifts it into a separate container that can be manually emptied. The heater keeps the grease waste from becoming too hard, which would hinder removal. Figure 3-18 shows a conceptual drawing of a grease recovery unit.

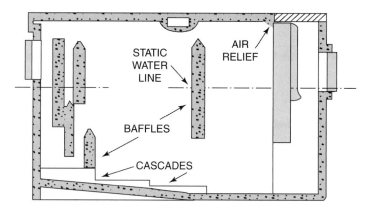

Figure 3-18 **Conceptual drawing of a grease recovery unit. Remember that grease and oils will float on top of the water.**

ICE-MAKING MACHINES

Commercial *ice-making machines* are used to make ice in quantities large enough to meet the demand of a public facility. The backflow protection to this equipment is often an atmospheric vacuum breaker. The waste outlet should be protected by air gap.

WALL HYDRANT

Wall hydrants or *hose bibbs* are found on just about any building. Mostly they are located outside, but in some cases they can be found inside the building. Typically they are constructed of brass and may be nickel- or chrome-plated. Figure 3-19 shows a recessed bronze hydrant equipped with an integral backflow preventer. Since the off-on working mechanism is inside the building, this device is considered freeze-proof provided the product is installed so it can drain. For added vandal protection, a loose key is provided. Integral backflow protection and self-draining features are preferred.

A relatively new product on the market is the hot-cold wall hydrant, which allows the user to get tempered water. Figure 3-20 shows one made by Moen that features frost-proof design with built in vacuum breaker.

Figure 3-19 **Typical recessed commercial wall hydrant.**

INDUSTRIAL FIXTURES AND APPLIANCES

Considerations for Industrial Installations

Since many fixtures already described for residential or commercial applications are also used in industrial settings, they will not be reviewed in this chapter. You will encounter special industrial equipment that requires careful plumbing installation. In each case, study the details of the fixture, the manufacturer's recommendations, and the needs of the job to ensure a proper installation.

Industrial establishments frequently generate unusual waste products that require special handling. Industrial waste is any liquid or liquid-borne waste resulting from the processes employed in industrial and commercial establishments. Such industrial wastes must be separated, collected, filtered, or neutralized before they are discharged into the drainage system. Special traps and interceptors are used to accomplish these purposes.

Figure 3-20 **This wall hydrant allows for the mixing of hot and cold water so the user can control the outlet water temperature. Photo courtesy of Moen Incorporated, 2007.**

Many industrial devices and configurations make cross-connections likely. To guard against the occurrence of such cross-connections, carefully survey each installation to be sure that the arrangement is safe and proper. A large variety of industrial wastes are encountered, but relatively few methods are available for handling them. The methods most common are separation, collection, filtration, or neutralization.

The requirements for a correct and properly maintained system are extensive. The industrial company is required by many governmental departments to maintain a healthful workplace for employees, produce safe products, and maintain a pure environment. Thus, any system or device for such a customer must be effective and properly installed.

FLOOR DRAIN

Floor drains are installed to collect any spilled water or other liquids, remove them from the floor surface, and deliver them to either a storm drain or a sanitary drain system. Floor drains are also used in areas where regular or frequent floor washings are required. Figure 3-21 shows a cross-section of a common floor drain.

Floor drains include a skid-resistant cover, large diameter inlet, and an outlet opening that may be caulk, no hub, threaded, or solvent weld-type. The housing may be made of cast iron, copper alloy, plastic, or other special materials. The basic requirements of a floor drain are the following:

- It must be sufficiently strong and be installed in such a manner to support the traffic passing over it.
- It must be equipped with a strainer (or debris bucket) appropriate to the waste entering the drain.
- It must afford the required drainage capacity for the application.
- If greasy or oily wastes are involved, a grease interceptor or recovery unit is used in conjunction with the floor drain.

A *deep seal trap* can be installed in areas where there is extended time between wash downs. A deep seal trap contains more water then a standard design trap so that evaporation will not cause a trap seal to fail and allow sewer gases to enter the building. In the lower example of Figure 3-21, a deep seal trap also contains a connection for a self-priming feature to prevent the loss of the seal. This is often referred to as a *trap primer*, a device that periodically supplies water to the trap.

This device must be protected against backflow in some areas based on code requirements.

ROOF DRAIN

Industrial buildings with large roof areas are usually provided with a flat roof. With the flat roof, drainage must be provided at reasonably close spacing to conduct rainwater off the roof to prevent a build-up of water that may become a structural problem. In addition, most roofing materials have a longer life if they are dry, so an important consideration of roof drain placement is to select locations that will leave a minimum of *ponding* on the roof after the rain ceases.

A *roof drain* is similar to a floor drain in general construction, consisting of a large diameter sump and a strainer top. While floor drains have a flat top, roof drains have a domed top. This feature presents a large area for flow and still prevents any large solids from entering the drain.

Roof drains should be accessible for cleaning. Traps are generally not used on roof drains, but may be required by local codes when used on drains that connect to a combined sewer and serve a roof that is used as a walkway.

Where excessive amounts of water could collect if the roof drain system were to fail, a secondary drainage system is required. The secondary roof drainage system

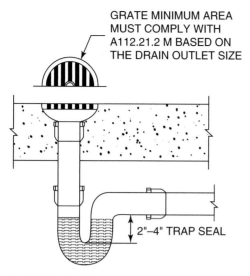

GRATE MINIMUM AREA
MUST COMPLY WITH
A112.21.2 M BASED ON
THE DRAIN OUTLET SIZE

2"–4" TRAP SEAL

A FLOOR DRAIN TRAP AND STRAINER

FLOOR DRAIN

4" DEEP SEAL

A DEEP SEAL TRAP

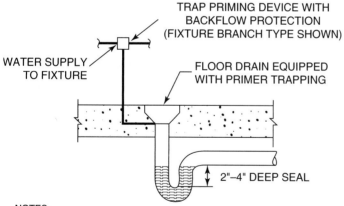

TRAP PRIMING DEVICE WITH
BACKFLOW PROTECTION
(FIXTURE BRANCH TYPE SHOWN)

WATER SUPPLY
TO FIXTURE

FLOOR DRAIN EQUIPPED
WITH PRIMER TRAPPING

2"–4" DEEP SEAL

NOTES:
1. A FIXTURE BRANCH TYPE TRAP PRIMER CAN PRIME MORE THAN ONE TRAP IF IT IS
FITTED WITH A MULTI-OUTLET MANIFOLD.

A FLOOR DRAIN WITH TRAP PRIMER

**Figure 3-21 Three possible floor drain configurations.
Courtesy of Plumbing-Heating-Cooling Contractors–National Association.**

should be completely separate from the primary roof drainage system. The drainage fixtures for the secondary system will have an elevated opening, sometimes referred to as a *dam*. This will prevent anything from entering the fixture until the water level has increased to at least 1-1/2" above the primary roof drain openings. Chapter 24 covers the different types of roof drains and how storm drainage systems work as a whole.

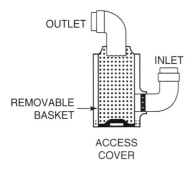

SOLIDS INTERCEPTOR

Figure 3-22 Solid waste interceptors are designed to trap any heavy solid materials to prevent them from entering the sewer system. Courtesy of Plumbing-Heating-Cooling Contractors–National Association.

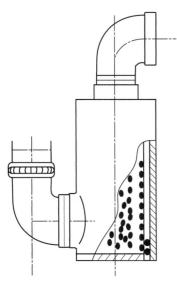

Figure 3-23 Hair interceptor typically found in barber shops.

SOLIDS INTERCEPTORS

Solids interceptors are used to collect materials (heavier than water) that are either valuable or hazardous to discharge into the waste system. Dental offices and photographic dark rooms are examples of applications where valuable solids need to be retrieved from the waste piping (see Figure 3-22).

The interceptor is either a cylindrical or box-shaped watertight container with a basket strainer. Waste liquids flow through the basket and solids are retained in the basket. When servicing, remove the cover and withdraw the strainer basket to recover the intercepted solids.

HAIR INTERCEPTORS

Hair interceptors are used in barber shops, beauty salons, pet stores, and veterinary clinics to keep hair from entering drain lines and forming stoppages. A basket holds any hair entering the device and a removable cover provides access for cleaning (see Figure 3-23).

SAND INTERCEPTORS

Sand interceptors are settling basins for sand, grit, and similar solids to settle out of sewage flow. These devices are usually large and similar in construction to a septic tank. Baffles may be used to encourage the solids to drop out of the flowing water. As with other interceptors, it is necessary to provide a means for periodic removal of the collected solids (see Figure 3-24).

OIL INTERCEPTORS

Oil interceptors are used in conjunction with sand interceptors to serve parking garages and service stations. These devices prevent lighter-than-water materials entering the drainage system and are built like grease traps, with baffles penetrating the surface of the liquid. The water flows below the baffles and the oils are held in the interceptor by the baffles (see Figure 3-25).

Since no congealing takes place in the oil interceptor (in contrast to the grease interceptor), the oil must be drawn off periodically or the material will eventually enter the drainage system.

NEUTRALIZING OR DILUTION TANK

For installations where concentrated aggressive chemicals are discharged, a *neutralizing tank* is the only satisfactory method of rendering the wastes suitable for the drainage system. A tank is installed in the branch or building drain and the waste

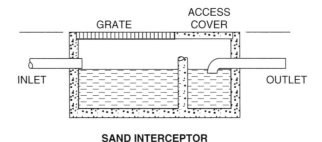

SAND INTERCEPTOR

Figure 3-24 Sand interceptor.
Courtesy of Plumbing-Heating-Cooling Contractors–National Association.

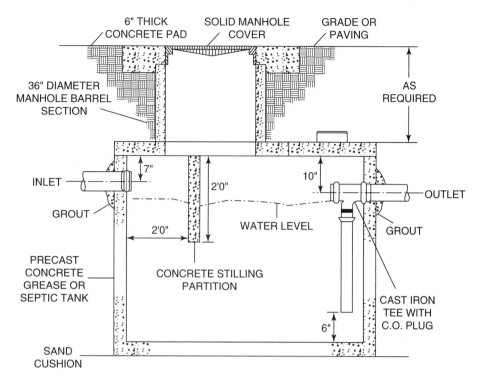

Figure 3-25 Oil interceptor.

is either diluted by water or exposed to a neutralizing chemical. The chemical must be replenished from time to time to maintain the effectiveness of the system.

Figure 3-26 shows a typical arrangement. Note that the cover of the tank must be accessible and removable in order to add the neutralizing chemical. A more sophisticated system uses pH sensors to control chemical feed pumps to deliver the proper amount and type of neutralizing chemical to the waste stream.

IN THE FIELD

2006 *National Standard Plumbing Code* states the following: "Vents for neutralizing or dilution tanks shall be constructed of acid-resistant piping and shall be independent from sanitary system vents."

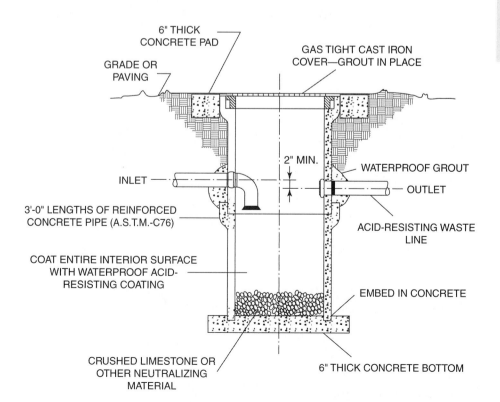

Figure 3-26 Dilution basin used to dilute chemicals before they enter the drainage system.

EMERGENCY EQUIPMENT

Eye Wash

At first glance, the fixture appears to be a double drinking fountain with a large handle, but it is actually specially designed to deliver a low-pressure stream of tempered water. It is available in wall, counter, or semi-recessed mounting styles (see Figure 3-27).

This fixture is installed wherever strong chemicals are stored or used. In the event of a chemical spill, any affected worker can immediately flood the eyes with tempered water to dilute and wash away the hazardous material. Since the first seconds are the most important in such accidents, these ready-to-use eye wash fixtures must be immediately available.

The affected person places his eyes near the two spouts and activates the handle to flush the eyes with large streams of water. The fixture is provided with a fine-mesh strainer so that contact lenses will not be lost with the water wash.

Figure 3-27 Eyewash station. These stations should be clearly marked and readily accessible.

Emergency Shower

For situations similar to those requiring the eyewash, emergency showers are installed for worker (or student) protection. These units have a large water flow that can dilute and wash off chemicals or extinguish fire from the clothing of a laboratory worker (see Figure 3-28).

A large handle within reach is located in the shower area so that the affected person can easily turn on the water. A heavy spray of water is delivered for maximum effectiveness in washing away the strong chemical or flame. These devices are vital protection against a serious hazard. Unfortunately, they are subject to abuse by horseplay, especially in a school laboratory setting. In such applications, consideration should be given to adequate drains and waterproof walls in the vicinity of the emergency shower.

INSTITUTIONAL FIXTURES AND APPLIANCES

There are three general categories of institutional fixtures to consider:

1. Handicap use (or physically challenged)
2. Medical
3. Vandal resistant

Figure 3-28 Emergency shower.

The extensive variety of products available and the types of institutions served require considerable thought in selecting the equipment for the task at hand. In new work, the designer makes the decisions, but for repair or remodeling projects, the plumbing contractor is usually involved in the selection process.

Fixtures and trim for the physically disabled or physically challenged person are designed to allow safe access and easy use. The fixtures (or settings) considered for handicap-use installations include the following:

- Water closet and compartment
- Lavatory
- Drinking fountain
- Shower
- Urinal

Medical fixtures include all those pertaining to medical care and treatment in hospitals and clinics. Such fixtures include the following:

- Water closet with bedpan accessory
- Various types of bathing devices
- Lavatory
- Various types of sinks
- Special trim products

Vandal-resistant fixtures are provided for institutions where security is a problem and vandalism must be minimized. The following are examples to consider:

- Water closet
- Lavatory
- Shower
- Cuspidor
- Drinking fountain

For the Physically Disabled (Handicap-Use)

NOTE: Design and installation of fixtures and fixture fittings for use by the physically challenged are governed by local codes. In the absence of local codes, consult the Americans with Disabilities Act (ADA), or ANSI Standard A117.1, Making Buildings Accessible to the Physically Impaired.

Water Closet

Water closets for the physically impaired are always the elongated version, designed and installed to be accessible. The bowl design is the same as conventional designs, but the top rim of the bowl is 16"–18" above the floor, rather than 14"–15" as in the regular design (ANSI A112.19-2M and ADA requirements). To meet ADA and ANSI requirements, the resulting height of the seat cannot exceed 19". If wall-hung bowls are used, the rough-in for the fixture can be placed as desired to have any rim height needed.

Substantial grab bars should be provided on the fixture or on the wall(s) adjacent to the water closet to permit the impaired person to transfer to and from the wheelchair to the bowl. In addition, sufficient clearance must be provided so that a wheelchair can be maneuvered to gain access to the water closet. The effect of the required rules is that the minimum stall size required for the handicap-use water closet installation is about 3'6" × 6'3". Figure 3-29 shows a typical handicap water closet stall. NOTE: Always refer to ADA requirements for the proper installation.

> **IN THE FIELD**
>
> Make sure to determine who is responsible for the installation of the grab bars and blocking. This should be discussed in the pre-planning meetings at the beginning of the job. In some cases, it may not be the plumber who is responsible for their installation.

Figure 3-29 Typical handicap toilet stall. Grab bars are securely attached to support the user's weight.

Lavatory

Lavatory installations for access by handicapped persons must provide legroom under the fixture. The clearances required by the standards usually exclude the channel for the overflow. All piping must be covered with a soft insulation to prevent bruising or burning the legs of a person in a wheelchair. Some persons in wheelchairs have no sensations in their legs, so they will not feel even a serious burn or bruise.

Most wall-hung lavatories can be installed to meet handicap applications, but there is a special line of lavatories that enhances the requirements for handicap applications. These fixtures require special mounting arms to obtain adequate installation strength. Other considerations include the following:

- Extend the fixture out from the wall a little more than usual (27" rather than 20" to 24").
- Make sure the fixture has a shallow front vertical dimension.
- Make the drain opening fairly close to the wall.
- Equip the fixture with blade-handle, single-lever, or automatic sensor faucets.

Drinking Fountain

Drinking fountains for easy access have the following characteristics:

- Semi-recessed or wall-mounted with leg clearance
- A shallow front vertical dimension
- A bubbler located close to the front
- A pressure plate, blade-handle, or sensor for operation

These units extend out from the wall far enough for the person in a wheelchair to get into a position to drink from the bubbler. Because they extend out from the wall, they could obstruct walking traffic, particularly the sight impaired; therefore, they should not be installed in heavy-traffic corridors.

Shower

Handicap showers may provide a seat for the bather or may be the roll-in type for the more severely restricted patient. Either type should be equipped with the following features for convenient bathing:

- Grab bars
- Water controls, including scald protection with a thermostatic balanced pressure or combination type-valve
- Soap trays
- Hand-held shower assemblies

Urinal

Conventional wall-hung urinals may be used, but the lip height should be no more than 14" above the floor. Extended lip urinals appear to provide the preferred access. Sensor-activated urinals provide the easiest fixture use.

Medical

Water Closet

Medical water closets are suitable for the able or disabled person, with the seat height usually 17"–19" high. The water closet may be equipped with bedpan lugs and bedpan cleaning accessories, including a spray nozzle, support hook, vacuum breaker, and control valve. Grab bars are usually provided for use by the handicapped or infirm. Figure 3-30 shows a flush valve with a bedpan washer.

Options available for the bedpan washer include warm water supply and/or germicide *aspirators*. An *aspirator* is a fitting or device that causes a vacuum when supplied with water or some other fluid, as it passes through an integral orifice or "constriction" under a positive pressure (see Figure 3-31).

The aspirator is used in morgues, operating rooms, and similar locations where a suction line is needed. Its assembly consists of an aspirator ejector nozzle for the vacuum hose, a control valve, a vacuum breaker, a 2" receptor, a 1-1/2" tailpiece, and the related chrome nipples and escutcheons.

No matter what options are used, the bedpan washer water supply must be protected against backflow. The backflow preventer must be installed to comply with local codes.

Figure 3-30 Flush valve with bedpan washer. Courtesy of Sloan Valve Company.

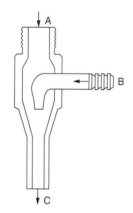

A - INLET FLUID FLOW
B - INDUCED VACUUM
C - COMBINED OUTLET FLOW

Figure 3-31 Aspirators are strictly regulated by code section 14.13. Courtesy of Plumbing-Heating-Cooling Contractors–National Association.

Bowl with Bedpan Lugs

This type of bowl is similar to a normal elongated bowl in style, trapway design, and operation. Installed in many patient-care facilities, it is also equipped with lugs on the rim to support a bedpan during the bedpan-cleaning operation.

Sitz Bath

Sitz bath fixtures are soaking tubs for soothing and treating the perineal section of the body. Tempered water is introduced into the fixture and the patient sits in the water. (Sitz is the German word for a *sitting.*) The units are made of vitreous china, with armchair sides, wide bathing well, and integral overflow. They may be floor- or wall-mounted.

Institutional Bath

These baths are somewhat oversized bathtubs mounted on a base above floor level for easier handling of patients. The drain end of the tub is placed against a wall, with the other three sides providing access for attendants to assist the patient.

Hand-held showers and scald-protection valves are commonly installed for these units. Grab bars can be used, but with three open sides, only the end wall is available for mounting them. The manufacturer may provide grab bars in the design of the tub for greater safety. Special sling chairs are available to raise and lower the patient into the bathing vessel.

Whirlpool Bath

Whirlpool baths provide oscillating or pulsating jets of water flow for body massage. Some of these fixtures are sized and shaped for certain body parts (an arm or leg) and others are for total body immersion. The tubs include an electric or air-operated pump that produces the flow as well as the external piping and nozzles necessary to direct the flow.

Patient's Lavatory

A patient's lavatory is installed in a hospital room to provide for the special needs of the location. The fixture is one-piece and includes a drain board and backsplash. Overflows are not built into these lavatories. Spout aerators are also not used for fear that air-borne germs may be drawn to the water supply. A corner installation model is available, so that cleaning the lavatory (and keeping it clean) is made easier. The faucets are blade-handle or automatic sensor-type.

Instrument Sink

The *instrument sink* is installed where medical equipment is cleaned or medical staff washes up. They are made of vitreous china and incorporate a drain board and backsplash. Overflows are not built into these fixtures and faucet aerators are not used. Knee, foot, or sensor-operated faucets are provided for wash-up locations, but blade-handle faucets are normal for general-purpose locations.

Scrub Sink

The *scrub sink* is used where staff personnel perform cleanup. Locations include surgery, emergency rooms, morgues, and treatment rooms. The fixture has a deep washing well and uses knee, foot, or sensor-control faucets with a gooseneck spout. The deep well minimizes splashing and the special controls maintain a sterile field on hands and forearms. No aerator is used on these faucet spouts. Figure 3-32 shows a scrub sink with a knee-operated soap dispenser.

Clinic Service Sink

Installed in utility rooms, emergency rooms, autopsy rooms, clean-up rooms, and soiled linen rooms, this sink receives liquid or solid wastes. It consists of a vitreous china bowl with flushing rim, integral trap (siphon jet or blowout), spud inlet, and

Figure 3-32 Scrub sink with knee control allows for a more sanitary condition.

waste outlet. It is either floor- or wall-mounted. Flushometer valve and bedpan washers are usually provided as well as stainless steel rim protectors for china bowls. It is usually mounted about 28" above the floor for the convenience of the staff.

Vandal Resistant

Water Closet

Water closets for prisons and some other applications must provide the required level of sanitation and be as resistant to vandalism as possible. In addition to being able to deliver a complete flush on each cycle, it must be unbreakable. A piece of a broken fixture could easily be used as a weapon and a broken-off or dislodged bowl could allow water to flood the cell. They must also be able to resist thermal stress (prisoners could build a fire within the bowl). They should be fixed-in-place (an escape route may be provided if access can be gained to the plumbing chase), and sound-deadened to eliminate noise propagation during disturbances. Accessories must also be secure to eliminate any hiding places for contraband. As an added measure, the seat is integral with the bowl.

Wall-hung styles are mounted with concealed bolts and the fixture is fabricated of stainless steel. The details of trap design are selected to deliver the greatest flushing capacity and to minimize any space where contraband can be hidden. Flushometer valves, piping, and cleanouts are located in plumbing chases located between back-to-back pairs of water closet installations, i.e., between adjacent cells. In this way, only a push-button operator is exposed in the prison cell. Figure 3-33 shows a flush valve designed to be installed in a plumbing chase so that only a push-button is exposed.

Many plumbing systems in modern penal facilities are connected to a vacuum waste system. Such systems are desirable as contraband is recoverable and the system is less likely to be tampered with by inmates.

Lavatory

Lavatories are installed in conjunction with the water closets discussed above to provide for personal hygiene (see Figure 3-34).

They are usually formed on the same mounting system that supports the water closet. Thus, installation costs and security problems are minimized. The water controls are push-button-operated, and the drain outlet is a fine-screen strainer assembly. The water supply, therefore, has a minimum exposure to tampering, and the fine-screen drain opening reduces the chance of concealing contraband. The lavatory trap is installed in the plumbing chase outside of the cell.

Handicap-Use Combination Unit

This device is a combination lavatory, soap dispenser, light, mirror, and towel holder designed in one packaged unit. It is mounted in the wall, making it accessible for the wheelchair patient. All dispensers, faucets, and similar accessories are easily reached. The mirror may be tilted for greater convenience by the user.

Securely bolted to the building structure, this fixture has self-closing or self-metering controlled faucets and a vandal-resistant spout. The drain contains a grid strainer. All piping is located behind the fixture in a plumbing chase. The fixture is usually made of stainless steel for maximum-security application. It may be vitreous china for minimum-security institutions or schools.

Gang (Multi-Person)

Installed in school or minimum-security facilities, this fixture can serve the hand-washing needs of more than one person at one time. The soap and tempered water supplies are often centrally distributed. The fixture is generally stainless steel or terrazzo in composition. Each 18" of space constitutes one lavatory equivalent. Such products are now equipped with electronically controlled faucets. Figure 3-35 shows a typical group-type wash fountain or wash sink.

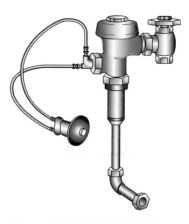

Figure 3-33 Pushbutton controlled flush valve allows for the valve and piping to be located in a piping chase.
Courtesy of Sloan Valve Company. All rights reserved.

Figure 3-34 Combination water closet and lavatory helps conserve space and offers less chance for vandalism. Courtesy of Acorn Engineering Co.

Figure 3-35 Gang or multi-person sinks are common in shops and schools.

Shower

The shower is intended to deliver sufficient volume of water at the proper temperature for body bathing. The design must provide a strong, tamper-resistant, contraband proof, and suicide-resistant installation. Such showers are often used in schools to provide the greatest protection against vandalism. The maximum-security styles use a push-button initiated, timed water control, and tapered showerhead. The minimum-security unit consists of a single water control and vandal-resistant showerhead, and it is available in single units for privacy or in multiple-head column style for use by several persons at the same time. Maximum-security units are single shower types. Maximum-security showers are supplied with tempered water and a single push-button operated control. The multiple column shower installation involves a minimum of equipment to keep clean and the lack of partitions enhances security.

Column Shower (Gang)

With the water, soap, and drainage lines stemming from the base of the column, this stainless steel fixture can accommodate up to six persons at one time. All faucets, soap dispensers, and showerheads are vandal resistant. The shower water temperature should be thermostatically or pressure-balance controlled.

Individual Wall Mount

Installed where column showers are not practical, these fixtures consist of pre-assembled units that must be securely attached to the building structure and then supplied with a tempered water supply. In maximum-security installations, push-button faucets and suicide-proof showerheads are common.

Cuspidor

These units are installed for expectorating (spitting) purposes and used considerably in the dental industry. Even though they are still available, most dental offices have replaced them with a vacuum system that allows for a quicker removal of fluids from the patient's mouth.

SPECIALTY TRIM ITEMS

Knee Faucet Controls

Either wall- or fixture-mounted, these knee-operated controls divert hot, cold, mixed, or tempered water to the faucet spout or pan washer device. They are commonly used on specialty sinks in the medical industry.

Plaster or Casting Interceptor

Installed directly below a plaster or casting sink this appurtenance is designed to prevent plaster or plastic particles from entering the sanitary drainage system. It must be readily accessible. When it is used, no additional trap is needed below the casting sink, as the interceptor provides the trap seal.

REVIEW QUESTIONS

1. What is an advantage of the flushometer valve over the flush tank?
2. Why are there no aerators located on some medical-related sinks?
3. What is an aspirator used for?
4. At what height is a handicap toilet usually placed?
5. Fixtures in maximum-security prisons are usually made of what material?

Installation Methods

The student will:

- Describe the installation steps and precautions necessary to install residential, commercial, industrial, and institutional fixtures.
- Summarize the important considerations for installing residential, commercial, industrial, and institutional fixtures.

KEY TO SUCCESSFUL INSTALLATION

A well-rounded plumber must be experienced in all phases of the plumbing trade, from rough-in to finish work. In the previous two chapters, residential, commercial, industrial, and institutional fixtures were discussed from a theoretical standpoint. In this lesson, the transition from theory to practical application will be examined for common plumbing fixtures. Because the fixture design can vary by manufacturer, it is important to read and follow the instructions supplied with the fixture.

IMPORTANT: The following information should be used for general guidelines and local codes should always take precedence. Whether installing residential, commercial, industrial, or institutional fixtures, the following guidelines should always be observed.

First, consider the installation from beginning to end. This gives the installer the opportunity to get organized and gather the safety equipment, tools, and supplies needed to complete the work safely and efficiently. A list of recommended tools for each installation job has been provided. A clean, well-organized jobsite will be safer and more productive. Second, conduct a visual inspection of the fixture to verify that it is the correct one for the job and free of visible defects and damage. Size, color, and style are also extremely important. Third, be certain that the installed fixtures are securely mounted and check the complete operation of the fixtures. It may be necessary to adjust water levels and stopper linkages. Once all piping is installed, it should be checked for leaks. Then remove any debris, packing materials, and other unused materials from the jobsite. Only fixture setting will be discussed in this chapter. Faucet and trim installation will be discussed in Chapters 5 and 6.

WATER CLOSET

Before removing the fixtures from their shipping cartons, confirm that the proper rough-in dimensions and correct models have been provided (see Figure 4-1). This information can be found on the manufacturer's rough-in sheet.

Check the water and waste connections to ensure that they match the rough-in dimensions for the fixture and have been installed properly.

Be sure that the floor or wall surface is smooth, flat, and satisfactory for the bowl type being used. The closet flange should be solidly attached to the floor. Inspect the fixture for shipping damage or manufacturing defects when it is removed from the shipping carton.

While the rough-in dimensions for a two-piece and a one-piece toilet may be different, the installation process is essentially the same once the tank has been mounted onto the bowl. The one-piece fixture is heavier and considerably harder to handle than the close-coupled fixture. Be sure to have help when you handle it so there is less risk of breakage or injury. One-piece units are much more expensive than two-piece designs, so breakage would be costly! When installing a two-piece toilet, the plumber must decide whether to install the bowl onto the closet flange first and then the tank onto the bowl, or to install the tank onto the bowl and then install them as one unit. There's no right or wrong way because each choice has its advantages and disadvantages. If the bowl is installed first, it is easier to handle, but you will need to reach between both the bowl and the wall in order to install the tank bolts and nuts. This can be accomplished by sitting on the bowl while facing the tank.

On the other hand, if the tank is installed onto the bowl before the bowl is secured in place, all of the mounting bolts are easily accessible. However, due to their combined weight, this means that it will be more difficult to lift both the tank and bowl at the same time. Once the tank and bowl have been joined together, never lift it by holding onto the tank because the bolts or bowl flange could easily be damaged by the weight of the bowl. Usually the center of gravity will be more toward the back, so grab the bowl as close to the tank as possible.

IN THE FIELD

National Standard Plumbing Code states that a minimum clearance of 21" must be maintained in front of any water closet. It also states that there must be at least 15" clearance between the centerline of the water closet and any side walls. NOTE: For specified bathrooms/fixtures, these minimum clearances may be higher according to the Americans with Disabilities Act (ADA).

KOHLER

MEMOIRS™
TOILETS

FEATURES

- *12" (30.5cm) rough-in*
- *Vitreous china*
- *9-3/4" (24.8cm) x 9" (22.9cm) water area*
- *Round front bowl*
- *With classic design (K-3452), or stately design (K-3462)*
- *1.6 gpf (6 lpf)*
- *Ingenium™ flushing system*
- *2" (5cm) glazed trapway*
- *Includes polished chrome trip lever*
- *Combination toilet*
- *Less seat and supply*
- *With Insuliner® insulated tank lining (-U)*

K-3452, K-3462

K-3452
CLASSIC DESIGN

K-3462
STATELY DESIGN

CODES/STANDARDS APPLICABLE

- *ASME/ANSI A112.19.2M*
- *ASME/ANSI A112.19.6M*
- *Energy Policy Act of 1992 (EPACT)*
- *IAPMO/UPC*
- *Canadian Standards Association (CSA)*
- *States of Massachusetts & Texas*
- *City of Los Angeles, CA*

COLORS/FINISHES

- 0   White
- Other   Refer to Fixtures Price Book for additional colors

Accessories:
- 0   White
- CP   Polished Chrome
- PB   Polished Brass
- Other   Refer to Faucets Price Book for additional finishes

SPECIFIED MODEL:

Model	Description	Colors/Finishes		
K-3452	Round front bowl toilet, classic design	☐0 White		☐Other_____
K-3452-U	Round front bowl toilet with insuliner, classic design	☐0 White		☐Other_____
K-3462	Round front bowl toilet, stately design	☐0 White		☐Other_____
K-3462-U	Round front bowl toilet with insuliner, stately design	☐0 White		☐Other_____
Recommended Accessories				
K-4662	Lustra™ seat with cover	☐0 White		☐Other_____
K-4662-A	Lustra™ seat with cover (includes anti-microbial agent)	☐0 White		
K-7637	Angle supply with stop	☐CP	☐PB	☐Other_____
Optional Accessories				
K-9439	Trip lever (non-CP)	☐PB		☐Other_____

PRODUCT SPECIFICATION:

The round front combination toilet shall be 12" (30.5cm) rough-in. Toilet shall be made of vitreous china. Toilet shall have 9-3/4" (24.8cm) x 9" (22.9cm) water area. Toilet shall be with classic design (K-3452), or stately design (K-3462). Toilet shall be 1.6 gpf (6 lpf) with Ingenium™ flushing system. Toilet shall have 2" (5cm) glazed trapway. Toilet shall include polished chrome trip lever. Toilet shall be less seat and supply. Toilet shall have Insuliner® insulated tank lining (-U). Toilet shall be Kohler Model K-_____-_____-_____.

We reserve the right to make revisions without notice in the design of fixtures or in packaging unless this right has specifically been waived at the time the order is accepted.

Page 1 of 2
115754-4-**DC**

Figure 4-1 Water closet rough-in sheet. Courtesy of Kohler.

MEMOIRS™

PRODUCT INFORMATION

Fixture:	
Configuration	2-piece, round front
Water per flush	1.6 gallons (6L)
Passageway	2" (5cm)
Water area	9-3/4" (24.8cm) x 9" (22.9cm)
Water depth from rim	5-3/8" (13.7cm)
Seat post hole centers	5-1/2" (14cm)
Included components:	
Bowl	K-4257
Tank with classic design	K-4454
Tank with stately design	K-4464
Tank cover, classic design	84406
Tank cover, stately design	84407
Trip lever	K-9439

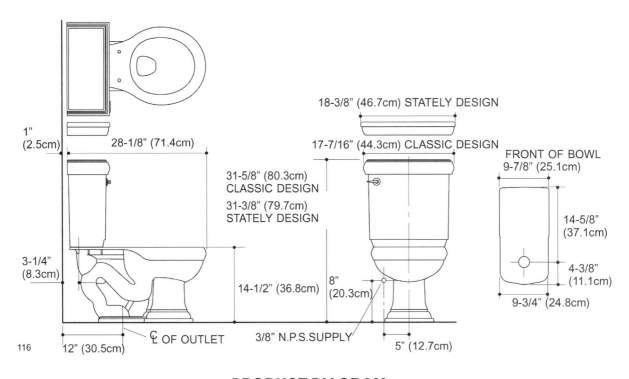

PRODUCT DIAGRAM

K-3452, K-3462 Memoirs ™ Toilets
Page 2 of 2
115754-4-**DC**

THE BOLD LOOK
OF **KOHLER**®

Figure 4-1 (Continued)

When mounting the tank to the bowl, first install a foam rubber gasket onto the threaded portion of the flush valve, which is located on the bottom of the tank. This will seal the opening between the tank and the bowl. There are usually two or three tank bolts with rubber gasket washers that seal the head of the bolt to the inside of the tank. The metal washer and nut go on the underside of the bowl and secure the two parts together. As you're doing this, tighten the nuts alternately and as evenly as possible. The flush tank should be bolted securely to the bowl, but do not over-tighten. Some flexibility in this connection is acceptable. Also, make sure to turn the nut and not the bolt. Turning the bolt increases the chance of cutting the rubber washer, which will result in a leak (see Figure 4-2).

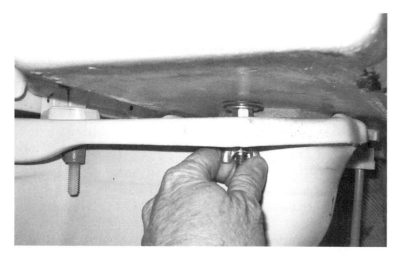

Figure 4-2 Tighten tank bolts as evenly as possible. Make sure to use your body weight to compress the wax ring seal.

When installing the toilet bowl to the closet flange, you'll need to install the 1/4" closet flange bolts and then the wax ring seal. The wax ring creates a watertight seal between the toilet bowl and the toilet flange on the floor (see Figure 4-3).

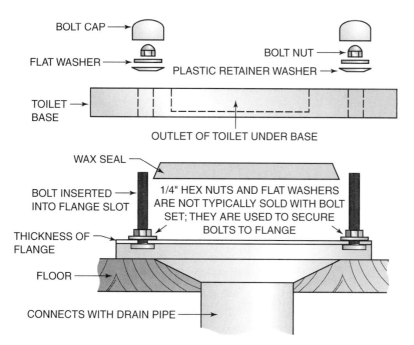

Figure 4-3 Typical components for a floor-mounted water closet. Bolts securing the closet flange to the floor are not shown.

Some plumbers will place the wax ring on the underside of the toilet and then place both the toilet and ring onto the closet flange. Sometimes it can be difficult to get the wax ring to stick to the bottom of the toilet, so warming it up and twisting it into place can help. Other plumbers place the wax ring onto the closet flange and then put the toilet in place. Just be careful not to cut the ring as you place the toilet on it. Check with your employer to see which installation method is preferred in your company.

When placing the toilet onto the closet flange, use your body weight to compress the wax ring. This will reduce your chances of cracking the bowl while securing it to the closet flange. Then install a plastic retainer washer, a metal washer followed by a hex nut. The plastic retainer washer is responsible for holding the decorative cap in place. The decorative cap covers both the hex nut and closet bolt. After installing the washers and hex nut, make sure that the toilet is level and the tank is parallel with the back wall. In some cases, such as a tile floor, you may need to place a few small plastic shims under the edge of the bowl base in order to keep it from rocking.

Next, tighten the closet flange bolts evenly with a 7/16" boxed-in wrench. This type of wrench fits nicely into tight spots and reduces the chances of over tightening or rounding the corners of the nut. Tighten the closet bolts alternately.

After the water closet is in place, install a water supply line to the flush valve opening, located on the bottom of the tank to the supply stop. This is usually accomplished with a 3/8" supply line that is either chrome-plated copper, plastic, or a braided stainless steel line. After the waterline is connected, turn on the water and check for leaks. After the tank is full, flush the water closet several times and check to make sure the water level in the flush tank is up to the level marked in the tank, or at least 1" below the overflow tube (see Figure 4-5).

Attach the toilet seat to the bowl by using the hinge bolts and nuts. Secure the nuts, but remember that you cannot get the seat bolts tight enough that the front of the seat won't have some lateral movement.

As a final operation, cut off excess length of closet bolts and cover with the decorative caps. Be careful not to scratch the fixture surface when cutting off the bolts. Some local codes require a moisture resistant caulk be applied around the base of the bowl where it meets the floor.

LAVATORY

No matter whether the lavatory is wall-hung, pedestal, or a vanity-type, it will usually have two 3/8" supply lines and a 1 1/4" trap. The rough-in drain line may be 1 1/2", so a 1 1/2" × 1 1/4" trap adapter may be necessary. Hand-tighten the drain assembly, then turn a quarter turn more with a wrench or adjustable pliers. Because there

Figure 4-4 Ratcheting wrenches allow the tightening of nuts to be done much faster, thereby saving time and money.

Figure 4-5 Water level in tank must be below the overflow tube. If not, it could indicate a leaking fill valve.

Table 4-1   Tools and Supplies for Water Closet Installation

Quantity	Tools/Supplies
1	Safety glasses
1	Medium flat head screwdriver
2	Small adjustable or appropriate size combination wrenches
1	Miniature hacksaw
1	Tubing cutter
1	Small level (torpedo or 2 ft.)
1	Rag
1	Tube of caulk (color matched)
1	Wax ring seal
1	Tongue-n-groove pliers
1	3/8" supply tubes

is limited space under a sink, install the fixture fitting and waste assemblies before making the final placement of the lavatory. Do not use pipe dope on plastic parts because the oil in the dope may attack the plastic material. When installing the drain line, use a smooth-jaw wrench for tightening plated parts with flats and a strap-wrench for plated parts that are tubular.

Wall-Hung

Wall-hung and pedestal lavatories are set on a wall bracket. Be certain that the bracket is level and securely mounted to backer boards (if installed) or wall studs with heavy screws. If it is designed to sit on legs or a pedestal, the height from the bracket to the finish floor is very important, so keep this in mind if the finished floor is not already installed. The joint where the lavatory meets the wall should be caulked to keep water from getting between the lavatory and the back wall (see Figure 4-6).

Vanity

When a lavatory is mounted on top of a cabinet, it is referred to as a vanity and usually installed as a single unit. The unit should be installed so that the top is level, which may mean that the unit will have to be shimmed from below. Once the shims have been installed, the cabinet can be securely mounted to backer boards (if installed) or wall studs with heavy screws through the horizontal backer board in the cabinet.

Sometimes the sink will have to be mounted into an opening in the countertop. If it was not specified that the countertop fabricator cut an opening for the lavatory, then the plumber may be required to do so. If this is the case, a template from the lavatory manufacturer should be used to mark the countertop for the proper location. If needed, the fixture itself can be used to make a template simply by turning the lavatory upside down and tracing the outline on the countertop. If you do this, then you must cut 1/2" to the inside of the mark to obtain the proper opening! Remember, you can always make the opening larger, so be conservative. Use a jigsaw blade with fine teeth to cut the top. The teeth should cut on the down stroke to avoid chipping the laminate on the countertop. Place tape on the bottom of the jigsaw base so that the metal base won't leave black marks on the countertop.

Figure 4-6 Caulking the back of the sink where it meets the wall prevents the wall from being damaged.

Table 4-2 **Tools and Supplies for Lavatory or Bidet Installations**

Quantity	Tools/Supplies
1	Safety glasses
1	Medium flat head screwdriver
2	Small adjustable or appropriate size combination wrenches
1	Miniature hacksaw
1	Tubing cutter
1	Small level (torpedo or 2 ft.)
1	Rag
1	Tube of caulk (color matched)
1	Basin wrench
1	Tongue-n-groove pliers
1	PVC glue/primer
1	1 1/4" P-trap and necessary adapters
2	3/8" supply tubes
1	Can of pipe dope or Teflon tape
1	Jigsaw with fine tooth blade (if countertop is to be cut)

When installing the fixture, place caulk between the lavatory and the top to form a watertight joint around the lip of the sink. Make sure that the caulk is approved for the countertop material.

BATHTUB

Bathtubs, showers, and whirlpools can be quite large and heavy, which means they are set and piped in the early stages of construction. Unfortunately, this also means that the fixture is exposed to damage for a longer period of time. After it as been delivered to the jobsite, it should be inspected carefully. When the rough-in is complete, apply a protective covering. Before the protective covering is placed on the fixture, most contractors have the building foreman sign that the tub is not damaged. Do not store materials or tools in the tub, and to avoid chipping the tub finish, do not drag it across the floor.

Recessed

Typically the bathtub is set and piped in the rough-in stage of the work. Be sure the carpenters know the rough-in dimensions so that the proper supports can be installed. Also be sure that the tub is supported so that it stays level and secure when the weight of the water is applied. NOTE: Some tubs may require special supports. Enameled cast-iron tubs are very heavy and will require several people to install. Enameled steel or plastic tubs are easier to handle and they are gaining in popularity. Plastic and fiberglass units are also easy to handle. It is usually easiest to install the bath waste and overflow before the tub is installed. Use putty to assemble the drain fittings in the waste and overflow (see Figure 4-7).

Place water in the tub and check the trip-lever assembly for operation and for leaks.

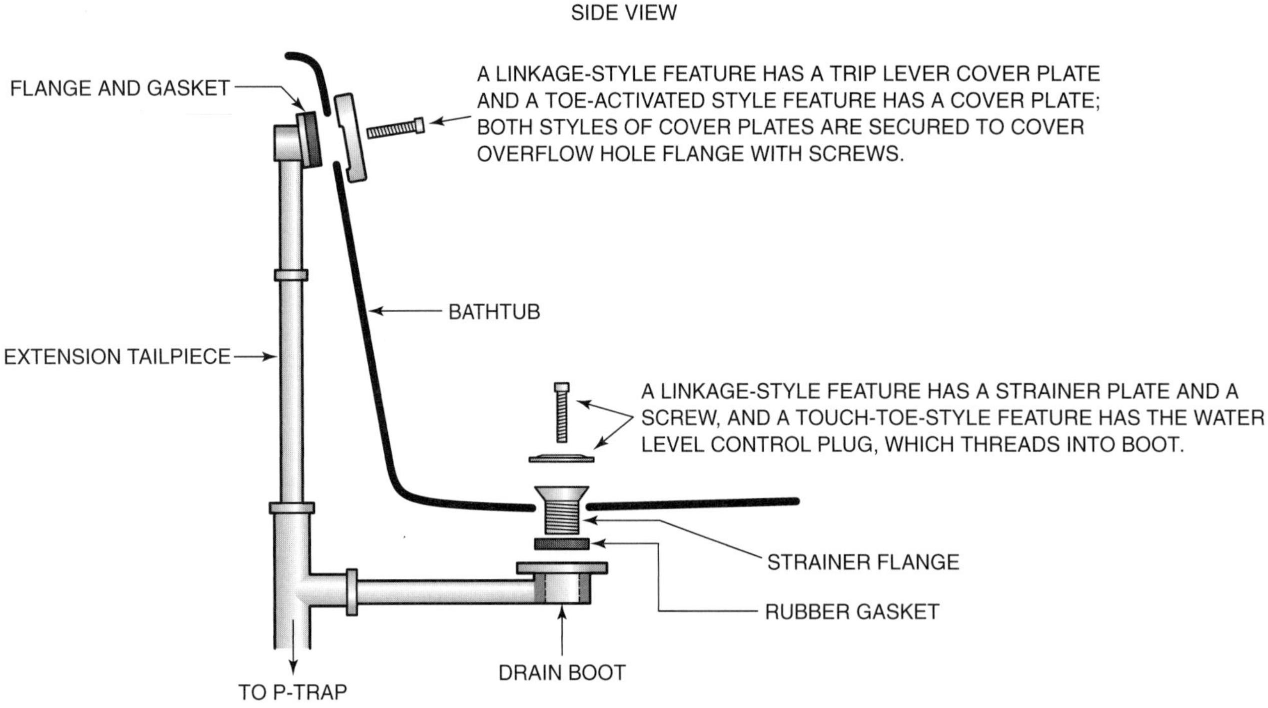

SIDE VIEW

FLANGE AND GASKET

A LINKAGE-STYLE FEATURE HAS A TRIP LEVER COVER PLATE AND A TOE-ACTIVATED STYLE FEATURE HAS A COVER PLATE; BOTH STYLES OF COVER PLATES ARE SECURED TO COVER OVERFLOW HOLE FLANGE WITH SCREWS.

BATHTUB

EXTENSION TAILPIECE

A LINKAGE-STYLE FEATURE HAS A STRAINER PLATE AND A SCREW, AND A TOUCH-TOE-STYLE FEATURE HAS THE WATER LEVEL CONTROL PLUG, WHICH THREADS INTO BOOT.

STRAINER FLANGE

RUBBER GASKET

TO P-TRAP

DRAIN BOOT

Figure 4-7 Bathtub waste and overflow showing a toe-operated drain plug.

BATH AND SHOWER

This unit can be a conventional bathtub and shower or a formed plastic tub and wall surrounds. These units also have to be installed in the rough-in stage, so substantial protection is required. One-piece units require large building openings and passageways to place the units in final position. The formed plastic-types must be installed in a recess that is within 1/4" of the required dimension to avoid distorting the assembled fixture. The base must be installed on a level floor with continuous support.

Use fine-tooth jigsaws or hole saws to cut the openings for the water controls and shower arm. To keep them secure and firm, the faucet and shower riser should be attached to the structure (see Figure 4-8).

Figure 4-8 Shower faucet without the trim plate.

Avoid making any solder joints next to the fixture. To prevent heat damage to the finish of the enclosure or tub, make up a sub-assembly on the bench and install it in the fixture.

WHIRLPOOL

The whirlpool tub should be installed with the same precautions used for conventional tubs. Never pick up the tub by the piping. Be sure that you do not disturb the circulating piping on the unit and be sure that the piping drains to the tub. Also, be sure that access is provided to the electric pump for service as required. Always water test the unit for proper function and leaks upon installation (see Figure 4-9).

SHOWER, FIELD FABRICATED

The field-fabricated shower consists of a tiled floor and tiled walls. Because water can work its way through the tile and mortar bed, it is necessary to install a waterproof membrane. The membrane can be made of sheet lead, copper, plastic, or composition

> **IN THE FIELD**
>
> Make sure to wear a dust mask and safety glasses when cutting fiberglass.

> **IN THE FIELD**
>
> The electric circuit to the pump must be protected with a ground-fault circuit interrupter breaker.

Figure 4-9 Whirlpool tub with rough-in piping. Work closely with the carpenters to make sure the mounting deck is the correct size.

material. According to the *International Plumbing Code,* sheet lead can not weigh less than 4 pounds per square foot (19.5 kg/m^2), coated with asphalt paint or other approved coating. The lead sheet must be insulated from conducting substances, other than the connecting drain, by 15–pound (6.80 kg) asphalt felt or its equivalent. Sheet lead must be joined by burning. The lead membrane must be covered with asphalt paint or the chemicals in the concrete will attack the lead. Plastic liners should be a minimum thickness of 40 mil (.040 inches). Be careful not to cover the weep holes in the drain body with the protective coating. Do not apply tar to any plastic membrane unless the manufacturer indicates that the tar will not affect the plastic. A plastic liner must be laid on a smooth surface so that it won't be punctured. Figure 4-10 shows a typical shower safety pan being installed over a wood subfloor.

The drain fitting is usually a three-piece assembly: one piece to fit above the tile, one piece to fit between the tile and the waterproof membrane, and one piece to fit below the membrane. There are weep holes in the middle section that allow any water that reaches the membrane to be directed back into the drain (see Figure 4-11).

Before placing the membrane, the shower area subfloor must be carefully cleaned. Any nail heads should be set down to the floor level. The membrane is cut to be 6" greater in all directions than the base. The corners are then folded so that the membrane extends up the sides at least 2" above the threshold. Provided the nails are only placed along the top edge of the material, the material can be nailed in place. After the drain and membrane are installed, the pan should be filled with water and tested for leaks. The inspector should verify the water level and be the witness to a watertight system. After the test is completed, loosely assemble the finished drain riser into place for final adjustment by the tile contractor. At this point, it is good practice to have the customer or building foreman sign that the shower installation is satisfactory. It is very important to protect the shower installation with substantial covering material so that it will not be damaged by subsequent operations. Table 4-3 is a list of tools and supplies required for bathtub, shower, and whirlpool installation.

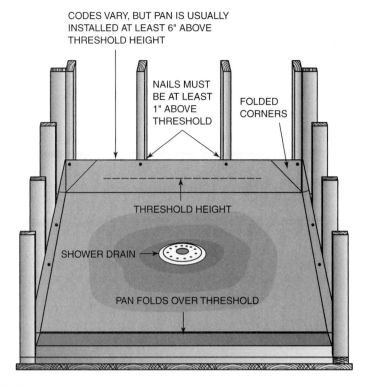

CODES VARY, BUT PAN IS USUALLY
INSTALLED AT LEAST 6" ABOVE
THRESHOLD HEIGHT

NAILS MUST
BE AT LEAST
1" ABOVE
THRESHOLD

FOLDED
CORNERS

THRESHOLD HEIGHT

SHOWER DRAIN

PAN FOLDS OVER THRESHOLD

Figure 4-10 The plumber must install the safety pan liner and drain and make certain it is water tight. Work closely with the tile installer for a smooth installation.

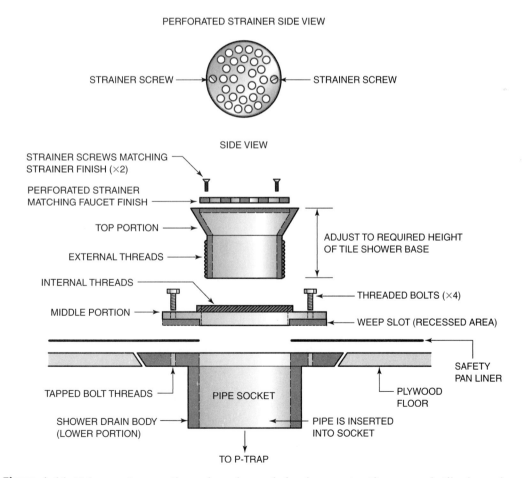

PERFORATED STRAINER SIDE VIEW

STRAINER SCREW   STRAINER SCREW

SIDE VIEW

STRAINER SCREWS MATCHING
STRAINER FINISH (×2)

PERFORATED STRAINER
MATCHING FAUCET FINISH

TOP PORTION

EXTERNAL THREADS

ADJUST TO REQUIRED HEIGHT
OF TILE SHOWER BASE

INTERNAL THREADS

THREADED BOLTS (×4)

MIDDLE PORTION

WEEP SLOT (RECESSED AREA)

SAFETY
PAN LINER

TAPPED BOLT THREADS

PIPE SOCKET

PLYWOOD
FLOOR

SHOWER DRAIN BODY
(LOWER PORTION)

PIPE IS INSERTED
INTO SOCKET

TO P-TRAP

Figure 4-11 Make sure to use a three-piece shower drain when constructing a ceramic tile shower base.

Table 4-3   Tools and Supplies for Bathtub and Shower Installations

Quantity	Tools/Supplies
1	Safety glasses
1	Medium flat head screwdriver
2	Small adjustable or appropriate size combination wrenches
1	Hacksaw
1	Tubing cutter
1	Level (2 ft. or 4 ft.)
1	Rag
1	Tube of caulk (color matched)
1	Can of pipe dope or Teflon tape
1	Tongue-n-groove pliers
1	PVC glue/ primer
1	2" P-trap and necessary adapters

SINK, KITCHEN

The kitchen sink can be either an undermount sink, which is usually installed by the countertop manufacturer, or a countertop sink. The plumber may be responsible for installing the above countertop mounts. If so, an opening must be cut in the countertop to receive the fixture. If available, use the template supplied with the fixture or the fixture itself to determine the cutout pattern. If the fixture is the template, cut 1/2" inside the mark to obtain the proper opening.

The faucet and basket strainer are installed using the same procedures as the lavatory. Once the faucet and basket strainer have been installed, the sink can be attached to the countertop. Apply a liberal amount of sealant between the sink and the countertop and tighten up the rim clamp screws (see Figure 4-12).

Use a damp rag to remove excess sealant for a professional looking job. Once water supply lines and drain lines are installed, test the assembly for leaks.

BIDET

The piping to a bidet is very similar to that of a lavatory; it typically uses two 3/8" supply lines and a 1-1/4" P-trap. The installation of the unit itself is like that of a water closet as it is bolted to the floor. Be careful not to mar polished trim pieces. Use special care to ensure that adequate backflow protection is provided. It may be an over-the-rim filler that provides the necessary air gap, a deck-mounted vacuum breaker, or a pipe-applied vacuum breaker. Because there are numerous designs, it is very important to follow the manufacturer's rough-in instructions. Do not over-tighten the fixture as it could break. Test the finished installation for proper operation and be sure there are no leaks.

LAUNDRY TRAY

The installation of a laundry sink or tray is very similar to that of a lavatory. The most common sink is one that is installed on four legs, either anchored to the floor or attached to the wall. The laundry tray should be installed level and firm for proper drainage. If the fixture can move or rock, the drain connection will eventually work loose. The faucet is usually a 4" spread faucet which may or may not have a hose-end spout. If a hose-end spout is used, some form of backflow protection other than an air gap must also be used (see Figure 4-13).

Table 4-4 Tools and Supplies for Kitchen Sink and Laundry Tray Installation

Quantity	Tools/Supplies
1	Safety glasses
1	Medium flat head screwdriver
2	Small adjustable or appropriate size combination wrenches
1	Hacksaw
1	Tubing cutter
1	Small level (torpedo or 2 ft.)
1	Rag
1	Tube of caulk (color matched)
1	Basin wrench
1	Tongue-n-groove pliers
1	PVC glue/primer
1	1 1/2" P-trap and necessary adapters
2	3/8" supply tubes
1	Can of pipe dope or Teflon tape
1	Jigsaw with fine tooth blade (if countertop is to be cut)

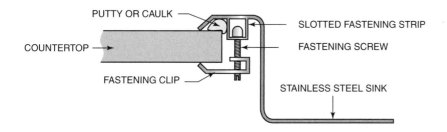

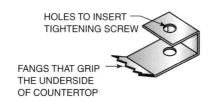

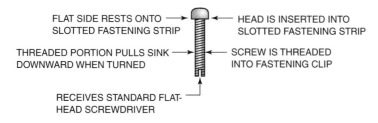

Figure 4-12 Fastening clips for a top-mounted stainless steel kitchen sink.

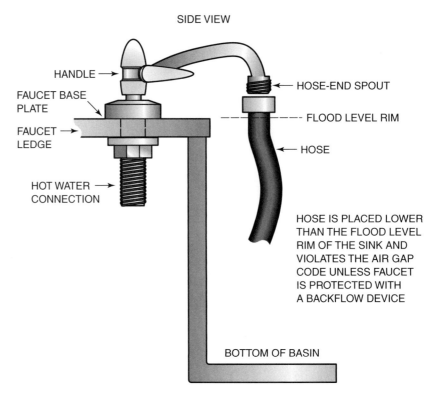

SIDE VIEW

HANDLE →

FAUCET BASE PLATE →

FAUCET LEDGE →

HOT WATER CONNECTION →

← HOSE-END SPOUT

---- FLOOD LEVEL RIM

← HOSE

HOSE IS PLACED LOWER THAN THE FLOOD LEVEL RIM OF THE SINK AND VIOLATES THE AIR GAP CODE UNLESS FAUCET IS PROTECTED WITH A BACKFLOW DEVICE

BOTTOM OF BASIN

Figure 4-13 Make sure that a vacuum breaker is installed on a faucet with a threaded spout.

The faucet is usually mounted on the back ledge of the sink, which is usually marked with knockouts for the plumber to remove prior to the faucet installation.

FOOD WASTE GRINDER

The food waste grinder is commonly called a *garbage disposal*. It is an electrically powered fixture that is installed to the underside of the sink. It comes with its own mounting bracket that replaces one of the typical basket strainers. Be sure the drain rough-in is low enough to accept the grinder discharge with a gravity waste line. The grinder should discharge into the sink trap, with the second compartment (when present) draining into the side branch of the continuous waste tee. If local codes allow the plumber to install the electrical connections, it is usually easier to do so before the disposal is installed under the sink. Be sure the unit is properly grounded. Like the basket strainer, the sink flange is installed from above, with a ring of plumber's putty compressed between it and the sink. A fiber gasket, backup collar, and snap ring pull down are installed on the sink flange from below. A rubber gasket seals the motor assembly and the mounting collar, while also acting as a vibration dampener. The motor assembly usually twists into place. Figure 4-14 shows how the garbage disposal attaches to the sink basin.

If the grinder is to receive the discharge of a dishwasher, be sure to remove the plug or blank in the food waste grinder housing. Ground-joint traps, rather than slip-joint types, provide better vibration resistance for this installation.

DISHWASHER

The dishwasher should be installed level and plumb and securely fastened in place (unless it is the portable type). Make certain that the fasteners used are not long enough to penetrate the surface of the countertop. When installing the supply line,

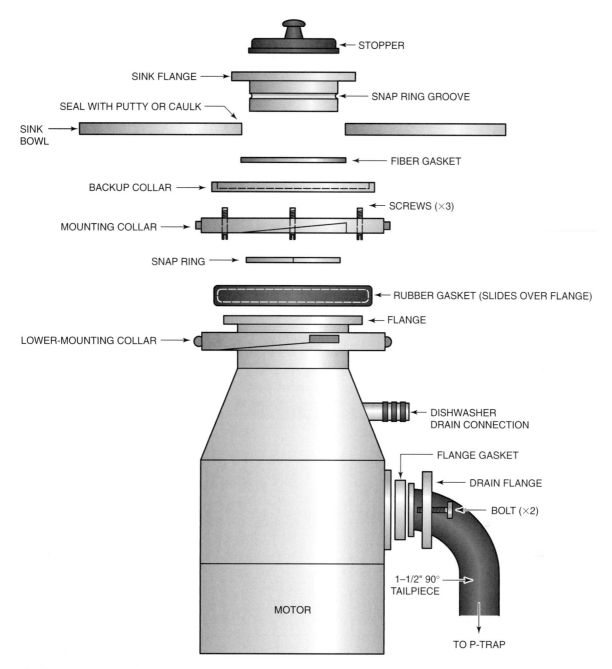

STOPPER

SINK FLANGE

SNAP RING GROOVE

SEAL WITH PUTTY OR CAULK

SINK BOWL

FIBER GASKET

BACKUP COLLAR

SCREWS (×3)

MOUNTING COLLAR

SNAP RING

RUBBER GASKET (SLIDES OVER FLANGE)

FLANGE

LOWER-MOUNTING COLLAR

DISHWASHER DRAIN CONNECTION

FLANGE GASKET

DRAIN FLANGE

BOLT (×2)

1–1/2" 90° TAILPIECE

MOTOR

TO P-TRAP

Figure 4-14 **Typical garbage disposal connection for a kitchen sink.**

do not make a solder joint at the dishwasher solenoid valve. The standard solenoid valve used by most manufacturers has a 3/8" female pipe thread. Normally a 90° brass elbow adapter is required, which has a 3/8" male pipe thread on one end and a 3/8" compression fitting on the other end. Flush the water supply line vigorously before making final connection to the dishwasher. It is good practice to install a water filter ahead of the dishwasher, especially in older buildings where scale is sometimes dislodged. The manufacturer usually builds water supply backflow protection into the unit.

Some local codes require an air gap fitting in the drain line. If so, the fitting is usually installed on the sink ledge or on the countertop. If your code does not mandate the air gap fitting, the discharge hose should be looped up and affixed to the underside of the counter before attaching to the food waste grinder connection. In this way, the sink would have to be nearly full to the flood level rim before sink wastes could flow back into the dishwasher (see Figure 4-15).

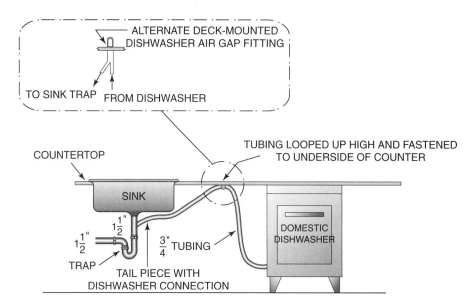

Figure 4-15 Typical connection for a residential dishwasher and sink. Check local codes for compliance.
Courtesy of Plumbing-Heating-Cooling Contractors–National Association.

Table 4-5   Tools and Supplies for Food Waste Grinder and Dishwasher Installation

Quantity	Tools/Supplies
1	Safety glasses
1	Medium flat head screwdriver
2	Small adjustable or appropriate size combination wrenches
1	Hacksaw
1	Tubing cutter
1	Can of pipe dope
1	Rag
1	Tube of caulk (color matched)
1	Basin wrench
1	Tongue-n-groove pliers
1	Voltmeter, wire nut, electrical tape (if allowed to make connections)
1	1 1/2" P-trap and necessary adapters

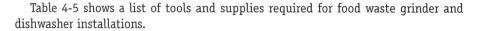

Table 4-5 shows a list of tools and supplies required for food waste grinder and dishwasher installations.

AUTOMATIC CLOTHES WASHER

The automatic clothes washer is not set and installed by the plumber in the same sense as the other fixtures described. The plumber normally provides a drain opening and hot and cold water supplies so the appliance can be connected by anyone.

A common device for new work is a receptor box which includes the incoming water connections and the waste opening. A ball valve-type washing machine assembly valve is an improved alternate to individual stop faucets. The box is roughed-in and a decorative frame is placed as a finish operation. Be sure that the standpipe length to the trap is from 18"–48" so that satisfactory operation is ensured. The standpipe and trap size should be 2" (see Figure 4-16).

Figure 4-16 New center mount washing machine box. This allows more space between wall studs.

No water supply backflow protection is required as the manufacturer builds it into the machine. The drain line from the machine is connected through an air break.

CAUTION: Because newer front-loading washing machines discharge water at such a high rate, it is important to use the drain hose clamp provided by the manufacturer. Otherwise, the hose can be pushed out of the drain box.

COMMERCIAL AND INDUSTRIAL INSTALLATION

Fixture installations in commercial or industrial buildings must be made as sturdy as possible. Carrier fittings and flushometer valves are frequently used in these non-residential applications. As with residential jobs, be sure that the fixtures are as specified for the job and check to see that you have the necessary accessories and tools to complete the job.

Carrier installation is a rough-in operation, but the essentials will be discussed here. Carriers are used to support the weight of the fixture without requiring any help from the wall itself (except that residential carriers do require wall support). Carriers for water closets and urinals also incorporate the drainage piping.

Figure 4-17 Typical water closet carrier.
Courtesy of Zurn Industries, LLC.

Water Closet

Toilet rooms with carrier-supported water closets are easier to rough-in because most of the work is above the floor, as only the waste stack penetrates the floor. The finished restroom is also easier to keep clean when the fixtures are wall-hung. Closet carriers can be used for siphon jet or blowout bowls. For most manufacturers, the structural part of either carrier is a common product—the only item that changes is the face plate that contains the bowl connection and fixture support bolts. Carriers are made for single fixture, back-to-back or battery installations, with either vertical or horizontal (left or right) drainage (see Figure 4-17).

To determine whether a fitting is described as left or right, face the stack wall in the position where the fixture is to set. Closets to the right of the stack use right-hand fittings. The same system can be used to order the stack discharge fitting. A vent connection is always provided for in these fittings.

Further variations include *short barrel* or *long barrel* discharge pipes and various opening options for different venting or fixture drain entries. Short barrel means that the discharge pipe is just a few inches long while long barrel means that the discharge pipe is about 3' long. Figure 4-18 shows an adjustable siphon jet water closet carrier.

The carrier fitting must be installed at the proper height to ensure that the water closet height will be correct. The face plate that attaches to the carrier is adjustable, so a single carrier does not involve critical placement. If a battery is to be installed, then the planning must be done more carefully. Figure 4-18 shows a preassembled battery of carriers.

The stack fitting is placed low enough so that the horizontal run to the battery carriers will have proper slope and all the carrier openings are within the adjusting range of the face plates. In this way, the final installation will have all the bowls at the required height (usually the same), and the piping will slope to provide proper drainage. The carrier face plates and feet are then installed and fastened into place.

Carriers for blowout bowls are mounted higher than those for siphon jet bowls because of the differences in the bowl discharge location. Blowout face plates are set up for three mounting bolts, whereas siphon jet plates provide four bolts. While these fittings are usually considered for commercial and institutional use, they are becoming popular in residential applications as well.

Figure 4-18 When fixtures are installed several in a row, a prefabricated carrier bank can be a great time-saver.
Courtesy of Marlin Mechanical Corporation, Phoenix, AZ.

IN THE FIELD

Some manufacturers are making prefabricated carrier assembles that greatly reduce the amount of on-site assembly. Refer back to Figure 4-19.

URINAL

Urinal carriers incorporate the urinal attachment and drain fitting on a framework that is attached to the floor (see Figure 4-19).

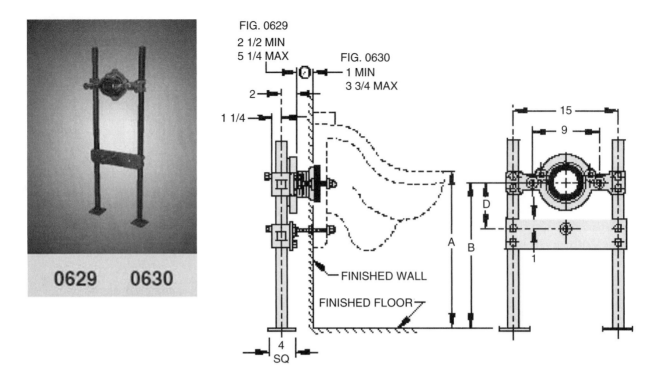

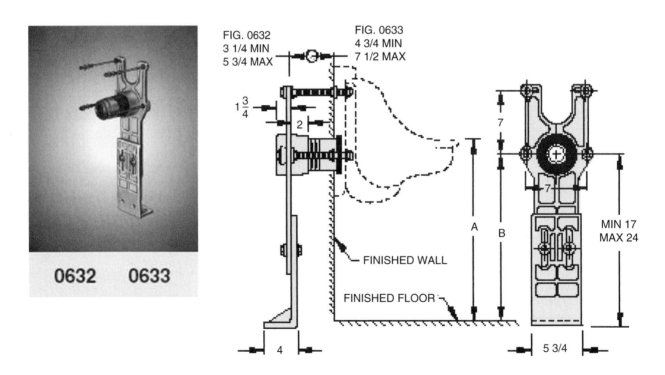

Figure 4-19 Carriers come in a variety of sizes and options. Always make sure all components are present before beginning the installation. Courtesy of Jay R. Smith Mfg Co.®

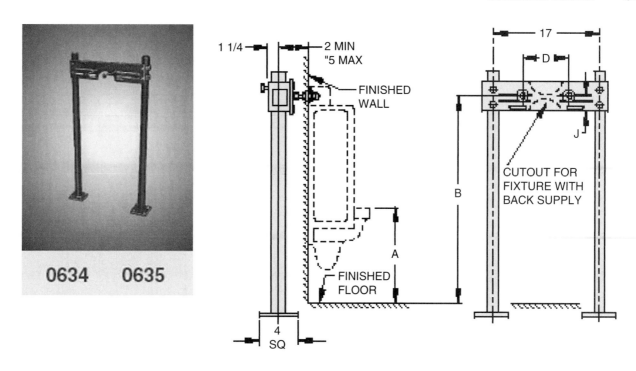

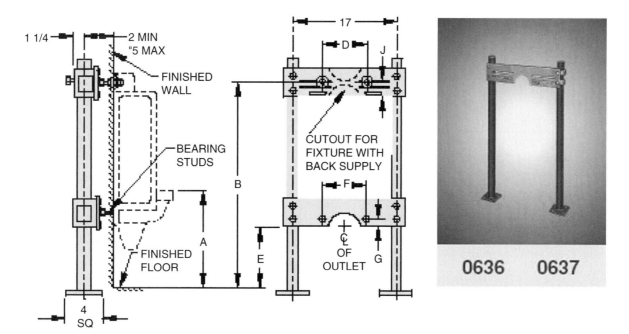

Figure 4-19 (Continued)

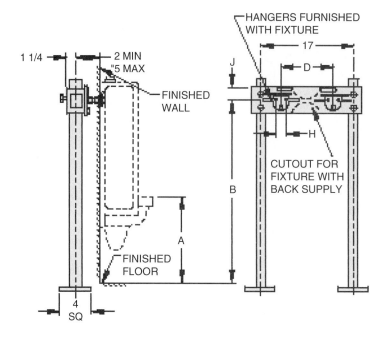

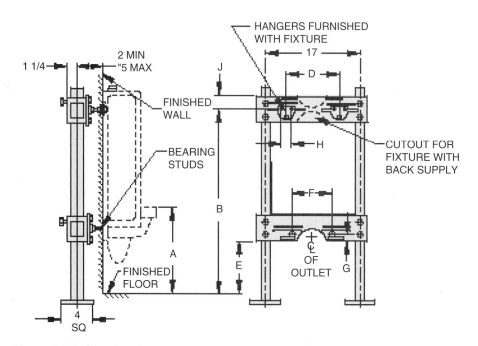

Figure 4-19 (Continued)

The top panel supports the top of the urinal. Setting wall-hung water closets or urinals requires that you adjust the waste nipple, gasket, and mounting studs to the thickness of the finished wall, set the fixture, and tighten the bolts. Then make the water connection and test the operation.

LAVATORY

Lavatory carriers do not incorporate the drain opening. They consist of a substantial frame in the wall and heavy arms that support the lavatory. Some lavatories contain the arms within the fixture so the carrier is said to be concealed. Exposed carriers consist of arms below the lavatory and extend approximately to the front of the fixture. The lavatory is attached to the arms for a solid installation. The lavatory carrier is attached to the floor, unless the wall structure is substantial enough to support the lavatory. In this case, the carrier consists only of a steel plate imbedded in the wall (usually masonry) which supports the arms for the lavatory. Figure 4-20 shows a wall-mounted lavatory carrier.

Figure 4-20 Wall-mounted carrier for lavatory. Courtesy of Zurn Industries, LLC.

KITCHEN EQUIPMENT

Kitchen equipment should be assembled levelled, and plumbed. Multiple compartment sinks used for food preparation must be indirect-wasted to the building drainage system. Bar sinks and culinary sinks should also be indirectly connected to the waste system. Grease interceptors or grease recovery devices are required for waste lines that have a heavy grease load. These units must be accessible for service. Flow controls may be required to prevent excessive flow rates that would break up the materials previously intercepted by the device.

SERVICE SINKS

Service sinks can be supported by a special trap called a *trap standard* or a *carrier*. Some carriers, such as the Zurn Z1218, can be used for both a urinal or a wall-mounted service sink. Floor sinks may be supported by mortar or concrete. The faucet serving such items must be solidly attached, especially if a bucket hook is part of the faucet spout.

DRINKING FOUNTAIN AND WATER COOLER

These fixtures are installed with the top surface level. Wall-hung models require substantial fasteners and wall elements to support the mounting bracket(s).

IN THE FIELD
Drinking fountains or water coolers may not be installed in restrooms.

WALL HYDRANTS

Non-freeze wall hydrants are installed where freeze protection is required. The actual shut-off mechanism is located inside the wall and the outlet end of the device is sloped to drain so that no water is present in the area where freezing temperatures can exist. Hoses cannot remain connected to the threaded outlet of older models because the automatic draining feature may be thwarted. More recent designs, however, will drain safely even if hoses are not removed. To minimize the exposure to vandalism, a loose key cover and loose key valve operator can be provided (see Figure 4-21).

If the protection is not built into the wall hydrant, always install a vacuum breaker on the hose outlet.

Figure 4-21 Recessed wall hydrant. Notice that a special key is required to open the fixture.

INDUSTRIAL DEVICES

Industrial devices include floor drains, roof drains, interceptors, and emergency aid fixtures.

Floor and Roof Drains

Floor and roof drains are installed according to manufacturer's instructions. Both drains are designed to receive water from either wash down or storms and introduce

Figure 4-22 Emergency stations should be clearly marked and readily accessible.

it into either a storm or combination drainage system. For a properly operating drain, the rough-in must be performed correctly, which means that the drain must be level or slightly recessed below the finished surface. While it is not always possible, try to install these devices in the low points of the floor or roof. Roof drains are covered in greater detail in Chapter 24.

Interceptors

Interceptors must be installed with the requirements of the job and the limitations of the interceptor in mind. The interceptor must be accessible for service with a moderate slope on the inlet piping. If necessary, equip the interceptor with flow control devices to prevent excessive velocity and inner turbulence. Set the interceptor on a firm base so the piping will not shift or settle. Vent the interceptor so it will receive waste freely and permit it to flow out readily.

Emergency Eye and Body Shower

These fixtures are installed wherever exposure to hazardous chemicals is possible. A large handle is installed within reach so that an affected person can operate the device easily. These fixtures must be solidly supported, as the injured person may act violently. A floor drain should be provided adjacent to the showers to convey the water away from the fixture. Give thought to the requirements of these safety devices. Guard the operation of the showers as much as possible from pranksters while providing the required protection for persons in the area. Figure 4-22 shows a typical emergency eye and body wash station.

INSTITUTIONAL INSTALLATION

Finish details that are special to institutional buildings pertain to handicap fixtures, medical applications, and vandal resistant installations; the related details that are required to obtain proper operation are discussed in the remainder of this chapter.

HANDICAP FIXTURES

Water Closet

Water closets are installed like those in conventional jobs except that the bowl must be solidly attached to the floor (or wall). If any rocking is possible, the bowl could break from the higher loads likely to be imposed on it. Rough-in heights may vary depending on the intended user. For example, water closets installed in daycare centers are intended for smaller children, so the bowl height may be around 9-1/2" to 13-1/2" above the floor while water closets intended for persons with disabilities are 17"–19" to the seat height, preferably at the low end (17"). Table 4-6, taken from the 2006 *National Standard Plumbing Code*, shows some allowable heights based on usage.

Table 4-6   Bowl Heights Based on Usage

Application	Age-Years	Height (above finished floor)
Children's use	5 and younger	9-1/2" to 10-1/2" to rim
Juvenile's (children) use	6–12	10-1/2" to 13-1/2" to rim
Accessible children's use	12 and younger	11" to 17" to top of seat
Accessible adult's use	13 and older	17"–19" to top of seat

Grab Bars

Grab bars are required in conjunction with the water closet, either mounted on the bowl seat or on the adjacent walls. The possible load on these bars is very high, so be sure they are attached to substantial backboards or other supports. The grab bars must be capable of sustaining a 300 pound load. Figure 4-23 shows a typical handicap stall with water closet and handrails.

Urinal

Urinals are also conventional, but they must be installed so that the lip of the fixture is not more than 14" above the floor. Extended lip models are usually selected for these applications, but should be located where they will not interfere with the walking pattern in the battery of fixtures.

Lavatory

Figure 4-23 A typical handicap toilet stall. Notice the extra room for wheelchair access.

Lavatories are installed high enough (about 1" higher than normal) to provide clearance below the fixture for wheelchair clearance. The lavatories are usually supported by carrier devices that are either concealed or bottom-mounted. So that a person without sensation in the legs will not be bruised or burned by contact with the piping, it is important to keep piping below the fixture, as far to the rear as possible, and well insulated. Sensor activated faucets provide for the greatest ease of fixture use.

Drinking Fountains

Drinking fountains are installed with the bubbler about 30" above the floor. The style is wall-mounted with an extension to provide wheelchair clearance. The fountain is operated with a pressure plate, lever handle, or sensor to control water flow.

Bathtub and Shower

Bathtubs and showers are conventional except that grab bars are required. The preparation and rough-in for these items has to be completed in the rough-in phase of the work. When the finish work is done, caulking should be used behind faucet trim, spout, and grab bar mountings to ensure that water cannot get into the wall.

MEDICAL INSTITUTION FIXTURES

In general, medical applications involve many different systems. Be sure to study the plans, specifications, and rough-in very carefully and be certain you apply the proper devices to the proper piping (see Figure 4-24).

Water Closet

Water closets may be installed on conventional rough-ins with conventional floor-mounted bowls or they may be handicap-style with a rim-height 3"–4" above the usual level. Wall-hung bowls may be set for conventional height or handicap height. The major modifications of the bowl are the presence of bedpan lugs or the incorporation of a bedpan washer connection on the flushometer. Backing support is required in the wall and knee or foot controls are often used. Vacuum breaker protection is included in the water supply. Take extra care not to scratch polished metal finishes or the fixtures themselves in any installation, but especially with medical applications; such scratches can harbor germ colonies.

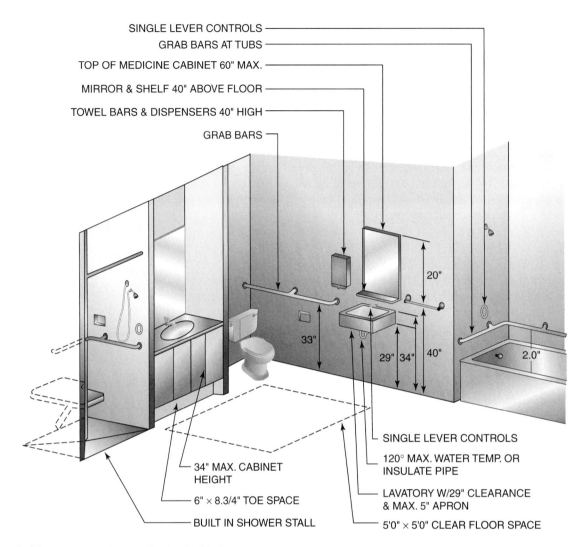

SINGLE LEVER CONTROLS
GRAB BARS AT TUBS
TOP OF MEDICINE CABINET 60" MAX.
MIRROR & SHELF 40" ABOVE FLOOR
TOWEL BARS & DISPENSERS 40" HIGH
GRAB BARS

20"
33"
29" 34" 40"
2.0"

34" MAX. CABINET HEIGHT
6" × 8.3/4" TOE SPACE
BUILT IN SHOWER STALL

SINGLE LEVER CONTROLS
120° MAX. WATER TEMP. OR INSULATE PIPE
LAVATORY W/29" CLEARANCE & MAX. 5" APRON
5'0" × 5'0" CLEAR FLOOR SPACE

Figure 4-24 Layout for the physically disabled.

Sitz Bath

Sitz bath trim is mounted on the fixture before it is set. After mounting on carrier arms, adjust the locking feature so the fixture is level, plumb, and rigidly attached. A thermostatic control or pressure-balanced valve is recommended to minimize the risk of scalding. Cover the fixture with substantial covering, such as a plastic protector or several layers of corrugated paper board, to protect the surface until the building is ready for final acceptance.

Bathtub

Institutional bathtubs are larger and deeper than residential fixtures and are installed for access from three sides. The tubs are installed in an elevated position so that hospital personnel can more easily handle the patient. Be sure to have enough manpower and equipment to lift the tub into position. Without adequate help, someone could be hurt or the fixture could be damaged. After the tub is set and trimmed out, cover the surfaces with substantial protection to ensure that the fixture is not marred during subsequent building operations. Store all handles, spouts, etc., in a marked box until needed at the time of final trim operations.

Whirlpool Bath

Whirlpool baths are available in a variety of sizes and for different applications. Study the manufacturer's instructions carefully for guidance in the required rough-in and finish methods to be used. Backflow protection must be provided. Full body whirlpool baths require special attention to support because they are large and can hold a large amount of water. If the tub is not properly and continuously supported, the weight of water could damage the fixture or building. Consult the manufacturer's recommendations for support. These recommendations vary with manufacturers and tub capacities. As with other tubs that have to be installed during rough-in, protect the finish with a covering that will reduce the chance of surface damage during the subsequent building operations.

Sink

Sinks in medical facilities must be solidly supported, preferably on a carrier or in a countertop. Knee or foot control faucets are frequently used for maximum cleanliness for the user. Interceptors should be used in situations where solids could enter the drain from the sink. The interceptor must be accessible for servicing.

VANDAL RESISTANT FIXTURES

Vandal resistant fixtures are desirable where large numbers of transient people may be present (schools, public buildings, places of assembly) or where the building occupants may be incompetent or antisocial (asylums, prisons, detention facilities).

Water Closet

Water closets for minimum-security buildings are vitreous china wall-hung with push-button flushometer valves. The wall-hung bowl can't be rocked loose like a floor-mounted bowl can, but the vitreous china fixture would not be secure enough in the living quarters of mental patients or prison inmates. The maximum-security water closet is wall-mounted, but the supporting bolts and nuts are in the pipe chase, not on the face of the bowl. The total fixture and mounting assembly is made of indestructible material (e.g., stainless steel). All piping is in the pipe chase and only the flushometer valve trip button is in the inmate's space. All piping and connections are inaccessible to the inmate and the fixture is ruggedly attached to the building. Water and waste connections are conventional in type; only the location is special for the prison application.

Shower

Showers may be arranged for one person only or several people at one time. The individual shower unit is set in a wall with all piping in a chase behind the wall. Shower columns (for up to six persons) are factory-assembled with water and soap connections, shower heads, water control valves, and drain provided. These fixtures are slender, vertical units that are frequently attached top and bottom to the building. They permit a clear view of the showering area for monitoring. In the case of showers especially, care must be taken to be sure that no protrusion on the column can be removed to be used as a weapon or used to support suicide. These facilities are built for troubled persons and to guard against any tragedies, the supervisors or other experienced individuals must be consulted for the best details for those installations.

REVIEW QUESTIONS

❶ Why is the bathtub the most likely fixture to be damaged?

❷ What is the minimum clearance required in front of a water closet?

❸ Give a disadvantage to installing the tank onto the bowl before installing a water closet onto the closet flange.

❹ True/False   A dishwasher has built-in backflow protection.

❺ True/False   An automatic clothes washer has built-in backflow protection.

Fixture Fittings and Trim

The student will:

- Describe the considerations required when selecting fixture fittings and trim for residential, commercial, industrial, and institutional installations.
- Identify the nomenclature and types of such trim.

RESIDENTIAL AND COMMERCIAL FIXTURE FITTINGS AND TRIM

When a plumber installs plumbing fixtures such as bathtubs or sinks, the installation is not complete until the fixture fittings and trim are installed. In many cases the general term *fitting* is used to describe any part of the plumbing system other than the pipe, elbow, and tees. For example, the term *fixture fittings* usually refers to products such as mixing valves, faucets, or other items that are needed for the operation of a particular plumbing device. The term *trim* is often used to refer to the items that are exposed or fill in the rough openings (e.g., traps or escutcheons). Trim may also be used loosely to describe all items accessory to the fixture.

When ordering fixture trim, you must consider and determine the following information:

- Quantity
- Type
- Style of the items
- Stock and/or model number or complete description of the items

It is a time-saving technique to keep a personal record of stock numbers or model numbers of items that are used regularly. It is also important to learn the meaning of terms such as *centerset faucet*, *8" faucet* (same as widespread), *basket strainer*, *waste-and-overflow*, *diverter*, *single-lever faucets*, and many, many more.

When an order is received, carefully check the shipment against the packing slip and against the original order. If there is any discrepancy, check with a supervisor at once. It is possible that a late change has been made that you should know about.

Water Closet

As described in Chapters 2 and 3, gravity flush tanks or flushometer flush tanks can be used in water closets for residential and commercial jobs. The *supply stop*—straight or angle—is the trim item that is used to control the supply water flow to the tank-type water closet. When ordering supply stops, the method used to open and close the valve must be decided. These options are loose key, screwdriver slot, or handle type. The environment usually dictates this choice. Handle type is mostly used when there is little chance of vandalism, while the others are used in more public or less controlled instances. The other option to consider is how it will be connected to the supply line. Typical choices are sweat or compression for copper tube, solvent cement, PEX connected, or pipe-thread supply nipple. Always make sure the escutcheon is installed before you install the angle stop or straight stop. Figure 5-1 shows an angled supply stop that was sweat to a copper stub out.

The connection from supply stop to tank is usually a compression joint on a plastic or chrome-plated copper tube. Plastic tubes with a braided stainless steel outer cover are also used. Brass ferrules are used for copper tube supplies and plastic ferrules are used with plastic tubes.

Flushometer valve closets are trimmed with an angle stop, adjustable tailpiece, flushometer valve, and vacuum breaker tailpiece. Polished escutcheons and polished supply nipple complete the trim package. Flushometer valves are available in a variety of finishes, with chrome being the most popular. The *angle stop* may be either handle- or screwdriver-type; the flushometer may be operated by a handle, push-button, sensor, or floor pedal. Figure 5-2 shows a sensor-operated flushometer valve with a screwdriver-type angle stop.

The connection to the bowl is made to a tapered brass spud, held fast in the top (or top rear) of the bowl by a gasket and locknut. The locknut pulls the tapered

Figure 5-1 Angled supply stop that was sweat to a copper stub out. Metal escutcheons have to be used due to the heat.

Figure 5-2 Sensor operated flushometer valve with a screwdriver type angle stop (cover removed for picture).

spud and gasket against the opening in the bowl, thus locking it in place. The gasket is a friction and interference fit; therefore, no lubricant is used on the washer-to-fixture seating surfaces.

Note that federal law now requires all residential tank-type water closets manufactured as of January 1, 1994, to be 1.6 gpf. Commercial units must be 1.6 gpf as of January 1, 1997.

Urinal

Urinals are operated by flushometer valves or high-flush tanks. Flush tanks are regulated by automatic means to minimize water usage. Water usage is generally in the 1.0–1.5 gallons per flush range. However, there are some newer models on the market that use 1/8 gallon per flush.

The flushing device is connected to the urinal by a spud, gasket, and nut, similar to the device used for water closets.

Bidet

Bidets are trimmed either with mixing faucets (preferably temperature controlled), or with a single-hole pop-up waste stopper control, with the wall-mounted faucet. The faucet may be finished in various ways (chrome or gold, polished or brush) and equipped with vacuum breaker, matching finish tube pipe covers, two handle controls, and inlet spuds. New style deck-mounted vacuum breakers are now available, with the critical level only 1" above the flood level rim of the bidet.

Lavatory

Lavatories are available in the greatest range of styles and varieties of trim of any fixture. Federal law requires that faucets manufactured after January 1, 1994, have a maximum flow rate of 2.5 gpm. Faucets include the following:

- 4" centerset (4" between hot and cold connections) with delivery spout in the center
- 8" widespread, where the control elements and spout are interconnected below the lavatory
- 12" widespread, where control valves and spout are also interconnected below the lavatory top
- Single bibb faucets, where individual hot and cold faucets with individual spouts are installed

Control variations are single lever, dual handle, self-closing, self-metering, and electronic sensor. The *self-closing faucet* is a single bibb faucet that closes after the handle (or push button) is released. *Self-metering faucets* are similar, except that the flow stops after a time delay, after the handle or push button is released.

Mounting styles include *deck mount* (faucet is installed on horizontal surface of fixture), *slant back* (faucet is mounted on a 45° angle panel on the fixture), and *shelf back* (faucets are mounted on the front of the rear vertical fixture face). Handles are made in many styles, such as single blade and cross handles, and many decorator types (usually made of acrylic plastic). Other options include shampoo hoses, special aerators, special drain assemblies, and ornate finishes.

Many manufacturers are currently offering a line of matching bathroom accessories such as towel holders, toilet paper holders, and water closet handles to match in color and style.

Lavatory supply stops and supply tubes complete the lavatory installation. In cases where the drain piping is exposed, a chrome or brass finish may be desired; if so, the installer should make every effort to protect the finish while performing the installation. For non-residential use, stops should be loose key or screwdriver-type.

Figure 5-3 Shower valve with rotation limiter.

Bathtub

Bathtub trim includes mixing valves for the incoming water and waste controls for the tub drain. NOTE: Since the bathtub waste and overflow are installed during the tub rough-in, you will find that addressed in Chapter 4 of this book. Water control options include single lever and two-handle valves. According to the 2006 *National Standard Plumbing Code,* "All showers and bath/shower combinations shall be protected with individual balanced pressure, thermostatic, or combination automatic compensating valves that comply with ASSE 1016 or ASME 112.18.1/CSA B125.1."

Such controls respond either to the leaving water temperature and reset the valves to attempt to maintain a constant temperature output; or, they respond to pressure changes at the inlet to the bathtub control valve and reset the valve mechanism to maintain a constant pressure ratio between hot and cold inlets. If a combination tub/shower arrangement is installed, a diverter spout is installed to direct the water to the tub spout or to the shower at the user's option. Each type valve helps to reduce the risk of serious injury, but only after a brief period of flow during which the change in flow conditions is sensed and operating elements within the valve reposition to compensate for either temperature or pressure change. Figure 5-3 shows a *rotation limiter* that comes standard on most faucets. Its purpose is to limit the amount of hot water so that the maximum temperature reached is 120° F.

Very large tubs may require more than one set of valves or a larger valve and spout to be able to fill the tub in a reasonably short time. Spouts may have a finished appearance of ceramic, chrome, brass, or gold. They may be provided with a lift gate to divert water to a showerhead and they may also include a threaded opening to receive a hand-held shower assembly. The hand-held shower option requires a vacuum breaker on the showerhead arm. Showerheads are available in models that deliver a fixed, adjustable, or pulsating spray pattern. The showerhead may be mounted on a ball and socket joint so that the direction of discharge is adjustable. The escutcheon, shower arm, and showerhead should be finished to match the tub spout and faucet handle appearance. The visible parts of the tub waste and overflow fittings should match the finished appearance of the faucets and spouts.

Shower

Showers use trim that is similar to that for tubs except that no tub spout is required or utilized. In many cases, it is the same valve body with the bottom discharge port plugged. Temperature control or balanced-pressure valves are required as described above to reduce the risk of scalding or thermal shock.

Further similarity with the tub-shower application is the choice of finishes for exposed faucet handles, shower arms, and showerheads. Please note that federal law now requires that water-conserving showerheads do not exceed 2.5 gpm flow rate. Sometimes more than one shower valve or head may be installed in certain applications. In those cases, consult your local code for specific requirements.

Sink

Culinary

Culinary sinks are usually equipped with at least one 8" (center-to-center) faucet, hose spray, swing spout with aerator, and 3 1/2" basket strainer. A culinary sink can be a three or four compartment sink. The water control for the faucet may be two handle or single handle. Figure 5-4 shows a popular residential fixture that combines the hand-held sprayer and the traditional spout.

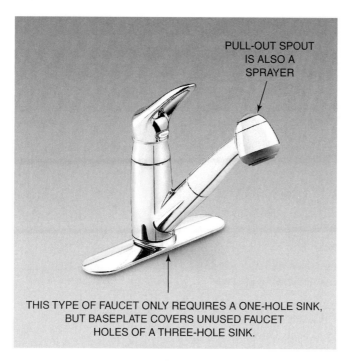

PULL-OUT SPOUT
IS ALSO A
SPRAYER

THIS TYPE OF FAUCET ONLY REQUIRES A ONE-HOLE SINK,
BUT BASEPLATE COVERS UNUSED FAUCET
HOLES OF A THREE-HOLE SINK.

Figure 5-4 Pull-out spout for residential kitchen sinks.

Bar

Bar sinks are equipped with a 4" center, gooseneck or other bar-type spout faucet, and a 1-1/2" or 2" waste outlet. Commercial applications may be equipped with a pop-up drain fitting with the control mounted above the sink rim. Bar sinks may require foot-operated valves in commercial applications.

Service

These faucets are mounted to the vertical back of the service sink or to the wall above the back. So that the arrangement can support the weight of a pail of water, a bucket hook is forged on the spout and a support rod is attached to the wall above. Usual construction materials are brass for faucets and brass or die-cast materials for drains. Stainless steel snap-on rim guards are usually provided for service sink rims.

Dishwasher

Dishwashers in residential locations may be equipped with air gap fittings in the discharge pipe. If installed, this device should be mounted on the sink or countertop. An alternate method is known as a *high loop,* covered in Chapter 4 under installation methods. Some areas require these units, while other areas prohibit them. Check your local code for guidance. Commercial dishwashers are connected indirectly. An air gap fitting is supplied integrally in the unit in order to protect the potable water supply.

Clothes Washer

Clothes washers are usually connected at a special washing machine rough-in box that includes hot and cold hose faucets and/or 2" standpipe opening for the drain. For a pleasing completed appearance, this box is provided with a cover frame trim. In cases where water hammer is a problem, boxes with built-in water hammer arrestors are available (see Figure 5-5).

Figure 5-5 Washer box with built-in water hammer arrestors. This box is mounted on either side of the wall stud.
Photo courtesy of Sioux Chief Mfg. Co.

INDUSTRIAL AND INSTITUTIONAL FIXTURE FITTINGS AND TRIM

When choosing trim for industrial and institutional fixtures, fashion is typically less important than cost, low maintenance, and simplistic design. The installer wants to use trim that is good quality (at a low cost) and can be quickly installed while the owner wants something that will be heavy-duty construction for long life, simple to clean, and vandal resistant (more so depending on the particular application).

Industrial

In the previous chapters, various considerations for industrial trim items have been discussed with the exception of the safety eye washes and emergency body showers. The choice of trim for these units is greatly affected by the location they're in and type of environment. In an environment like a school lab or manufacturing clean room, more attention will be paid to the effect of discharge water on surrounding surfaces whereas one installed in a chemical factory production area may be exposed to a harsher environment and, therefore, would require a more durable finish. Other options available that might be considered include the following:

- Floor or wall mounting
- Regular or frost-proof design
- Horizontal or vertical supply connection
- Various surface finishes
- Alarm devices linked to the shower-operating mechanism
- Fine-screen strainers to catch contact lenses

Institutional

Institutional trim is available for handicap, medical, and vandal-resistant applications.

Handicap

Handicap trim has three basic requirements: accessibility, ease of operation, and rugged construction and installation. Thus, faucets use lever or blade-handles within close reach and single-lever faucets or automatic-sensor designs are favored. Substantial grab bars are also provided. Piping and other trim below lavatories are insulated for thermal burn and bruise prevention. Always make sure you are current on local code requirements.

Medical

Medical applications call for fixture fittings that are easy to use and made for long life. Sanitary installations are important for these fixtures, both for patient and staff safety, so blade-handle faucets and mountings free of crevices and recesses are used.

Special items peculiar to medical applications include the following:

1. *Bedpan washers* consisting of a hose and control device, spray nozzle, vacuum breaker, and wall hook. Usually they consist of brass parts with heavy chrome plating on visible components.
2. Flushometer valves equipped with a bedpan washer tailpiece and a soap dispenser. The washer accessory is pivoted out of the way when not in use.
3. Lavatory faucets equipped with blade-handles and lavatory drains that have grid strainer assemblies.
4. Surgeon sink faucets equipped with gooseneck spouts and knee or foot controls.
5. Aspirator devices used in surgery rooms and morgues, consisting of a supply valve, vacuum breaker, aspirator nozzle and ejector fitting, hose or tubing, tubing receptor with air gap, and the tailpiece. When water is passed through the aspirator, a vacuum is produced that is used to withdraw fluids from the body. The extracted fluids and the flowing water are discharged into the air gap fitting.
6. Bath trim, shower trim, and fixture traps are similar to these devices for other applications except that heavy duty, heavily plated products are used. Bath and shower trim should be thermostatically controlled, pressure-balance, or combination-type valves.

Maximum-Security Institutions

Maximum-security installations require fixture fittings that are inaccessible to the inmate except to initiate the operation of the device. For example, push-button controls will trip a flushometer valve and begin the flow of a self-metering lavatory faucet or drinking fountain valve.

The flushometers or faucets are *non-hold-open types*, i.e., once the button is pushed only a certain amount of water will flow before the water flow is stopped. This operating mode is **required** so that the inmate cannot flood the fixture by blocking in the push-button with a broomstick or similar device. In addition, flushometers must be equipped with vacuum breakers. Lavatory faucets may be single or dual control. The lavatory drain should be either a grid strainer or a non-removable stopper.

Maximum-security showers use push-button controls and vandal resistant, suicide-proof showerheads. Single units are arranged for wall mounting. Multi-use showers are made up in column assemblies for up to six persons at one time.

REVIEW QUESTIONS

❶ What is an angle stop?

❷ Give a location that might warrant installing a screwdriver-operated angle stop.

❸ Shower discharge is limited to a maximum of _____ gpm?

❹ What is an advantage to using a sensor-operated faucet on a lavatory?

❺ Give an example of when a vacuum breaker may be required on a bathtub.

CHAPTER

6

Trim Installation

LEARNING OBJECTIVES

The student will:

⊗ Describe the method used to install fixture trim for residential, commercial, industrial, and institutional fixtures.

⊗ List effective ways to clean fixture trim.

RESIDENTIAL AND COMMERCIAL TRIM INSTALLATION

In general, to achieve a satisfactorily completed project, the fixture trim must be appropriate to the application and must be installed properly. This will ensure that the fixture will function well, require minimal maintenance, and have the appearance of a professional installation. In many cases, the small details, such as a clean-looking project and a friendly, knowledgeable worker, will determine the amount of repeat business the customer gives. Table 6-1 gives a recommended checklist to ensure that the job will be satisfactory.

Table 6-1   Fixture Trim Checklist

- ☑ Check the fixture trim to be sure it is compatible with the fixture with which it will be used.
- ☑ Check the equipment received against the packing list received with the shipment.
- ☑ Inspect the trim carefully for defects. Do not install trim that is marred or damaged.
- ☑ Protect finished surfaces when installing fixture trim. Use strap wrenches, smooth jawed wrenches, or pieces of inner tube to accomplish this protection.
- ☑ Protect surrounding areas from damage during finishing operations.
- ☑ The best practice is to install fixture trim before setting the fixture.
- ☑ Do not perform any finishing operation before the jobsite is secure from theft or vandalism.
- ☑ After installing fixture fittings and trim, test the complete assembly. Faucet aerators should be removed until the piping and faucets are thoroughly flushed.
- ☑ With tank-type water closets, flush several times and adjust the water level in the tank in accordance with manufacturer's recommendations to deliver the best flush performance.
- ☑ Remove labels, adhesives, excess putty, or other sealants from exposed surfaces. Caution: Do not use any harsh chemicals or scrapers that can damage the finish.
- ☑ Shine all polished surfaces with a clean, dry cotton cloth. Do not use abrasive cleaners. Many non-abrasive commercial products are available that will clean off dirt and foreign matter. Hot water is usually effective with this cleaning.

Water Closet

The water supply is connected to the tank. The water rough-in out of the wall (or floor) may be sweat, threaded, solvent cemented, or compression joint connected to the stop valve. If the rough-in is properly located, the final connection may be made quickly. Use the proper stop (angle or straight) with the appropriate connection for the supply pipe. Be sure to install the escutcheon at the wall joint.

Turn off the water to the area and install the stop. If the system was pressure tested with water, a small pail is handy to catch any water when the stub out is cut open. The connection from the stop to the tank can then be made (see Figure 6-1).

This connection can be made most readily with a flexible supply tube, but some water closets require a larger supply pipe size to deliver a larger flow. If straight pipe is to be used, alignment of the stop and tank supply is important to be able to make the connection quickly. If a metal supply tube and cone washer is used on the tank connection, the end of the tube should be flared slightly to prevent the cone washer from coming off the end of the tube.

Remember to use only 100% Teflon tape or approved thread compound on plastic parts. Oil-based thread compounds could react with the plastic and produce a delayed joint failure. Tighten the connection on the ballcock shank first and then tighten the compression joint on the stop (this joint has some tolerance for the position of the tube when it is tightened).

After the supply connections are made, turn on the water and check the fixture for operation. Check all the connections for leaks. Many professionals use a flashlight

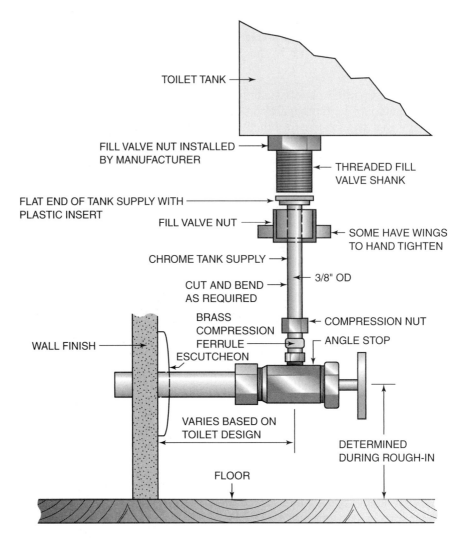

Figure 6-1 Angle stop for flush tank water closet. Some supply tubes use a cone-shaped end or cone washer to connect to fill valve. Remember to use plastic ferrules with plastic supply tubes.

to aid in this part of start-up. The light helps detect minor drips the eye might not be able to see in natural light.

For flushometer valve installations, install the angle stop on the supply pipe so that the center of the outlet of the stop is the same distance from the wall as the top inlet to the bowl. Then, install the flushometer valve, vacuum breaker, and tailpiece. Be sure you put the escutcheons on at the wall and at the bowl. When all joints are completed, test the bowl for proper operation and check for leaks.

Urinal

Urinal trim is installed using the same methods as for flushometer valves on water closets. High flush tanks should be installed according to the manufacturer's recommendations.

Bidet

The water supply valves (trim) can be either wall-mounted or fixture-mounted. For the fixture-mounted option, install the trim before setting the bidet. The pop-up drain assembly and faucets are similar to lavatory trim (see Figure 6-2).

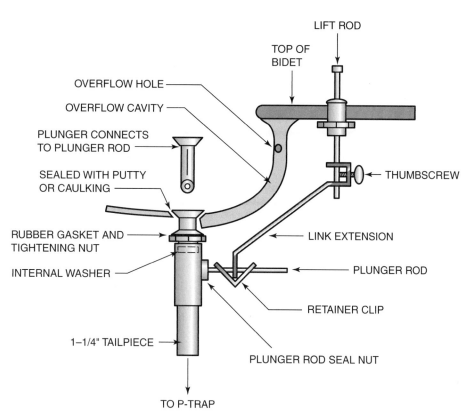

SIDE VIEW

LIFT ROD

TOP OF BIDET

OVERFLOW HOLE

OVERFLOW CAVITY

PLUNGER CONNECTS TO PLUNGER ROD

SEALED WITH PUTTY OR CAULKING

THUMBSCREW

RUBBER GASKET AND TIGHTENING NUT

INTERNAL WASHER

LINK EXTENSION

PLUNGER ROD

RETAINER CLIP

1–1/4" TAILPIECE

PLUNGER ROD SEAL NUT

TO P-TRAP

Figure 6-2 Bidet drain assemblies are very similar to lavatory drain assemblies. Remember to install the drain and linkage before installing the bidet.

The faucet outlet is equipped with a diverter which either sends water to the flushing rim or to the spray nozzle. When submerged inlets are part of the design, do not forget the vacuum breaker. After the bidet is set, the drain line and water supplies are connected. The drain outlet is sealed with slip nut and washer. The water connections are made with supplies and stops as described for water closet tanks. The bidet is fastened to the floor to complete a solid, substantial installation. The fixture should then be tested for operation and for leaks.

Lavatory

The trim is mounted on the lavatory and the lavatory is placed in position. Do not use putty on marble or cultured marble lavatories because these porous surfaces may absorb the oil and discolor the fixture—use appropriate caulking only.

When installing the faucet, tighten the two sides alternately and be careful not to over-tighten the locknuts. Remove all excess putty or sealant before it hardens.

The water stops and supply tubes are connected as described for water closets. Install the lavatory trap and waste pipe to the wall opening. Be sure to install the waste pipe escutcheon (see Figure 6-3). Test the completed installation for satisfactory performance.

Bathtub

Bathtubs require little work at the time of finishing. The spout, shower arm (if used), showerhead, and faucet handles are installed to complete the operation. Be sure that the rough-in procedures properly positioned the faucet body, spout and shower openings, and grab bar backing boards.

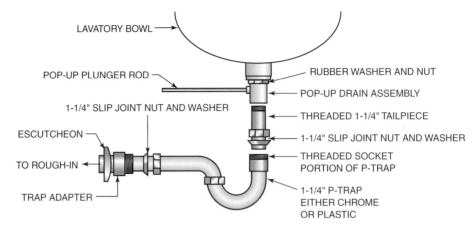

Figure 6-3 Lavatory drain assembly. Notice how the bevel on the 1-1/4" slip joint washer points into the fitting and not the nut.

Figure 6-4 Rough-in mistake. Notice how the valve body is not high enough. Now the tub spout will interfere with the shower trim plate.

Putty or caulking must be used between the finished wall and final trim components to ensure that water splashing against the wall cannot pass into or behind the wall structure. Double-check the settings on the thermostatic or pressure-balancing high-limit stops.

As a final test, fill the bathtub above the overflow and check for leaks.

Shower

Showers are trimmed out using the techniques discussed for bathtubs.

Sink

Kitchen

Sinks are trimmed out in a similar fashion as lavatories. Faucets are sealed to the sink with putty or sealant and locknuts pulled up securely. Drain openings are trimmed with basket strainers (or a food waste grinder flange assembly).

The water supplies are connected to stops at the wall and the waste opening(s) is (are) equipped with a trap and connected to the drain opening at the wall. If two compartments are involved, a continuous waste is used to connect the two compartments together and then to the single trap.

If a dishwasher is involved, the dishwasher drain can be connected to the food waste grinder head. If no food grinder is involved, connect the drain to a branch tailpiece above the trap.

If a food waste grinder is used with a continuous waste, arrange the grinder discharge to flow straight through the continuous waste tee, with the other compartment drain entering the side of the tee. Remember that some three-compartment sinks in food preparation areas need to be separately wasted to the drain line by an air gap or air break.

If connecting to a multiple bowl sink, completely fill all bowls with water and then drain simultaneously to ensure that they drain properly and have no leaks.

Service and Laundry

Service and laundry sinks are trimmed with special faucets and drains intended for these applications. Faucets must be solidly supported to provide long, satisfactory service. Service sinks are usually supported on trap stands while laundry trays are supported on stands or by countertops. If either of the faucet types terminates with a hose thread, it must be protected with a vacuum breaker.

Dishwasher

Residential dishwashers need to be installed level and in line with adjoining cabinets. Size the water supply line to be adequate for the required flow rate. Typically a 3/8" supply tube is connected to the 3/8" NPT female threads of the control valve with a 90° brass adapter. The most common adapter is 3/8" compression to 3/8" male thread. Dishwashers include an air gap in the incoming water connection, so an external protective device is not required. Local codes usually require either an air gap or a high loop in the drainage discharge. Be sure to check your locality. If no air gap is used, the discharge tubing should be fastened to the underside of the countertop. This can minimize the risk of backflow into the dishwasher in the event of sink drain stoppage. After installation, run through a complete wash cycle and check for leaks.

Clothes Washer

Residential clothes washers are trimmed with two hoses attached to the faucets in the laundry trim box and the discharge hose placed in the 2" drain standpipe. The power cord is plugged into the electric outlet in the box. The hoses should be installed without kinks. Air gaps are built into the appliance; therefore, additional backflow preventers in the water lines are not necessary. After installation, run through a complete wash cycle and check for leaks.

INDUSTRIAL AND INSTITUTIONAL TRIM INSTALLATION

Information about industrial and institutional trim installations that are special to these applications is included below.

Water Closet and Urinal

Flushometer valves with special operating devices are used in handicap, medical, or vandal-resistant settings. Foot pedal, press plates, push-button, or automatic sensor-operated valves are available options in these cases. Locate the piping and operating mechanism to accomplish the purpose intended. Visible elements should be heavily plated, so use care with assembly methods to avoid marring the finishes.

Install the spud (water closet or urinal) so it is secure in the fixture and complete the piping. Even if the spud is factory-assembled, be sure to check for tightness. Adjust and test for proper operation and check the installation for leaks.

Lavatory

Lavatories are usually trimmed with self-closing, self-metering, or automatic sensor faucets. These are installed like any other faucet—with putty or sealant, tube supplies, and stops.

Emergency Eyewash and Emergency Drench Shower

Emergency eyewash and drench showers are a vital safety device wherever hazardous chemicals are present. It is essential that these devices be located as close as possible to the area of hazard exposure and readily activated by large, accessible handles. Because emergency use could be forceful, the emergency fixtures must be substantially attached to the building structure. Remember, if an emergency occurs, these fixtures must work!

As with other fixtures, a proper rough-in ensures an easy, efficient finish operation. Be sure to flush the supply lines before connecting the fixtures. After installation, test the units for proper operation. Advise the permanent building personnel that these devices must be tested on a regular basis. Remember, ANSI code requires that tepid water be used for eyewashes and drench showers.

Handicap

Handicap bathrooms in institutional settings use conventional installation methods and equipment except for the addition of grab bars. Backing boards, installed during the rough-in stage, are required for proper installation. Adequate wood screws (#12 or #14 × 1-1/2" or 2") are used to develop the 250-pound strength needed for the grab bars. Blade handles or pressure plates should easily operate all faucets and other devices.

Medical

Medical applications use bedpan washers, blade-handle faucets, and similar sanitary or easy-to-operate devices. Special fixtures (e.g., whirlpool baths) must be installed per manufacturer's recommendations. Flush all rough-in piping before connecting finish items, then test and check the completed installation.

Vandal Resistant

To achieve the required level of security, vandal-resistant trim requires careful attention to all details of the installation. If any screw heads must be in the inmate space, use Allen head, tamper-proof, or soldered screw heads. Any trim used must be non-removable. All accessories should be an integral part of the main units or solidly attached to the structure.

Push-button flushometer valves or faucets must be solidly attached to the structure so that abuse or high forces to the button will not damage, distort, or move the main device. Whenever possible, no exposed fasteners of any kind should be used.

The descriptions given in this chapter are not meant to be all-encompassing. To have a completely satisfactory installation for any fixture fitting and fixture trim, follow the installation instructions carefully.

REVIEW QUESTIONS

1 A material that would not be appropriate in setting a fixture onto a cultured marble vanity top is _____.

2 What size screws are typically recommended for attaching a grab bar?

3 What size pipe threads are used on residential dishwashing machines?

4 A(n) _____ is a piece of trim that covers the hole where the pipe penetrates the wall.

5 Use a (n) _____ wrench to protect finished surfaces when installing chrome or brass-plated drain piping.

Blueprints and Specifications

The student will:

- Describe the general range and extent of building plans.
- Summarize the principle uses for specifications.

Chapters 7, 8, and 9 discuss blueprint reading and are backed by partial sets of plans for two projects: a new restaurant job and a job remodeling an office in an existing building. Specifications are included in Appendix C for the office remodeling job.

The office job is from the state of Indiana while the restaurant job is from the state of Florida. Please note that each state has its own codes and may therefore not be compliant with codes in your state. In addition, while the plans may have conformed to their home state codes when prepared, it is possible that they may no longer conform to the applicable codes for those states (or any other) at the time of this printing or any time there after.

Refer to the plans included with this book whenever you see the following icon in the margin:

OVERVIEW

Blueprint is a term that describes an older method of reproducing an architect's or engineer's original drawing. They were typically white lines on a blue background. Modern prints are usually drawn with black or blue lines on a white background. Nowadays, all drawings are constructed through some form of *computer-aided design (CAD)* package and then sent to a printer or copy machine. These computer-based designs are more easily stored, reproduced, or altered than the previous hand-drawn versions.

A set of blueprints comprises the drawing instructions of how to build a building or execute some particular project. Blueprints, or plans, become part of the contract between the owner and builder and make the scope of the job clearer. In order to prepare quotations from the plans and make sure a project is built correctly, understanding the language of blueprints is required. A term commonly used in today's construction industry is *plan*. The word *plan* actually has two meanings as applied to drawings: (1) as a general term, used for a set (or a single sheet of a set) of drawings for a project; and (2) in reference to any drawing that shows a horizontal view of the building or project as seen from an overhead position.

Plan Preparation

The following steps are typical for the development of a set of drawings:

1. The owner engages the services of a design professional who develops a written program to fulfill the owner's needs. The owner will usually describe how the building should be situated on the land and what function the building will serve.
2. The design professional will make a preliminary set of drawings that attempts to meet the requirements of the owner as well as the local building codes and ordinances.
3. The design professional will present the owner a conceptual set of drawings that describe the major points of the project and provide a basic feel for the layout of the building. After discussions over the owner's essential needs, the final set of blueprints is developed as a guide for the builder to perform the work.

Figure 7-1 shows two possible chain-of-command flow paths, or the interrelationships that are typical for most building projects.

PLAN TYPES

The following drawing types are usually needed to describe any project in complete detail.

Grade Plan

Prepared by a civil engineer, the *grade plan* shows cut and fill data, finish elevations, surface drainage, and contours. The information presented helps to lay out sewer and water connections.

Plot Plan

Prepared by the architect, the *plot plan* shows building location; sidewalks, curbs, and drives; mechanical items and service locations; and relationships to adjoining streets.

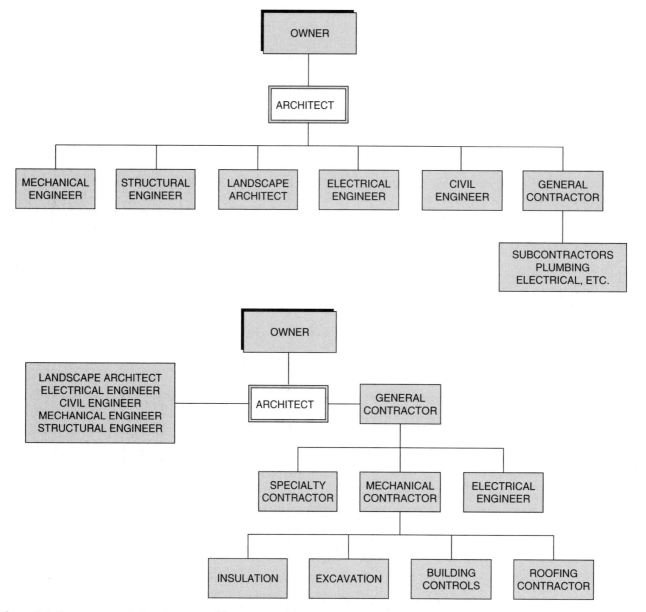

Figure 7-1 Two common chains of command for a construction project.

Landscaping Plan

Prepared by the architect or landscape planner, the *landscaping plan* shows the plantings and site improvements for the project.

Foundation Plan

Prepared by the architect or structural engineer, the *foundation plan* shows footings, foundation walls, and any items that support the structure.

Soil Reports

A civil engineer or soil specialist will prepare a *soil report* so that the architect or structural engineer can determine the dimensions necessary for the footings and foundation. Since the type of soil can have a great effect on the methods used for trenching, soil testing is discussed in Chapter 18 as well. The soil report may also be used along with a percolation test to determine the size and location of any septic field.

Floor Plan

Prepared by the architect, the *floor plan* shows the room layout for a floor. A plan is required for each floor of the building.

Framing Plan

Prepared by the architect or structural engineer, the *framing plan* shows the building elements that form the supporting structure of the project. A plan is required for each floor and the roof.

Roof Plan

Prepared by the architect, the *roof plan* shows details of roof construction and location of devices on the roof.

Details

Prepared by the architect, *detail* plans show the precise components and arrangements of parts of various building elements. Details are shown in sectional views, exploded views, or cutaway views to present the specific building requirements.

Mechanical and Electrical Plans

Prepared by the mechanical and electrical engineers, *mechanical and electrical plans* show the details of plumbing, heating-ventilating-air conditioning, electrical, fire protection and life safety, fuel gas, and any other special mechanical or electrical systems needed for the project. In many cases, a particular type of fire-stop caulking may be specified for the different types of piping materials, especially in fire-rated walls.

Equipment Schedules

Prepared by the mechanical engineer, the *equipment schedule* will list the size, power requirements, and other relevant specifications of the equipment being installed in the building. Equipment such as boilers, air-conditioning equipment, exhaust fans, and air compressors are examples of items that might be listed.

Special Equipment Plans

Prepared by specialized designers, the *special equipment plans* show the dimensions, locations, and requirements for special equipment. An example of a specialized piece of equipment could be a humidification unit that would require both an electrical power supply and a water supply.

Plan Notes

Typically there are two types of plan notes, *general and referenced*. The *general notes* apply to all trades and are used to express information not normally found on the graphics. An example of a general note is found on page A3 of the enclosed prints: "Keep areas under construction secure and clear of debris." An example of a *reference note* can be found on that same page. It specifies what size quarry tile to be used. Of course, any specification should always be cross-referenced to what is indicated in the specifications.

Addenda

Addenda are graphical or written instructions that are used to revise or clarify information to the original bidding documents. The changes could require adding or deleting information, such as material substitutions or product options. Addenda could also be used to clarify conflicts between drawings and specifications or other unclear details.

Addenda should be issued to all prospective bidders who are listed as having received a complete set of bidding documents. There is usually a specified deadline as to when addenda can be issued; generally, it is not within five days of the bid deadline unless its purpose is to extend the bidding deadline. Addenda should be listed on the proposal and numbered.

Scope Letters/Proposals

The *scope letter* specifically states the extent of the work to be performed and basically provides a general narrative of the work to be performed. It should state the sections of the specifications that will be included and those that will be excluded in the proposal submitted by the contractor.

Design Build

The *design-build delivery method* is when one company or firm is contracted by the owner to serve as the architect, engineer, and builder of the project. By placing all of the responsibility on one company, the project can typically flow more easily because of better internal communication and less duplication of work. Since all work will reflect back on the company, there is usually a greater focus on quality and performance. In many cases, this can translate into a savings of time and money. Because of the internal communications, there is a tendency for the drawings to be less than complete. More information can be found on the Design-Build Institute of America's website (http://www.dbia.org).

BLUEPRINT REVIEW

Reviewing a set of blueprints entails a study of the relationships among the various special crafts to be sure that all the equipment is adequately described, functioning, and detailed. The reviewer should be able to cross-check the various drawings to ensure that the equipment will be installed properly. The following are some typical comparisons that a builder would look for when beginning to bid a job:

1. The foundation plan should be compared with the mechanical plan to be sure that sub-soil drainage is properly described.
2. Floor plans should be checked for proper location of thimbles and sleeves.
3. Plot plans should be checked for interceptors, catch basins, sewer and water service routings, as well as any setback requirements.
4. Roof plans should show the locations of drain and vent openings.
5. The framing plan should indicate where obstructions occur that require offsets in the piping or air-conditioning ductwork.
6. Detail plans should be referred to whenever a possible conflict is encountered.
7. Kitchen and specialty plans should be studied for any plumbing connection requirements.
8. The architectural plans must be checked to ensure that the mechanical details conform to the general floor plan.

If any discrepancy is noted in any of the above comparisons, it should be checked out and approved by the job superintendent or your supervisor. Any discrepancies must be resolved and solutions approved, in writing, by the person in charge, which is usually the general contractor, architect, or owner.

When plans are not to code, notify your supervisor. The supervisor can then notify the general contractor, architect, and owner. Going above and beyond the code requirement is acceptable.

SPECIFICATIONS

It is possible to describe a building project completely on the set of drawings, but it is usually more convenient to present additional information in a book of *specifications*. This book contains a sample construction contract, a substantial area of ground rules for the requirements to be met by all contractors, and a listing of particular equipment to be provided by each specialty contractor. The combination of specifications and building drawings describes the job better than can be done by the drawings alone. Usually, in case of contradictory provisions, specifications take precedence over drawings. A statement saying so is often found in the specification text.

Specifications should be custom written, using a standard outline for each building project; but occasionally a stock specification is used without adaptation to the project. Because they are so generalized, it is not apparent how stock specifications fit the current situation, making them more difficult for bidders to deal with.

Specifications are usually presented in a format developed by the American Institute of Architects (AIA) called *Master Spec,* or by the Construction Specifications Institute (CSI), and called *Master Format*. The outline table of contents of the CSI format is shown in Appendix A. It should be noted that these specifications are constantly under review and will change over time.

Certain divisions such as Division 0 (Bidding and Contract Requirements) and Division 1 (General Requirements) pertain to all contractors; while others, such as Division 3 (Concrete), are more focused to one trade. Division 15 (Mechanical) applies to all the parts of the work that fall under the general heading of "Mechanical." Mechanical focuses on piping issues such as vibration isolation, piping insulation, and plumbing equipment. It also covers the heating, air conditioning, and refrigeration components such as fans, ductwork, and controls.

A contractor should read these parts carefully and analyze the job needs. Certain terms such as "shall" are used to mean that the following condition is mandatory whereas "may" and "should" are permissive terms. It is a good idea to read the General Requirements section first. Prepare an outline of your responsibilities, the work to be performed, and major equipment to be furnished and/or installed. Also, note other crafts that are required to be involved in the installation of equipment. Other crafts might be involved in work pertaining to HVAC, electrical, insulation, fuel gas, fire protection, irrigation, and process piping.

There are areas of work for which the limits of responsibility are unclear. These include the following areas:

1. Site Work
   Who replaces trees, shrubs, and sod? Who installs sewers, water mains, or storm water disposal?
2. Hoisting and Rigging
   Does each contractor provide his own or is there a service on the site? If a service is available, when is it accessible to specialty contractors and at what cost?
3. Cutting and Patching
   Who is responsible and who pays for this work? Under what conditions, specifically?
4. Related Accessories
   Who provides toilet and kitchen accessories?
5. Sub-soil Drainage
   Who is responsible and who performs site regrading?
6. Irrigation Systems
   Who is responsible? Who provides water supply connection?
7. Large Construction Equipment
   Who provides compressors, cranes, trenchers, and other special equipment?
8. Trash Removal, Security, Storage
   Who is responsible? How frequently?

9. Existing Facilities

Is sewer, water, heat, and electricity available to contractors? Toilet rooms?

10. New Projects

Who provides toilets, electricity, water, and temporary heat?

After considering the general information, the contractor should look at the more specific sections to get a clear picture of who is to perform which parts of the job. Note if any part or aspect of the job is unclear or note if any conflicts or gaps occur. Get clarification in writing of any such points.

Search out any alternate construction methods or materials called for in the specification so that you can prepare the alternate proposal. An *alternate proposal* is when the bidding contractor submits to the design professional a different method or material to perform a specific task. For example, a plumbing contractor may want to use PVC pipe on the job, but the specifications call for ABS. In many cases, alternate proposals should be submitted well before the bidding deadline. Upon completion of the first reading, read the specifications through again to be sure of your first analysis.

You should then have the following information in summary form:

1. General requirements and technical specifications lists
2. List of responsibilities
3. List of materials and applications
4. List of discrepancies, vague requirements, special details, omissions, or conflicts

If you have only partial plans and partial specifications upon which to base your bid, be sure you limit your proposal (in writing) to the material covered by the documents you have studied. Be sure to list the exact date and plan revision on which your bid is based.

OUTLINE FOR SPECIFICATIONS

Table 7-1 shows a sample specification analysis sheet to help you organize the information and point out any conflicts.

Table 7-1   Specification Analysis Sheet

Job Name				Reviewer	
Job Location				Date	
Page and Article No.	Division No.	Item of Responsibility	Requirement	Other Trades Required	Remarks

The following is an explanation of how it should be filled out.

1. The first column shows the page number of the specification book and the article number (if the spec is organized that way).
2. The second column identifies the division of the specification such as General Conditions, Plumbing, Ductwork, etc.

3. The third column shows the items you are responsible for, such as toilets or water heaters.
4. The fourth column lists the specific requirement, such as size or model of equipment, efficiency ratings, etc.
5. The fifth column lists notes detailing any requirements for coordination with other trades.
6. The last column is available for any remarks.

To prepare a complete analysis, practice reading the specifications and reviewing both Division 0 and Division 1 as well. It should be noted that when the specifications are placed on the drawing pages, the materials usually presented in Division 0 and 1 are mostly or entirely missing. Thus, to achieve a mutual understanding, the owner or architect must convey this information in some other way.

Once all available specifications have been reviewed, a review of the drawings should be made to see how all the elements come together. Usually, some questions will remain that need to be clarified either by more study of all documents or by inquiry to the designer.

Cross-check the completeness of drawing requirements versus specification requirements. It is not unusual for an item to be called out in the specification that is not actually needed in the project. Do not be dismayed, but also check very carefully to confirm your judgment that the product is indeed not required.

Example 1

Use the specifications in Appendix A to fill out Table 7-2 for the Suite 108B remodel job.

Table 7-2 Specification Analysis Sheet

Job Name: *Suite 108B*				Reviewer	
Job Location: *The Commerce Center*				Date	
Page and Article No.	Division No.	Item of Responsibility	Requirement	Other Trades Required	Remarks
29	15400	Eljer water closet	Model No. 111-1215	Masonry?	Thickness of floor
29	15400	Eljer lavatories	Model No. 213-0138 052-0164 552-1400		Who cuts countertop?
29	15400	water heater	Sears lowboy	Electrician	Voltage? Amperage?
29	15400	gas piping	What component?		5-ton heat pump, so where is gas?

SUMMARY

Specifications are written to fulfill three project needs:

1. Display the form of the contract
2. Present general and special conditions for the project
3. Provide a detailed list of the responsibilities for each party to the job

The specifications supplement the working drawings with the technical information for each construction specialty. Along with the drawings, they provide a legal format for the construction project.

For negotiated jobs or projects with a limited list of bidders, an alternate way of operating is sometimes used. The architect places sufficient notes on the drawings to completely describe the technical requirements. The Division 0 and Division 1 information is understood between the parties and is not formally set out.

Note that with the two projects included in these chapters: (1) the new restaurant job contains all of the specifications on the drawings, and (2) the office-remodeling project has the specifications in book form (Appendix C).

Study the plumbing specification Division 15400 to see how specifications are worded and presented, although this list is not exhaustive. Later lessons will continue the review of these plans and specifications.

Note that the CSI *MasterFormat*™ is always under review to include new facets of the building industry. Therefore, some sections listed will be eliminated and others included.

REVIEW QUESTIONS

1 What does CAD refer to when discussing blueprints?

2 On blueprint AP-1, what size water heater is required?

3 What sheet would have an elevation view of the water closet?

4 Who usually has direct communications with the owner?

5 According to Table 7-1, what division does plumbing fall under?

CHAPTER

8

Drawing Types, Floor Plans, and Site Plans

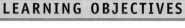

LEARNING OBJECTIVES

The student will:

- Describe the information presented on floor and site plans.
- Interpret elevation, plan, and sectional views.
- Recognize symbols used to depict typical materials on prints.

Chapters 7, 8, and 9 discuss blueprint reading and are backed by partial sets of plans for two projects: a new restaurant job and a job remodeling an office in an existing building. Specifications are included in Appendix C for the office remodeling job.

The office job is from the state of Indiana while the restaurant job is from the state of Florida. Please note that each state has its own codes and may therefore not be compliant with codes in your state. In addition, while the plans may have conformed to their home state codes when prepared, it is possible that they may no longer conform to the applicable codes for those states (or any other) at the time of this printing or any time there after.

Refer to the plans included with this book whenever you see the following icon in the margin:

FLOOR PLANS

A *floor plan* is an overhead view of a specific floor or portion of a building. The size and shape of the rooms for that particular area are shown, as well as any other architectural features. Figure 8-1 shows the floor plan of a typical residential home. This layout should contain enough information to give the viewer an overall feel of the building, but not so much that it becomes confusing.

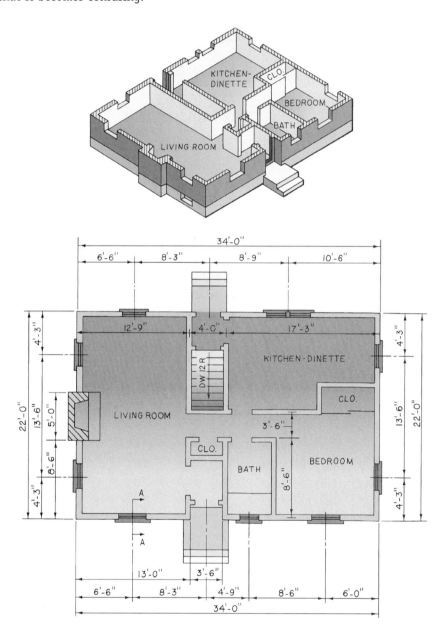

Figure 8-1 The floor plan should help the viewer locate key items in the building.

Separate views will be prepared for each floor of a multi-story building unless a plan can represent more than one floor. Upper floors of many commercial buildings are the same or nearly so; therefore, a single print sheet can show the features and details of all floors that are the same. Such a plan is indicated as *typical* and will be identified as to the floors it represents.

The usual floor plan is conceived as if a cutting plane were passed through the building horizontally 5' above the finished floor. More simply put, it's as if someone measured up 5' from the floor, cut through every wall, and then removed the top portion of the building. Any objects in the space below that cutting plane will show up as visible objects while anything above that level will not. Sometimes a hidden line may be used to represent some details above the cutting plane such as wall cabinets in a kitchen or workroom. This reminds the reader that there will be something located above.

Waitress Station #1 on blueprint A-1 of the restaurant has a hidden line representing the cabinets above the cutting plane used to create the floor plan. Also see Sectional View K on blueprint A-3 of the same job for clarification. Ceiling features, for example, would not be shown. The ceiling is detailed by imagining that the floor is a mirror and the ceiling is reflected in the mirror. This drawing is called a *reflected ceiling plan*.

The architectural floor plan is the beginning point for the building design. To complete the building it is usually necessary to prepare several additional floor plans to show all the parts for all the crafts required. Most CAD packages have a feature called *layers* which allows the draftsman the ability to use the same floor plan and then show or hide items for each trade separately or together.

The architectural floor plan shows the dimensions of the building, wall and partition locations, and other building details. Heavy object lines are used for outside walls; other components of the building, like interior walls, doors, and cabinets are sometimes shown by somewhat lighter lines. Dimension lines are used in conjunction with extension lines to show the view of how long a particular feature in the building is. For example, the bottom most line in Figure 8-1 shows the viewer that the building is approximately 34'-0" long. Remember these dimensions are to the finished framing, not finished walls. This can be critical when locating the rough-in plumbing. If there is any question about a distance, refer back to the architectural drawings; using a scale on mechanical drawings can lead to mistakes. Dimensions may be shown in one of the ways seen in Figure 8-2.

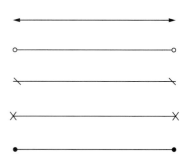

Figure 8-2 Ways to represent dimension lines.

Study the dimension marks carefully to determine whether the dimensions are to wall centerlines, rough face, or finish face. The specialty contractor must be able to interpret these aspects, determine the best way to plan the installation, and select the best tooling to accomplish the job.

Architectural features (pipe chases, furred-out areas, etc.) required for mechanical features will be shown on the architectural plans. Check these items carefully to ensure that the equipment and material will fit. For example, a contractor needs to know if the architect has located a 3" vent pipe in the middle of a 2" × 4" wall; in case of doubt, contact your supervisor immediately. The mechanical details themselves are shown on mechanical floor plans. For many jobs, these special blueprints include floor plans for plumbing, heating-ventilation-air conditioning, electrical power, electrical lighting, etc.

In many cases, the architect presents the beginnings of the architectural floor plans (plans without dimensions or general construction notes) to each specialty designer. These designers then place the information pertinent to their specialty on a floor plan devoted to that trade.

Table 8-1 is an example of a form that allows the contractor to make notes of how his work will tie into the building as a whole. Notice the comments such as, "VTRs must be 10' from any air intake, coordinate with AC contractor"(found on blueprint P-1).

Note that there are many references to notes and sections on other pages of the drawing sets. Compare the similarities and differences between the floor plan and other specialty drawings. Remodeling jobs include a demolition floor plan; they frequently require this type of drawing so that the contractor will remove only the material not needed for the final execution of the work.

DRAWING TYPES

There are three general types of views for showing how a building is to be built: *elevation views*, *plan views*, and *section views*.

Table 8-1 Restaurant Job

Job Name:			Reviewer:		
Job Location			Date		
Building dimensions			Exterior wall construction		
Room	Wall finish	Ceiling finish	Special notes	Other details and special plans to reference (equipment, drainage . . .)	
				Item	Other plan (Page No.)
Kitchen	Tile		3 comp, pot sink		

Elevation Views

Elevation views show the following:

1. Vertical design and appearance of the building
2. Height of floors, roof, window, and door heads
3. Facing materials of exterior walls (see Figure 8-3)

Usually, elevation views are made for the exterior faces of a building, but they are also drawn to show an inside wall when the details cannot be described adequately with notes on the floor plan. In some cases, both a portion of the exterior and a portion of the interior will be shown. This allows the architect to give more information in a single drawing.

Plan Views

Plan views are very similar to floor plans except that they can typically extend to other parts of the project such as grade or landscaping. Floor plans are focused solely on the building itself. As discussed earlier, floor plans are drawn as seen by an observer looking down from a high viewpoint, similar to what a person would see from a low flying plane.

The building construction details may be shown on the plan view. Exterior details are indicated on site plans, which may be subdivided into the following parts:

1. Grade plans
2. Landscaping plans
3. Plot plans (also called site plans) (see Figure 8-4)

When possible, all the information will be combined onto one or two sheets.

Sectional Views

Sectional views represent cutaway views of major building components so that the details of construction may be visualized. Foundation plans, framing plans, and wall, window, and door details are frequently supplemented by section views (see Figure 8-5).

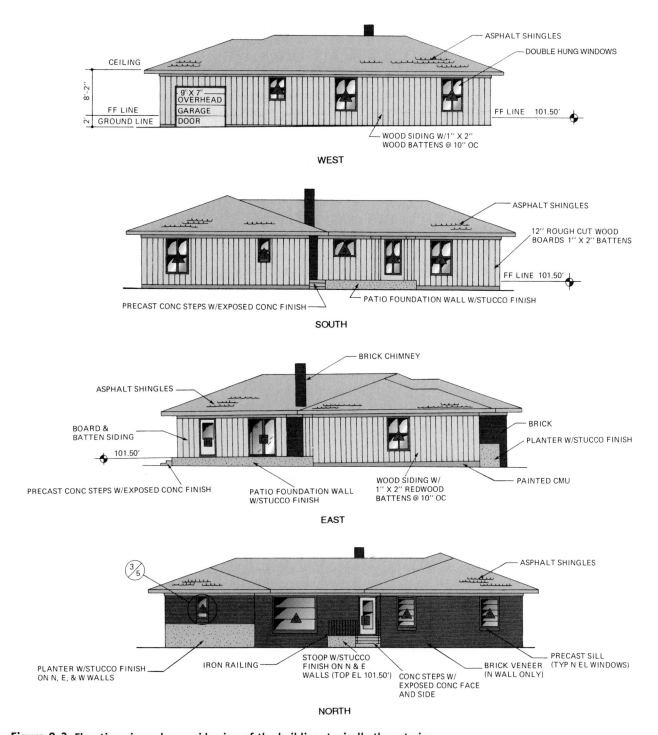

Figure 8-3 Elevation views show a side view of the building, typically the exterior.

Analyzing a set of blueprints means that you must study all the pages or risk leaving something out in your takeoff. Major specialty portions of the work, like plumbing, will probably be indicated on more than one drawing page. Check the drawings to be sure the work elements fit together and verify that there are no gaps or missing elements in the work. Any discrepancies should be resolved as soon as possible. For example, if the architectural plan for a basement shows a sump for a sump pump, the plumbing plan should show the pump and related piping, and the electrical plan should show the necessary electrical work to power the pump.

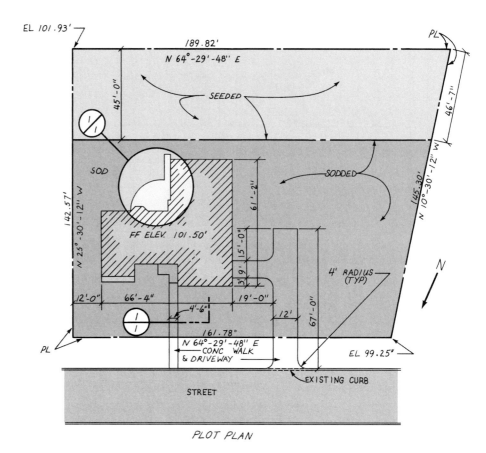

EL 101.93'

189.82'

N 64°-29'-48" E

45'-0"

SEEDED

SOD

SODDED

FF ELEV. 101.50'

142.57'

N 25°-30'-12" W

61'-2"

15'-0"

3' 9'

4' RADIUS
(TYP)

N

PL

46'-7"

N 10°-30'-12" W

145.30'

12'-0"

66'-4"

4'-6"

19'-0"

12'

67'-0"

PL

161.78"

N 64°-29'-48" E
CONC WALK
& DRIVEWAY

EL 99.25"

EXISTING CURB

STREET

PLOT PLAN

Figure 8-4 Plan views show construction and site details from an elevated view. Sometimes they are specialized to one trade such as landscaping; other times they may show all trades in one view.

On smaller or less complicated jobs, the plumbing will be shown on a mechanical plan, which may also include heating, ventilating, and air-conditioning equipment, ductwork, gas piping, venting, and any special mechanical work that is required. Such multiple application plans can be difficult to use for estimating or construction because it is usually difficult to distinguish clearly between the different trades. Most estimators and mechanics use colored pencils to mark out on the drawing the parts of the job that have been finished. In this way, the finished/unfinished work is graphically apparent.

Make a shop drawing of the plumbing layout using the architectural floor plan as the starting point. Use the shop drawing in the field as the basis for the plumbing system installation. The shop drawing should highlight potential conflicts with other trades. Such conflicts, if recognized early, can lead to solutions for the problems before they develop into costly delays or emergencies.

Symbols

Blueprints use generally standardized symbols for usual construction elements. The use of such symbols aids in understanding the details of the structure and simplifies the drawing. The architect should include a table of symbols used for the job to make clear what is being represented. There are commonly used symbols for typical building materials (see Figure 8-6) that were developed by organizations such as ANSI, AIA, Department of Defense, etc.

Using these standard symbols helps to reduce the chance for error. Be sure to check any set of drawings for the precise symbols used. Do not assume that the

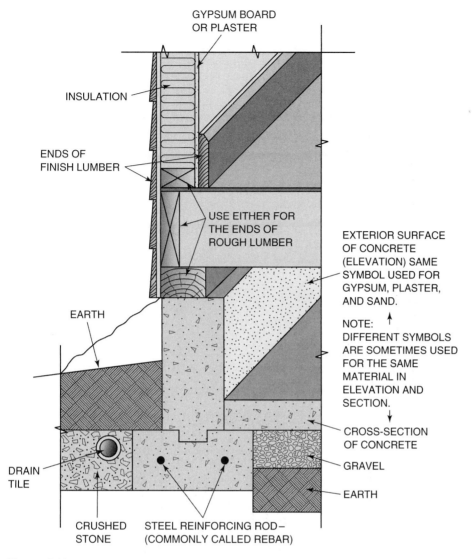

Figure 8-5 Sectional views help show how the construction materials connect to each other. Symbols can be used to show what type of material is being used.

architect used the standard symbols. Recognizing the building symbols will help you to select the following:

1. Pipe, fixture supports, and hangers appropriate to the structural components
2. Sleeves and thimbles, depending on building materials
3. Tooling to make or repair holes for the type of materials used

In an example of symbol usage that involves a sewage pump in the basement of a building, the plumbing plan should show a symbol for the following:

1. Sewage pump
2. Fixtures that drain into the sump
3. Valve details for the installation
4. Drain lines to the sump

The electrical plan would show the pump with power and control wiring including timers, operating switches, and alarms (see Figure 8-7).

Figure 8-7 shows the details in the immediate vicinity of a sewage pump. Note that the standard symbol shown is for concrete. Upper floor plans should show the

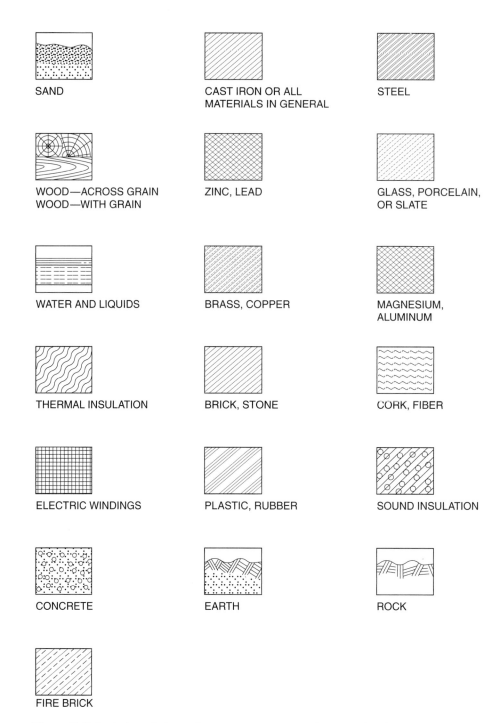

Figure 8-6 Typical symbols used to represent construction materials in sectional views.

 separate vent serving this sump. Note that the remodeling project includes standard symbols on blueprint T-l, but notes are still required to identify certain items (see blueprint A-3).

The restaurant project does not include a depiction of standard symbols, but the detail uses the conventional symbols in most of the views. See the note on blueprint A-5 that identifies a certain symbol as representing wood-framing construction.

Site Plans

Frequently, the first drawing of a set of plans describing a building is called the *title page*. It contains a list of all drawings, standard symbols used, and a plan view of

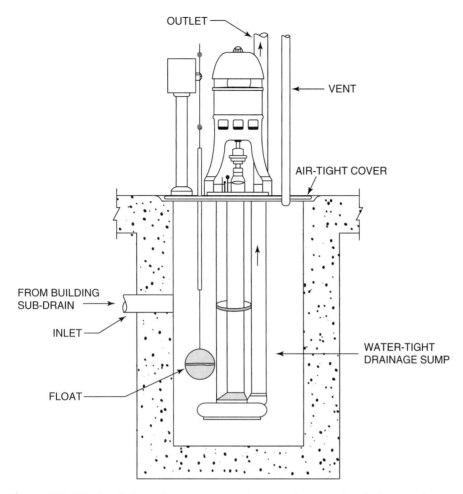

OUTLET

VENT

AIR-TIGHT COVER

FROM BUILDING
SUB-DRAIN

INLET

FLOAT

WATER-TIGHT
DRAINAGE SUMP

Figure 8-7 **This detail view of a sewage pump does not show any electrical connections, so it would be important to check the electrical prints to verify how power will be supplied.**

the building as it relates to the real estate on which it is to be built (refer back to Figure 8-4). This plan is referred to as the *site plan*, and for large projects it might require several drawings to completely describe the placement of the building on the real property and the details of the property development.

The principle information can be shown on a topographical plan, which is a scale drawing of the site and includes the following:

1. Size of the lot
2. Placement of the buildings and other major features
3. Location of streets, drives, and sidewalks
4. Relationship of the lot to true north
5. Location in the city, town, or county

Property lines are shown as well as benchmark locations (if any). Also shown are existing and final contour lines, water and stream locations, and locations of existing utility services. Elevational changes in the land prior to rough grading are usually depicted by dashed contour lines, while solid lines present the finish grade. Contour lines are typically spaced out at 1' intervals (vertically). This means that if contour lines are spaced closely together, then the ground elevation changes drastically in that area. The farther apart the contour lines the more gradual the slope (see Figure 8-8).

Elevations are shown for building floors, curbs, utilities, and surrounding grades. Also included are test bore results, which detail the soil conditions at the site. If available, landscape plantings and other features are also provided.

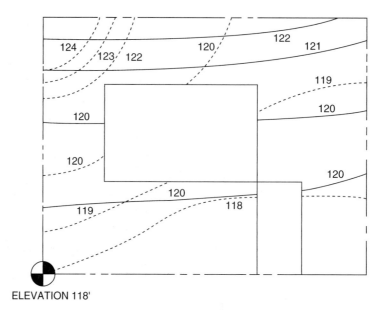

ELEVATION 118'

Figure 8-8 Dashed lines are elevations before construction (natural grade); solid lines are after construction (finished grade). The lot is more level from left to right after construction.

Symbols used on site plans include contour lines, hydrants, irrigation system details, trees, shrubbery, fountains, curb boxes, meter boxes, backflow devices, manholes, cleanouts, head walls, sewer flow lines, and benchmarks. Smaller projects may have all the above-described features shown on a single drawing, usually called a *plot plan*. Larger or more complicated jobs may require several sheets to present all the required information. Separate drawings may be titled *grade plans*, *landscape plans*, and *site plans*. In any case, while the range of needed information is the same, the architect or engineer may decide that the needs of the owner can be served best by presenting more than one drawing of the building setting. The analysis of these features is independent of how many drawing pages are required. The following issues must be covered and noted in your review of this aspect of the job:

1. Drawing scale to help ensure accuracy of takeoffs
2. Locations of existing obstacles such as sidewalks, trees, shrubs, streets, driveways, etc.
3. Responsibilities for excavation, shoring, and regrading and replacement of same
4. Location of property lines
5. Location of benchmarks
6. Grade and contour lines
7. Water and stream locations
8. Locations of existing utilities
9. Testing bore data, if available
10. Landscaping information, which might involve special plumbing requirements

 The building plumbing floor plan should be compared to the site plan(s) to be sure that all the elements tie together. While the restaurant project does not have a site plan, there is a *key plan* on blueprint A-l and a note on the *plumbing plan* (blueprint P-l) to see *Landlord's Drawing for Continuation* to continue the building sewer. The remodeling project is contained within an existing building, so no plot plan information is needed.

Table 8-2 shows a form that can be used to record your analysis of typical site plans.

Table 8-2   Plot Plan Analysis

Job Name:		Reviewer:	
Job Location:		Date:	
Benchmark location:		Scale of drawing:	
Dimensions of property:		Dimensions of the building:	
Grade elevation value(s) at the building			
NE corner:	NW corner:	SE corner:	SW corner:
High and low contour grade values:			
Obstacles that may be encountered in the installation of plumbing (e.g., storm sewer, sanitary sewer, water gas or other utility lines). Include locations from the benchmark or a common corner.			
Elevations and locations (from benchmark) of existing utilities:			
Gas			
Water			
Sanitary sewer			
Storm sewer			
Electrical			
Other			
Test data			
Other			
Special landscaping requirements			
Interrelationships with other trades			
Other information in need of clarification			

REVIEW QUESTIONS

1. What can be found on the floor plans for a building?
2. What does *typical* mean on a floor plan?
3. At what height is the cutting plan passed to develop a floor plan?
4. What kind of feature on a site plan may be of importance to a plumber?
5. What is the symbol for concrete?

Structural, Plumbing, Electrical, HVAC, and Detail Plans

The student will:

- Interpret information presented on structural and elevation plans.
- Identify possible conflicts between different trade installations.
- Compare plans for each trade and find interferences.
- Interpret items on HVAC plans and determine plumbing requirements, if any.
- Employ detail drawings to guide installation requirements.

Chapters 7, 8, and 9 discuss blueprint reading and are backed by partial sets of plans for two projects: a new restaurant job and a job remodeling an office in an existing building. Specifications are included in Appendix C for the office remodeling job.

The office job is from the state of Indiana while the restaurant job is from the state of Florida. Please note that each state has its own codes and may therefore not be compliant with codes in your state. In addition, while the plans may have conformed to their home state codes when prepared, it is possible that they may no longer conform to the applicable codes for those states (or any other) at the time of this printing or any time there after.

Refer to the plans included with this book whenever you see the following icon in the margin:

CONSTRUCTION TYPES

The type of structure chosen has a significant effect on the methods used to install the plumbing and it also impacts the cost of the plumbing installation. Regional or local conditions, economics and building codes affect how a building is to be constructed; therefore not all the following construction types may be performed in your area.

Frame construction uses dimensional lumber for the load-carrying members of the building. *Platform*, *balloon*, and *braced* are the three most common variations of frame construction recognized. Platform construction consists of one story being erected at a time. First, a platform is constructed and then the walls. To add a second story, another platform will be constructed on top of the lower walls. This is the most popular type of residential construction because it is quick, easy, and less costly (see Figure 9-1).

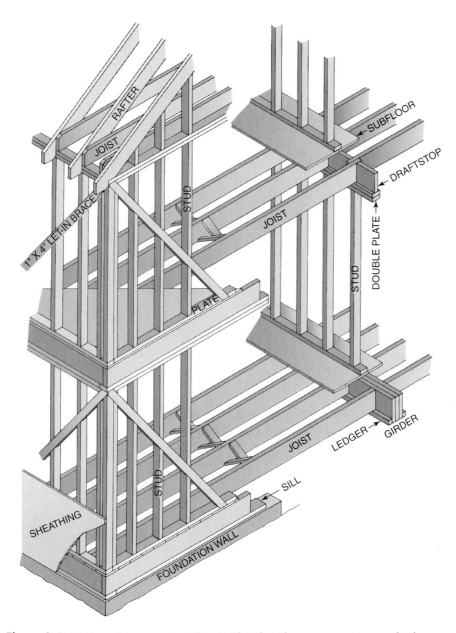

Figure 9-1 Platform frame construction. Notice that the corner posts are only the length of each floor.

Balloon construction consists of studs that extend from the foundation plate to the highest ceiling, with intermediate floors supported on cut-in members called *ribbons*. Because long framing components were needed, this type of construction is not as common in new construction (see Figure 9-2).

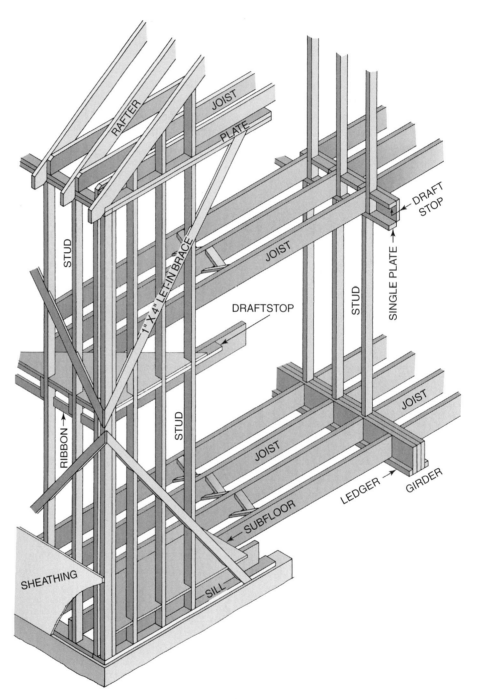

Figure 9-2 Balloon frame construction. Notice that the corner post is equal to the height of the building.

Braced or post and beam construction is similar to platform except that larger timbers are used and spaced at greater intervals (see Figure 9-3).

In any of these variations, exterior walls are completed with sheathing and then some form of siding. The interior walls are then covered with gypsum board, lath and plaster, or panel board.

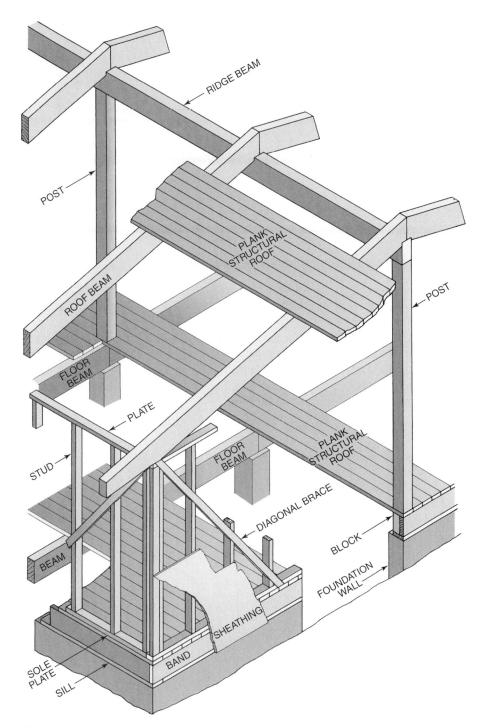

Figure 9-3 Braced or post and beam construction. Notice the larger framing members and wider center distances.

Masonry

Masonry uses brick, concrete block, stone, or a combination of these materials for the walls. Floors are constructed of lumber, steel framing, or reinforced concrete. Special chases are required to provide space for plumbing and air-conditioning work.

Masonry Veneer

Masonry veneer consists of a single course of masonry brick tied to the outside of any of the above-mentioned framed structures. The foundation wall must extend

Another type of construction gaining in popularity is the use of *insulated concrete forms (ICF)*. In this type of construction, a polystyrene form is constructed of smaller blocks. The two outer layers of polystyrene are held apart with plastic webbing. Once the forms are in place, the center is filled with concrete and rebar. In one step, the exterior walls will be framed, insulated, and wrapped. This makes for an energy-efficient, fire-resistant, and stronger building. With the increasing cost of lumber, these ICF cost only 4% more than a traditional wood-framed home. Because the walls consist of approximately 8" of concrete, all pipe-wall penetrations should be performed before the walls are poured; otherwise, this work will be very labor intensive. Figure 9-4 shows an ICF form being filled with concrete. Always remember to install a sleeve in the walls before the concrete is poured. The sleeve allows piping to be installed through the wall without having to bore through concrete.

Figure 9-4 ICF being filled with concrete for basement walls.

past the wood frame construction enough to allow a ledge to support the masonry brick wall (see Figure 9-5).

One end of a metal tie is embedded into the mortar joints of the masonry wall while the other end is fastened to the wall studs. The air gap that separates the two different materials allows for a limited amount of movement and the escape of any trapped moisture. The moisture will collect at the base of the wall and escape through weep holes in vertical mortar joints.

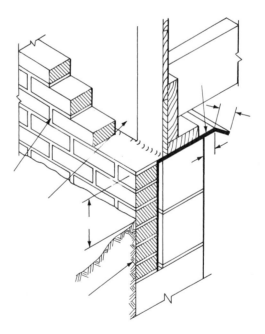

Figure 9-5 Masonry veneer along traditional wood framing.

Steel Frame

Steel frame uses steel structural elements for wall and floor supports, usually with poured concrete floor surfaces. Outside wall surfaces may be selected from a wide variety of options. This type of construction is more critical for the design of components, so the structural engineer must approve floor penetration, locations, and sizes. Some metal wall studs are constructed to the same cross-sectional size as dimensional lumber and are available in load-bearing and non-load-bearing types. The load-bearing studs are constructed of heavy gauge metal while the non-load-bearing ones are of lighter material. Special care should be taken so that plastic piping doesn't come into contact with a sharp metal edge. NOTE: Steel structural beams cannot be cut, so careful planning is essential.

Reinforced Concrete

Reinforced concrete uses metal rods, wire mesh, and fiberglass flake or metal cables to increase the strength in the concrete for columns and floors. Concrete is very strong in compression (when being squeezed), but very weak in tension (when being pulled apart). For example, when a beam is supported on both ends and a weight is placed in the middle, it would tend to bow down in the middle (see Figure 9-6).

The bottom portion of the beam is being pulled apart while the top portion is being pushed together. To better understand this, place your fingers together, as in Figure 9-6, and then bend them down in the middle. The top fingers are pushed together, while the bottom ones are pulled apart; this is compression and tension.

Placing steel rods or cables in the portion of the columns, beams, and floors increases the tensile strength required in the members. It is much less of a structural problem to develop penetrations in the original construction process than to attempt to cut them in after the concrete is poured. Either way, the structural engineer must be involved directly in selecting locations and sizes of openings in the floors.

Prestressed Concrete

In order to eliminate some of the tension forces in the members, the metal rods or cables can be stretched so that they are in tension. In *prestressed concrete*, the metal rods are stretched and then the concrete is poured. Once the concrete is fully cured, it will have bonded to the steel; so when the device stretching the rods is removed and the rods try to return to their original length, the concrete is squeezed (put in compression). When making penetrations, it is very important not to cut these tensioning members.

Tilt-up Construction

Tilt-up construction is a form of pre-cast construction gaining in popularity, especially on the West Coast. Typically after the floor slab has been poured and allowed to cure, it will serve as the bottom of a form for the wall construction. This way the walls can be cast horizontally. The form will consist of the outer perimeter of the wall and any openings for windows, doors, and sleeved openings for pipe penetrations. A release agent is then applied to the concrete floor slab so that the wet concrete for the wall does not stick to it. Once the wall casting is cured to full strength, it can be lifted in place through the use of a crane and spreader bar. The spreader bar allows the crane to lift the wall from several points, which reduces lifting stresses. This method also reduces the amount of material needed to make a form and eliminates the need for drilling holes for pipe penetrations.

FOUNDATIONS

Foundations for any building must be suitable to carry the loads imposed. The dimensional requirement of foundations is partially based on the soil conditions and the weight of the building to be constructed. Most foundations are either poured concrete or concrete blocks placed on adequate-sized footings. If there is

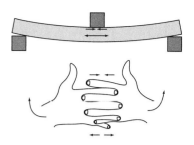

Figure 9-6 A beam supported on both ends will bow down in the middle if a large weight is placed on top. Notice how the bottom fingers are being pulled apart. This force is tension.

any possibility of water being present around the footings, footing drains should be installed (see Chapter 24 for subsoil drains).

CONSTRUCTION DRAWINGS

The drawings for the building must indicate the details of the construction. To the skilled craftsman, notes, sections, special views, and elevations convey how the building is to be built. Based on the structural and elevation blueprints, a skilled person should be able to determine the methods required and routings possible to install his or her work. To avoid surprises, it is very important that the estimator thoroughly inspect this material before beginning any construction.

Wall dimensions, wall surface materials, roof details, room dimensions, foundation details, and similar information are all needed to enable the mechanical installer to plan the work properly.

The analysis of these drawings should include the following items:

1. Type of hangers and supports needed and any special tooling required
2. Pertinent dimensions and other job details that affect your work
3. Pertinent roof structure notes that affect your work
4. Any requirement for sub-surface drainage and sub-surface pumping equipment, including venting
5. Floor and wall penetrations
6. Any offsets required
7. Pipe protection requirements—insulation and tracers, special mounting details
8. Fixture supports that may require backing boards, carriers, or other special supports
9. Any special foundation considerations

IMPORTANT: Any discrepancies or uncertain areas should be checked out with the architect or engineer before plumbers are on the project.

Typical Views

Usually, *typical views* (those that are the same throughout the building) are shown of all building components such as walls, doorjambs, window frames, etc. This prevents the architect from having to duplicate the same information. Enough different views of sections, elevations, and plans are shown to completely describe the item.

Elevation Drawings

Exterior elevation drawings show the exterior appearance of a building. They are usually presented showing the four faces of a building and the levels of floors, rooflines, and other major features. The elevations are usually named for the direction that the surface is facing. More simply put, if the observer were standing in the middle of the structure with a compass, the walls would be in the same position as the compass headings (see Figure 9-7).

Floor plans and elevation views can be used during the pre-planning stage of a project to determine if there are any discrepancies.

Table 9-1 lists some common items that should be considered when looking at a structural plan.

Note that neither of the sample jobs that accompany this textbook includes elevation views or fully developed structural plans. A large note on blueprint A-l of the restaurant project indicates that the outside walls are by the landlord. The restaurant blueprints include a ceiling plan and ceiling-framing plan.

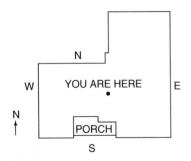

Figure 9-7 Notice how the elevations are named after the compass headings. Caution: North is not always toward the top of the paper.

Table 9-1   Structural Analysis Form

Job Name:	Reviewer:
Job Location:	Date:

Type of structural construction (frame, veneer, etc.)	
Type of foundation:	
Wall construction:	
Floor construction Basement floor	
Other floors	
Roof construction	
Special hangers or supports needed Type	
Location	
Type of pipe flashing required	
Number, size, and location of pipe thimbles or sleeves required. Label according to: stack A, riser A, vent A, etc. Floor 1	
Floor 2	
Floor 3	
Special types of pipe protection required (kickplates, insulation, heat tape, etc.)	
Fixture supports required (carriers, supports, backing boards, etc.)	
Special offsets or structural problems which must be noted or clarified	
Special notes or problems associated with the elevation drawings	

PLUMBING PLANS

The designer may show the plumbing work on a separate plan for plumbing only. For less complicated projects, the plumbing may be included on mechanical plans that show heating and air conditioning and, in some cases, even electrical work. For large jobs, the following elements of specialty work need to be shown:

1. Plumbing—drainage, waste, and vent (sanitary)
   a. Plumbing—water
   b. Plumbing—drainage (storm)
2. Heating, ventilating, air conditioning
3. Electrical—lighting
   a. Electrical—power

4. Fire protection, sprinkler, and/or life safety
5. Lawn sprinkler or irrigation
6. Fuel-gas piping
7. Vacuum, compressed air, special gases, and refrigeration piping
8. Any other special chemical or process piping

Symbols are used to identify particular piping and fixtures so that all the above systems can be visualized on the plans. Appendix B lists some of the most common of the customary fixture, piping, and appurtenance symbols used by many designers.

Analysis of Plumbing Plans

In the process of reviewing or bidding a job, the following points of work are typical issues to analyze on the plumbing plan:

1. Condensate piping from air conditioners
2. Water supply connections to special equipment, including backflow provisions
3. Water connections to fire standpipes
4. Any connections to special equipment, including refrigeration and food service
5. Location and possible interactions with HVAC components and HVAC flues and vents
6. Connections to solar equipment
7. Wiring of plumbing equipment
8. Foundation drainage
9. Check that where sewer line is going out of building matches where sewer line is coming into building

Also check for special hangers or equipment to any of the above equipment such as special purpose pumps, softeners, roof penetrations, backflow devices, special solutions, special metering or gauges, special-order devices, and any item whatsoever that needs plumbing connections to operate. Verify that each of these special items is to be installed in such a way that sufficient access is provided for service and maintenance. Review the implications for special tooling or critical tasks such as setting sleeves in concrete construction, etc. Be alert for conflicts with obstructions or insufficient available fall for required drain slopes.

Two Different Uses for Plans

There are at least two different purposes that a plumbing plan must serve: (1) to estimate the job and (2) to construct the job. The plan is used primarily to build the building, although we can also estimate from the plan.

Much of the fine detail shown is not necessary for estimating. Considerable detail indicated for all trades is not necessary for any one craft to build that portion of the project. As much as we would like to have perfect plans, it is extremely rare. However, every worker on a project can have all the purposes and limitations of the project in mind and still be aware that his work must coordinate with the work of others. If something is wrong, the architect or engineer should be notified as soon as possible so that corrections can be made.

Sample Plans

The remodeling job involves only one sink and a restroom with little detail shown. The estimator is required to check the job conditions, visualize the equipment connections, and prepare a quotation. The same process would be necessary to develop shop drawings for the installation. The work to be done is covered by Notes 1-5 on the drawing and by the specifications. Note that while the bar sink is not mentioned in the specifications, it is still a requirement of the job. This is an example of when clarifications need to be made with the architect or engineer and modifications made.

The restaurant piping is not marked very often and there is at least one error. If the incoming line is 1 1/2", then 3/4", and then 1 1/4", then the 3/4" size must be wrong! Also, note that there are some 1/4" branches to fixtures. In many jurisdictions, the smallest branch size is 1/2". Note that some piping is outdoors. While this may be satisfactory in warm regions, such an installation would lead to disaster in northern parts of the country!

Table 9-2 can be used to help organize the information that should be considered when dealing with the appliances on the job.

Table 9-2 **Plumbing Appliance Analysis**

Job Name:		Reviewer:			
Job Location:		Date:			
Appliance	Manufacturer	Size	Type	Input	Remarks: energy into appliance, flue piping, supply piping, backflow devices, special fittings, safety equipment, etc.

ELECTRICAL PLANS

Electricians and plumbers frequently work on the same equipment and install their apparatus in the same spaces. As a result, there are many circumstances where interference can occur. To minimize the chances of such interference, the electrical plans should be studied to discover where conflicts are likely to occur.

Electrical plans are working drawings that show the installation and material details of the electrical portion of a building project. The architect's floor plans are the beginning documents for these drawings. As in other specialty drawings, electrical drawings utilize symbols to represent switches, receptacles, circuit breakers and panels, and related wiring. Figure 9-8 shows typical symbols for electrical plans. It should be noted that these symbols will be different than those used on schematics for troubleshooting appliances, which are shown in Figure 29-2.

Analysis of Electrical Plans

Review electrical plans with the following points in mind:

1. Study the Symbols Table for the electrical devices.
2. Find the location of the electrical panel boards, usually a circuit breaker panel, or other sources that supply power to equipment you must furnish (pumps, appliances, and appurtenances).
3. Assure yourself that the proper power characteristic (voltage, 1 phase or 3 phase) is available at the panel board.
4. Check for conflicts between your piping and electrical conduits and raceways.
5. Determine who is responsible for the cost of electrical connections to mechanical equipment—the electrical contractor, mechanical contractor, or others. Example: Who supplies the disconnects for a water heater?

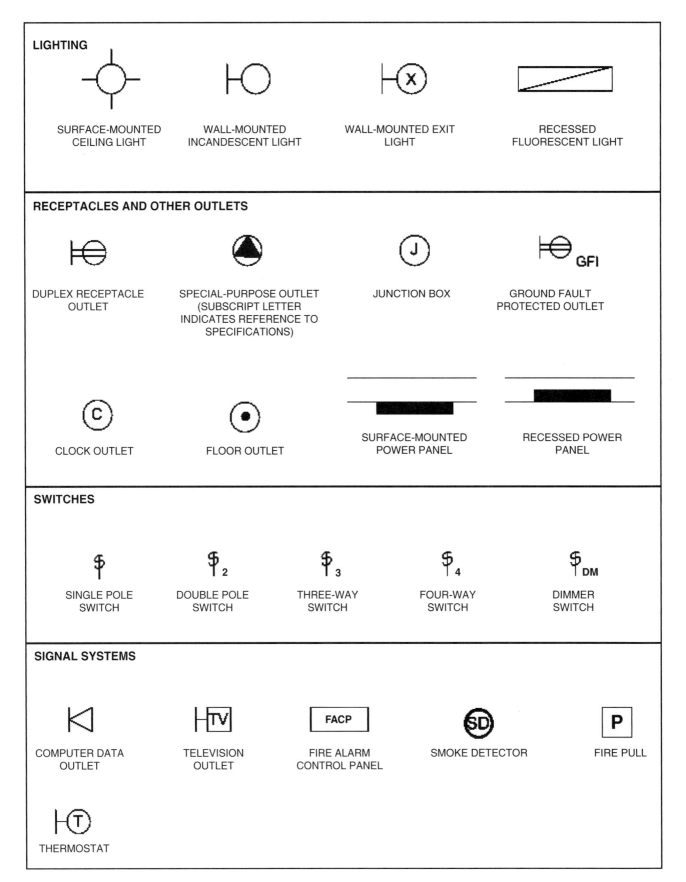

Figure 9-8 Electrical symbols found on architectural plans.

A side-by-side analysis of the plumbing and electrical drawings should include the following elements (see Table 9-3):

1. A list of plumbing equipment that requires electrical connections including item name, number and location, and electrical requirements
2. A list of conflicts or interferences
3. A list of special requirements for electrical information or instructions required to be furnished by the plumbing contractor

Table 9-3 Electrical Plan Analysis

Job Name:		Reviewer:		
Job Location:		Date:		
Item	Location of reference number	Voltage	Amperage	Phase
Interferences that occur in plumbing and electrical piping or equipment				
Location		Name of interference		
Special responsibilities of plumbing mechanic relating to the electrical system. For example, provide wiring schematics for the following equipment.				
Problems of discrepancies relating to the electrical system that must be resolved.				

HVAC PLANS

Designers will usually include as many aspects of the mechanical systems on a single drawing as can be shown without excessive complications. It is desirable from the installing technician's point of view to have several systems shown because it highlights conflicts and interferences. At some point, however, the confusion and difficulty in following one system when several are shown outweighs the advantage of showing correlations. Thus, on major projects, each mechanical subsystem is usually depicted on separate drawings which should be done before going to the jobsite.

Heating, ventilating, and air-conditioning functions are often provided by a single system. There are times when separate exhaust air systems are needed, and even though they require separate systems and equipment, they are almost always included on the HVAC plans. The air-conditioning ductwork is almost always the most difficult job component to be modified or rerouted. Therefore, the HVAC plans should be studied to see if the plumbing piping would have to be modified to avoid conflicts. Check for HVAC equipment that requires plumbing connections or has special components (such as flues or chimney piping). Make sure to coordinate with the HVAC contractor before you start running pipe or ductwork!

Analysis of HVAC Plans

The HVAC plan uses special symbols that are placed on the architect's floor plan to represent their location and size. These plans will show the following:

1. The ductwork layout
2. Equipment locations
3. Location of fuel piping, controls, flues, and other special apparatus required for the HVAC system
4. Water and drain piping
5. If running gas to rooftop unit, check where to stub gas line so it is located close to the HVAC unit

Typical HVAC symbols are shown in Figure 9-9.

The plumbing technician should develop a list showing any plumbing requirements or obligations for the HVAC equipment. These requirements often include water for humidifiers, drainage of condensate, or fuel gas supply for heaters. The technician should also list any interferences or conflicts that must be resolved. On industrial or large buildings, other possible mechanical systems might include vacuum systems, compressed air piping, fire protection systems, and irrigation systems. Figure 9-10 shows some common line type abbreviations for these other mechanical systems.

If the job involves any of these additional systems, it will be necessary to review each one for conflicts and required actions (if any) by the plumbing contractor. There are usually large fines for causing a false alarm that results in the local fire department responding, so take special precaution when working around an existing fire protection system. A list should be developed of special equipment or tools required to make the plumbing connections to the HVAC equipment. This list should include cross-connection protective devices, welding equipment, access, and hangers and supports. Table 9-4 lists some of the more common characteristics to look at when preparing a job.

DETAIL PLANS

In order to produce the designer's intentions, many aspects of a building require large-scale drawings that show the precise arrangement of parts. These large-scale drawings are frequently called *detail drawings* or simply *details*. Such drawings are developed to show the construction of door and window openings, wall construction, structural connections, and installation of accessory equipment of all sorts.

Plumbing and mechanical drawings usually have many features that require detail drawings. Hangers, roof openings, and equipment mounts are also typically redrawn in a larger scale so that the installer can see how the components will be integrated into the building.

In some cases, the contractor or equipment supplier furnishes the detail drawings to the architect/engineer for that professional to review the intended method of installation. Thus, the shop drawings are a version of detail drawings. Detail

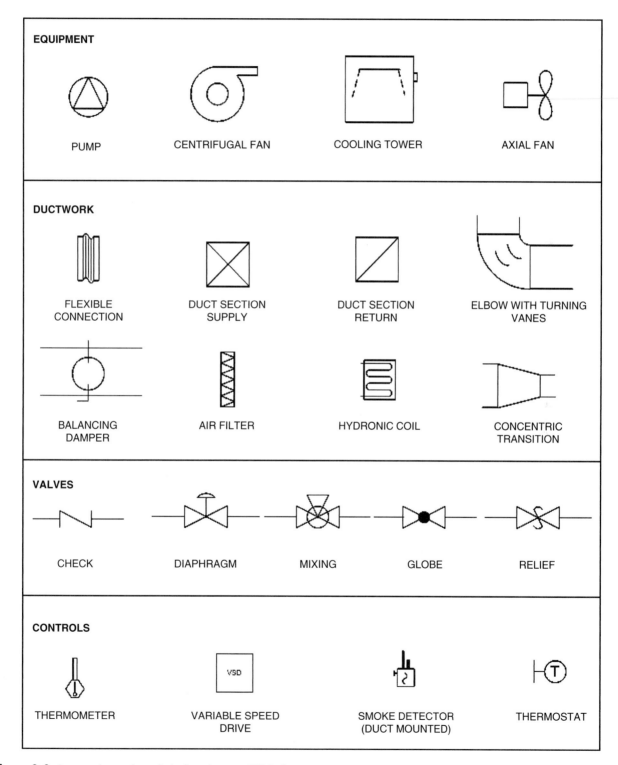

Figure 9-9 Commonly used symbols found on an HVAC plan.

AIR-CONDITIONING PIPING

COMPRESSED AIR	——A——
HOT WATER SUPPLY	——HWS——
HOT WATER RETURN	——HWR——
LOW PRESSURE STEAM	——RL——
CONDENSATE RETURN	——CR——
CHILLED WATER SUPPLY	——CHWS——
CHILLED WATER RETURN	——CHWR——
REFRIGERANT LINE	——RL——
REFRIGERANT SUCTION	——RS——
DRAIN	——D——
DOMESTIC HOT WATER SUPPLY	——DHWS——
DOMESTIC HOT WATER RETURN	——DHWR——

NOTE: DOMESTIC WATER IS COMMONLY USED TO REFER
TO POTABLE WATER LINES.

Figure 9-10 Common abbreviations for piping found on an HVAC or plumbing plan.

Table 9-4 HVAC Plan Analysis

Job Name:			Reviewer:		
Job Location:			Date:		
Type of equipment	Location of identifying #	Type of piping (water, DWV, fuel gas)	Material	Remarks: interferences between plumbing and HVAC ductwork and/or equipment	

drawings usually take precedence over other project documents, but in case of conflict, get clarification from the job superintendent.

Floor plans and other usual scale drawings indicate the portion of the building illustrated in details. Blueprint A-1 of the restaurant job shows a detailed cross-sectional view (I) of how a wall is to be constructed and anchored to the floor. The indicator is a symbol showing the detail number and the drawing page where it is drawn. Table 9-5 lists some of the more common characteristics of a detail drawing for a piece of mechanical equipment that might be reviewed when preparing a job.

Table 9-5 Detail Drawing Analysis Form

Job Name:					Reviewer:		
Job Location:					Date:		
Ratings of the appliance (only where applicable)							
Appliance	Electrical	Gallons capacity	Pumping capacity	Piping diameter inlet	Piping diameter outlet	Alternate operating supply needed	Remarks: (fasteners, hangers, backflow, preventers)

REVIEW QUESTIONS

1 How are elevation plans labeled?

2 What are the three most common types of frame construction?

3 What would be one reason a plumber might need to look at the structural plans?

4 What component of the HVAC plans usually causes the most interference?

5 Give an example of a piece of equipment that may need to have water supplied to it.

CHAPTER

10

National Fuel Gas Code, Materials, and Types of Fuel Gases

LEARNING OBJECTIVES

The student will:

- Describe suitable materials for gas piping systems.
- Describe proper techniques for working with gas piping systems.
- Summarize the physical and chemical properties of common fuel gases.

NATIONAL FUEL GAS CODE

uel gas codes came into existence for the same reasons that produced plumbing codes. Considerations of safety, adequate service life, efficient installations, and predictable system capacities were best addressed by having interested persons combine their experience and knowledge to develop what we now know as the *National Fuel Gas Code.*

A properly designed and installed fuel gas system provides efficient and safe use of the gas for applications such as cooking, water heating, space heating, and industrial uses. Licensing, permits, and inspection processes are set in place to ensure that work has been performed in a proper manner.

The principal hazards of improper or inadequate fuel gas installations include poor equipment operation, fire, explosion, and asphyxiation. To guard against these problems, codes specify minimum material standards, installation methods, sizing requirements, and testing procedures. It should be pointed out that the code provisions are the minimum requirements that will ensure safe, efficient operation of the equipment. The knowledgeable installer should suggest to customers that going beyond the code requirements is often a good idea. Increased pipe sizing, for example, makes future additions easy to accommodate. Additional valves make servicing equipment less disruptive.

The fuel gas code that is in the widest use in the United States is the *National Fuel Gas Code ANSI Z223.1-2006 NFPA 54-2006.* The most recent edition is dated 2006, but code revisions are being considered almost continuously. The table of contents shows the following major topics:

- Chapter 1 **Administration,** describes what systems are covered by the code and how it will be enforced.
- Chapter 2 **Referenced Publications,** (Reserved).
- Chapter 3 **Definitions,** describes the terms used in the code book.
- Chapter 4 **General,** describes how the work will be performed and potential ignition sources.
- Chapter 5 **Gas Piping System Design, Materials, and Components,** describes piping requirements.
- Chapter 6 **Gas Piping Installation,** covers the details of piping installation.
- Chapter 7 **Inspection, Testing, and Purging,** lists the how and where of these activities.
- Chapter 8 **Equipment Installation,** covers the design of equipment application.
- Chapter 9 **Installation of Specific Equipment,** lists the details for many types of specific appliances.
- Chapter 10 **Venting of Equipment,** covers safe venting procedures.
- Chapter 11 **Procedures to be Followed to Place Equipment in Operation,** describes the necessary actions for equipment start-up.
- Chapter 12 **Pipe Sizing,** describes the sizing and capacities of gas piping.
- Chapter 13 **Sizing of Category I Venting Systems,** describes piping requirements for venting equipment.
- Chapter 14 **Referenced Publications,** lists documents that have been referenced throughout the code book.

The following twelve appendices from the *National Fuel Gas Code* contain information supplemental or helpful to code applications:

- Appendix A: Explanatory Material
- Appendix B: Coordination of Gas Utilization Equipment Design, Construction, and Maintenance
- Appendix C: Sizing and Capacities of Gas Piping
- Appendix D: Suggested Method for Checking for Leakage
- Appendix E: Suggested Emergency Procedure for Gas Leaks

- Appendix F: Flow of Gas Through Fixed Orifices
- Appendix G: Sizing of Venting Systems Serving Appliances Equipped With Draft Hoods, Category I Appliances, and Appliances Listed for Use With Type B Vents
- Appendix H: Recommended Procedure for Safety Inspection of an Existing Appliance Installation
- Appendix I: Indoor Combustion Air Calculation Examples
- Appendix J: Example of Combination of Indoor and Outdoor Combustion and Ventilation Opening Design
- Appendix K: Other Useful (Industry) Definitions
- Appendix L: Informational References

Common Errors

Some common errors that occur in the gas piping industry are the result of either poor planning or poor installation. The installing technician should think about not only how to install the job as efficiently as possible, but also how to make it last. If the installer were to place aluminum pipe or tubing in contact with concrete, or place a non-coated ferrous pipe in an underground installation, the life span of the materials may be relatively short. The piping must be protected from any materials or atmosphere exerting a corrosive action. Consider that the material easiest to install may not be the best for the job requirements.

The other common error arises from faulty workmanship. The installer should not only purchase good quality material, but also should make every effort to install it properly. For example, the failure to hang pipe properly or sleeve a foundation wall could cause failure of fittings, valve bodies, or joints. Errors are also made by placing the equipment or piping in areas that could lead to damage, such as placing a gas meter near a driveway without proper impact protection. This could lead to a gas leak or explosion. Other problems, like insufficient ventilation for fuel-burning equipment, could cause failure or personal/property damage.

Proper Techniques

When working with gas piping or equipment, the technician should always think of safety, not only for herself, but also for those around her. The following are some proper techniques that should be observed at all times:

1. Never work on a fuel gas system when the gas is on.
2. Notify all residents of a building when alterations are to be made.
3. Turn off each appliance before turning off the gas supply.
4. Relight each pilot burner when changes are completed.
5. Use an approved gas detector, non-corrosive leak detection fluid, or other safe leak test material to check for leaks.
6. *Do not* activate any electrical switches—either on or off—if you suspect a leak. Use an only flashlights for illumination.
7. Do not smoke where gas piping is being installed, repaired, or serviced.
8. Test and purge any gas piping before placing in service.
9. Develop a complete design before starting a piping installation or repair.

ALLOWABLE MATERIALS FOR FUEL GAS PIPING

Materials used for gas piping should be clean of foreign materials and free of defects. All burrs from cutting and threading should be removed as well as any chips or scale. No defective piping should be repaired. The following is a list of the different types of materials approved by the *National Fuel Gas Code:*

Steel or Iron

- Steel pipe, conforming to ASTM A53, black or galvanized, is used for most installations.
- Standard malleable pipefittings rated 125 psi working pressure are used with steel pipe.
- Joints may be welded, flanged, or threaded. Threaded fittings are malleable, not cast. Special-purpose compression joint fittings are available for steel pipe.
- If installed underground, steel pipe should be wrapped or coated to prevent corrosion. Cathode protection, which is covered in Chapter 14, is also used on some systems.
- Ductile iron pipe (3" minimum size) may be used underground, but for outdoor use only.
- Corrugated stainless steel tubing with specially designed fittings can be used in fuel gas systems that operate at 5 psig. Fittings and pipe must be from the same manufacturer. Note that they may appear to be the same, but they are not. Some utilities may require that CSST be grounded and/or bonded. Consult with your local building official for requirements in your area.

Copper, Brass, or Aluminum

- Copper, brass, or aluminum pipe may be used if the gas is not corrosive to the pipe material. All fittings should be compatible for use with the pipe. For copper or brass tubing, the gas cannot contain more than 0.3 grains of hydrogen sulfide per 100 scf of gas.
- Aluminum pipe must be coated to protect against corrosion and should not be installed where it may be subjected to repeated wetting. Aluminum must not be used outdoors, nor in contact with concrete or mortar.
- Joints may be threaded, flanged, welded, or joined by slipnuts that develop high compressive forces on the exterior of the pipe wall. Such compression joints and conventional unions are required to be accessible.
- Brazing may be used if the filler metal has a melting point above 1000° F.

Metallic Tubing

- Metallic tubing, joints, and fittings may be copper, aluminum, Corrugated Stainless Steel [CSST], or steel, provided the gas is not corrosive to the tubing material.
- Aluminum tubing has the same restrictions as aluminum pipe.
- Approved fittings must be used.
- Brazing may be used if the melting point of the filler metal is above 1000° F.
- Joints in tubing (except brazed) are not permitted in concealed spaces.

Plastic Pipe, Tubing, and Fittings

- Plastic pipe, tubing, and fittings, which conform to ASTM D 2513, Standard Specification for Thermoplastic Gas Pressure Pipe, may be used outside, underground only, and in accordance with the manufacturer's instructions.
- Solvent cementing, adhesives, heat-fusion, compression couplings, or flanges may be used to make joints in plastic pipe.

Valves

Valves for gas systems must meet standards for gas service and may include gate, ball, and plug valves. Lever-handle plug cocks are preferred over square-head types because operation can be accomplished more quickly in an emergency. Many new gas valves have a small 1/8" NPT test port on them so that gas pressures can be checked. Gas valve trim shall include non-drying bonnet packing and substantial lubrication for larger plug-type cocks. Plug valves shall be designed and constructed so that the plug is retained in the body even if the packing collar is removed. Valves should be installed in locations that are readily accessible to facilitate operation in an emergency.

If a valve is located in a pipeline (or otherwise) so that the section served is not apparent, provide a tag on the valve that describes the area or devices that are controlled by the valve.

Always verify that the materials being used on a project are in conformance with job requirements and applicable gas code. Conventional ball and gate valves are permitted to be used for gas service. However, listed gas valves are usually less expensive, easier working, and known to be adequate for gas applications.

Table 10-1 is a list of standards for the major requirements for gas piping materials.

Table 10-1   **Major Standards for Approved Materials According to the** *National Fuel Gas Code,* 2002

Steel pipe	ANSI/ASTM A53. Standard Specification for Pipe, Steel, Black and Hot-Dipped Zinc-Coated, Welded, and Seamless
Aluminum alloy pipe	ASTM B241. Specification for Aluminum-Alloy Seamless Pipe and Seamless Extruded Tube
Copper tubing	ASTM B88. Specification for Seamless Copper Water Tube Type K or L
	ASTM B280. Specification for Seamless Copper Tube for Air Conditioning and Refrigeration Field Service
Steel tubing	ANSI/ASTM A539. Standard Specification for Electric-Resistance-Welded Coiled Steel Tubing for Gas and Fuel Oil Lines
	ANSI/ASTM A254. Standard Specification for Copper Brazed Steel Tubing
Aluminum alloy tubing	ASTM B210. Specification for Aluminum-Alloy Drawn Seamless Tubes
	ASTM B241. Specification for Aluminum-Alloy Seamless Pipe and Seamless Extruded Tube
Corrugated Stainless Steel	ANSI LC 1/CSA 6.26. Fuel Gas Piping Systems Using Corrugated Stainless Steel Tubing
Plastic pipe and tubing	ASTM D2513. Specification for Thermoplastic Gas Pressure Pipe, Tubing, and Fittings
	ASTM D2517. Specification for Reinforced Epoxy Resin Gas Pressure Pipe and Fittings
Metallic Threads	ANSI/ASME BI.20.1. Standard for Pipe Threads, General Purpose (Inch)

TYPES OF FUEL GASES AND THEIR PROPERTIES

A gas is a fluid that will expand to completely fill the vessel in which it is contained. The particles of a gas will easily move and change relative positions without a separation of the mass and easily yield to pressure. Fuel gases fall under one of the following four categories:

1. Natural gas
2. Liquefied petroleum (LP) gas
3. Manufactured gas
4. Special-purpose gas

Manufactured, LP, and special-purpose gases are usually distributed in pressurized tanks and transported by truck to the point-of-use. Natural gas is most often delivered by pipeline to the point-of-use. LP gas is used for portable appliances and as an alternative to natural gas in areas where natural gas is not distributed.

Fuel Gas Properties

Two characteristics of fuel gases are used to define their properties: *specific gravity* and *heating value* (or heat content).

Specific Gravity

Specific gravity is the ratio of the density of the gas to the density of air. A value greater than one means that the gas is heavier than air; a value less than one means that the gas is lighter than air. More energy would be required to move a heavier gas through a pipe than would be required to move a lighter gas. This energy difference will show up in the form of a pressure drop. Or more simply put, if the same quantities of gas were to move through a pipe, the heavier the gas, the more the pressure will drop.

Heating Value

The heating value of a fuel gas is the amount of heat that is released when the fuel is completely burned. The unit of measure in the United States is the British Thermal Unit (BTU). A BTU is the amount of heat required to raise the temperature of one pound of water one degree Fahrenheit. Table 10-2 shows the specific gravity and heating value of three common fuel gases.

Table 10-2 **Properties of Fuel Gases**

Fuel Gas	Formula	Specific Gravity (Air = 1.0)	Heating Value BTU/SCF*
Natural (Methane)	CH_4	0.60	1000
Propane	C_3H_8	1.52	2516
Butane	C_4H_{10}	2.01	3280

*SCF = Standard Cubic Foot

All these gases–natural, propane, and butane–are odorless. An odorant chemical is added to them to act as a telltale of a gas leak occurrence. Natural gas is a mixture that varies somewhat; specific gravity will vary from 0.60 to 0.65 and heating value will vary from 1000 to 1060 BTU/ft^3. The local jurisdiction gas supplier should be consulted for specific gravity and heat values. It should be pointed out that natural gas is much lighter than air, so when released into the atmosphere, it will tend to rise and disperse rapidly. The LP gases, on the other hand, are heavier than air and will sink and collect in low spots in the vicinity of the release point.

Natural Gas

Natural gas is a mixture of gases, principally methane and small amounts of other gases. Methane is the lightest product in a chemical family that extends through LP gases, gasoline, diesel fuel, and heating oils. Volume meters record natural gas volume so that the utility can charge the user for the fuel used. The usual basis of charge is the *therm*, the amount of gas that delivers 100,000 BTU.

Combustion

When combined with oxygen, methane (CH_4) produces carbon dioxide, water vapor, and heat. It should be pointed out that when air is used as the source of oxygen for the combustion process, a large amount of nitrogen (air is approximately 20% oxygen and 80% nitrogen) passes through the fire without participating in the chemical reaction. Even though it does not change the amount of heat released in the combustion process, the presence of the nitrogen reduces the net heat available to the material being heated because of the heat taken up by the nitrogen.

LP Gas

The LP gases are propane (C_3H_8) and butane (C_4H_{10}). They produce identical products of combustion, but have higher heating values per cubic foot of gas. The LP gases liquefy at relatively low pressures (in the range of 100 psi at normal room temperature), so very large amounts of energy can be contained in a small volume. When this pressure is relieved at usual temperatures, these products become gases. Thus, they are stored as liquids and used as gases.

Due to its high specific gravity, any leaked LP gas will tend to stay low to the ground and not disperse like natural gas. Because of this, LP gas installations must be handled with considerable care and all equipment should be guarded against damage.

Manufactured Gases

Manufactured gases are developed for special purposes or as a by-product of some other process. Manufactured gas, produced by the non-destructive breakdown of heated coal, was the principal fuel gas used in most of the United States in years past, but now it is used only in special cases.

Special-Purpose Gases

Special purpose gases include acetylene, hydrogen, MAPP™, and any other gas used for a heating purpose. Most often, these gases are used at a single burner and are hose-connected to a tank, but occasionally extensive piping systems are employed for industrial applications such as gas welding.

ACKNOWLEDGEMENT

Materials in Chapters 10 through 17 rely extensively on material from the *National Fuel Gas Code*. This material has been reprinted with permission as follows:

Reprinted with permission from NFPA 54-2006, *National Fuel Gas Code* Copyright © 2006, National Fire Protection Association, Quincy, MA 02169. This reprinted material is not the complete and official position of the National Fire Protection Association on the referenced subject, which can only be represented by the standard in its entirety.

REVIEW QUESTIONS

1   Where could the heating value of the natural gas be found?

2   What is a BTU?

3   Where can plastic gas piping be installed?

4   If a gas were to have a specific gravity of 0.55, would it be heavier or lighter than air?

5   What is a "therm"?

CHAPTER 11

Fuel Gas Pipe Sizing

FUEL GAS PIPE SIZING

This chapter will discuss using the longest run method for pipe sizing of small and large systems. Most residential and small commercial jobs will be low-pressure systems that can easily be sized by the longest run method which is the most conservative method to choose from.

In order to properly lay out a gas system, the following pieces of information will be required:

- Type of gas and its specific gravity
- Maximum gas flow required at each appliance
- Length of piping to each appliance
- Type of pipe to be used
- Gas pressure in the system
- Diversity factor (permitted for special systems)
- Allowable pressure drop from the meter or regulator to the farthest appliance

In most cases, the type of gas, amount of gas per appliance, and distance from the meter to the appliance have already been determined, so it's up to the technician to decide what type of pipe and what pressure can be used. The *system pressure* is the operating pressure that can be found from the gas meter to the gas valve on the appliance. Most gas valves used on natural gas-fired water heaters and furnaces are limited to a maximum of 1/2 psi or 14" W.C. The gas valve at the appliance will reduce the gas pressure down to 3"-3-1/2" of water column at the burner manifold.

Most residential and similar small commercial systems operate at a pressure equal to 6" of water column, usually written 6" H_2O or 6" W.C., which is equivalent to 0.217 psi or about 1/4 psi (one foot of water column is equal to 0.434 psi). Table 11-1, Gas Flow Capacity for Schedule 40 Pipe, is a presentation of the capacity of various pipe sizes for several different lengths. The table is valid for any piping system where the inlet pressure is less then 2 psi and the gas has a specific gravity of 0.60. The chart also gives the maximum amount of gas that can be expected to flow with a pressure drop of only 0.3" W.C. This table covers nearly all residential natural gas applications.

Example 1

For an appliance that requires 100 cfh (cubic feet per hour), what size pipe could be used for the following distances? Verify the following values from Table 11-2.

In order to properly use Table 11-1, the technician needs to know how much gas the appliance is expected to burn per hour of run time. This can be determined by dividing the firing rate of the appliance by the heating value of the fuel gas. The next example demonstrates how this is done for a common water heater.

Example 2

If a 50 gallon water heater requires 40,000 BTU/hr input, what gas volume is required if the heating value of the gas is 1000 BTU/ft^3?

$$\text{Volume of gas required} = \frac{\text{Firing rate of appliance}}{\text{heating value of gas}}$$

$$\text{Volume of gas required} = \frac{40,000 \text{ BTU/hr}}{1000 \text{ BTU/ft}^3}$$

$$\text{Volume of gas required} = 40 \text{ cfh}$$

Table 11-1 Gas Flow Capacity for Schedule 40 Pipe

Gas:													Natural	
Inlet Pressure:													Less than 2 psi	
Pressure Drop:													0.3 in. w.c.	
Specific Gravity:													0.60	

Pipe Size (in.)

Nominal:	1/2	3/4	1	1 1/4	1 1/2	2	2 1/2	3	4	5	6	8	10	12
Actual ID:	0.622	0.824	1.049	1.380	1.610	2.067	2.469	3.068	4.026	5.047	6.065	7.981	10.020	11.938
Length (ft)	\multicolumn{14}{Capacity in Cubic Feet of Gas per Hour}													
10	131	273	514	1,060	1,580	3,050	4,860	8,580	17,500	31,700	51,300	105,000	191,000	303,000
20	90	188	353	726	1,090	2,090	3,340	5,900	12,000	21,800	35,300	72,400	132,000	208,000
30	72	151	284	583	873	1,680	2,680	4,740	9,660	17,500	28,300	58,200	106,000	167,000
40	62	129	243	499	747	1,440	2,290	4,050	8,270	15,000	24,200	49,800	90,400	143,000
50	55	114	215	442	662	1,280	2,030	3,590	7,330	13,300	21,500	44,100	80,100	127,000
60	50	104	195	400	600	1,160	1,840	3,260	6,640	12,000	19,500	40,000	72,600	115,000
70	46	95	179	368	552	1,060	1,690	3,000	6,110	11,100	17,900	36,800	66,800	106,000
80	42	89	167	343	514	989	1,580	2,790	5,680	10,300	16,700	34,200	62,100	98,400
90	40	83	157	322	482	928	1,480	2,610	5,330	9,650	15,600	32,100	58,300	92,300
100	38	79	148	304	455	877	1,400	2,470	5,040	9,110	14,800	30,300	55,100	87,200
125	33	70	131	269	403	777	1,240	2,190	4,460	8,080	13,100	26,900	48,800	77,300
150	30	63	119	244	366	704	1,120	1,980	4,050	7,320	11,900	24,300	44,200	70,000
175	28	58	109	224	336	648	1,030	1,820	3,720	6,730	10,900	22,400	40,700	64,400
200	26	54	102	209	313	602	960	1,700	3,460	6,260	10,100	20,800	37,900	59,900
250	23	48	90	185	277	534	851	1,500	3,070	5,550	8,990	18,500	33,500	53,100
300	21	43	82	168	251	484	771	1,360	2,780	5,030	8,150	16,700	30,400	48,100
350	19	40	75	154	231	445	709	1,250	2,560	4,630	7,490	15,400	28,000	44,300
400	18	37	70	143	215	414	660	1,170	2,380	4,310	6,970	14,300	26,000	41,200
450	17	35	66	135	202	389	619	1,090	2,230	4,040	6,540	13,400	24,400	38,600
500	16	33	62	127	191	367	585	1,030	2,110	3,820	6,180	12,700	23,100	36,500

(continued)

Table 11-1 (Continued)

						Pipe Size (in.)						Gas:	Natural	
												Inlet Pressure:	Less than 2 psi	
												Pressure Drop:	0.3 in. w.c.	
												Specific Gravity:	0.60	
Nominal:	1/2	3/4	1	1 1/4	1 1/2	2	2 1/2	3	4	5	6	8	10	12
Actual ID:	0.622	0.824	1.049	1.380	1.610	2.067	2.469	3.068	4.026	5.047	6.065	7.981	10.020	11.938
Length (ft)						Capacity in Cubic Feet of Gas per Hour								
550	15	31	59	121	181	349	556	982	2,000	3,620	5,870	12,100	21,900	34,700
600	14	30	56	115	173	333	530	937	1,910	3,460	5,600	11,500	20,900	33,100
650	14	29	54	110	165	318	508	897	1,830	3,310	5,360	11,000	20,000	31,700
700	13	27	52	106	159	306	488	862	1,760	3,180	5,150	10,600	19,200	30,400
750	13	26	50	102	153	295	470	830	1,690	3,060	4,960	10,200	18,500	29,300
800	12	26	48	99	148	285	454	802	1,640	2,960	4,790	9,840	17,900	28,300
850	12	25	46	95	143	275	439	776	1,580	2,860	4,640	9,530	17,300	27,400
900	11	24	45	93	139	267	426	752	1,530	2,780	4,500	9,240	16,800	26,600
950	11	23	44	90	135	259	413	731	1,490	2,700	4,370	8,970	16,300	25,800
1,000	11	23	43	87	131	252	402	711	1,450	2,620	4,250	8,720	15,800	25,100
1,100	10	21	40	83	124	240	382	675	1,380	2,490	4,030	8,290	15,100	23,800
1,200	NA	20	39	79	119	229	364	644	1,310	2,380	3,850	7,910	14,400	22,700
1,300	NA	20	37	76	114	219	349	617	1,260	2,280	3,680	7,570	13,700	21,800
1,400	NA	19	35	73	109	210	335	592	1,210	2,190	3,540	7,270	13,200	20,900
1,500	NA	18	34	70	105	203	323	571	1,160	2,110	3,410	7,010	12,700	20,100
1,600	NA	18	33	68	102	196	312	551	1,120	2,030	3,290	6,770	12,300	19,500
1,700	NA	17	32	66	98	189	302	533	1,090	1,970	3,190	6,550	11,900	18,800
1,800	NA	16	31	64	95	184	293	517	1,050	1,910	3,090	6,350	11,500	18,300
1,900	NA	16	30	62	93	178	284	502	1,020	1,850	3,000	6,170	11,200	17,700
2,000	NA	16	29	60	90	173	276	488	1,000	1,800	2,920	6,000	10,900	17,200

NA means a flow of less than 10 cfh.
Note: All table entries are rounded to 3 significant digits.

Source: Reprinted with permission from NFPA 54, *National Fuel Gas Code*, Copyright © 2006, National Fire Protection Association, Quincy MA 02169. This reprinted material is not the complete and official position of the National Fire Protection Association on the referenced subject which is represented only by the standard in its entirety.

Table 11-2 Pipe Sizes for 100 Cubic Feet per Hour with a Maximum of 0.3" W.C. Pressure Drop

Length of Pipe (feet)	Pipe Size (inches)
10	1/2
60	3/4
200	1

The following is a more involved problem:

Example 3

What pipe size is required to accommodate a 55,000 BTU/hr heater located 150 feet from the meter if natural gas (1000 BTU/ft^3 heating value) is used?

$$\text{Volume of gas required} = \frac{55,000}{1000} = 55 \text{ cfh}$$

Table 11-1 shows that at 150' length, 1/2" pipe is only rated for 30 cfh while 3/4" is rated for 63 cfh. Therefore, the minimum pipe size is 3/4".

Longest Run Method

When there is more than one appliance in the building, the gas system must be able to supply all of the appliances with the required amount of gas at the appropriate pressure. The worst case scenario would be when all appliances are firing at the same time. The *longest run method* looks for the appliance that is located the farthest from the meter and bases all pipe capacities on this length. The method also starts at the farthest appliance and works back to the meter. As each additional appliance is connected to the system, the required capacity of the piping must be increased. Example 4 shows a typical residential layout with three common household appliances.

Example 4

Figure 11-1 shows a piping layout involving three gas appliances served by a 6" W.C. gas system. The maximum pressure drop is 0.3" W.C., specific gravity is 0.60, and the heating value is 1000 BTU/ft^3. Piping system is constructed of Schedule 40 pipe. Length to the farthest appliance = (30 + 15 + 30) = 75'. The table value of 80' must be used.

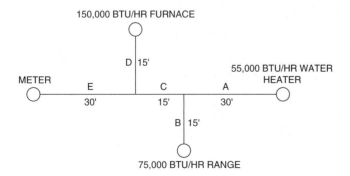

Figure 11-1 Residential layout using standard natural gas.

Follow these steps:

1. Indicate on the piping sketch the gas flow (in cubic feet per hour) needed at each appliance.
2. Determine the length of pipe from the farthest appliance to the meter.
3. Select the proper table from the code (Table 11-1 is the correct one in this case) and the correct column. Use the column for this farthest distance to size all piping.
4. If the actual calculated length is not listed in the table, select the column for the next greater length.

Table 11-3 shows the solutions to this problem.

Table 11-3   Example 4

Pipe Section	Gas Demand (cfh)	Pipe Size
A	55	3/4"
B	75	3/4"
C	55 + 75 = 130	1"
D	150	1"
E	130 + 150 = 280	1-1/4"

If the gas supplied by the local gas company were at a different specific gravity than the standard 0.60, the figures in Table 11-1 would have to be altered to account for the difference. Table 11-4 shows that as the gas becomes denser, or has a greater specific gravity, the less volume can be moved through the pipes at the same pressure drop.

Table 11-4   Multipliers Used with Capacities Shown in Gas Tables When Specific Gravity of Gas Is Other Than 0.60

Specific Gravity	Multiplier	Specific Gravity	Multiplier
.40	1.23	1.00	.78
.50	1.10	1.10	.74
.60	1.00	1.20	.71
.70	.93	1.30	.68
.80	.87	1.40	.66
.90	.82	1.50	.63

For a more complete chart see Table C.3.4 of the 2006 *National Fuel Gas Code*.

In the next example, the specific gravity has been increased to 1.5, which means this particular gas is much heavier compared to the standard natural gas.

Example 5

Solve Example 4 when the specific gravity of the fuel gas is 1.5. See Table 11-4 for the multiplier to use for this density. Table 11-5 shows how the capacity of each pipe size has been reduced because of the heavier gas. Figure 11-2 shows the pipe layout and length as well as the equipment firing rate.

Table 11-5   Pipe Capacities Modified for Heavier Gas

	Pipe Size (in.)					
Nominal	1/2"	3/4"	1"	1-1/4"	1-1/2"	2"
80' (0.60 sp. gr.)	42	89	167	343	514	989
80' (1.50 sp. gr.)	27	56	105	216	324	623

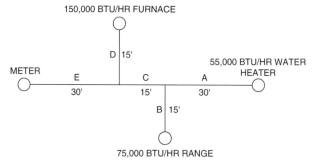

Figure 11-2 Residential layout using heavier gas.

Table 11-6 shows the modification of Table 11-3, reflecting the larger pipe sizes due to the heavier gas.

Table 11-6 Example 5

Pipe Section	Gas Demand (cfh)	Pipe Size (reduced capacity)
A	55	3/4"
B	75	1"
C	55 + 75 = 130	1-1/4"
D	150	1-1/4"
E	130 + 150 = 280	1-1/2"

NOTE: No actual gas has a heating value of 1000 BTU/ft^3 and a specific gravity of 1.5.

FUEL GAS PIPE SIZING FOR LARGER SYSTEMS

The sizing of large industrial gas piping systems is performed in a similar fashion to the methods used for smaller systems except that the equivalent length of piping from the meter to the farthest appliance is increased by 50% as an allowance for fittings in the pipe run or the equivalent length of each fitting is added in the pipe run.

Since a wider range of initial gas pressures and pressure drops is often encountered, various capacity tables have been developed as shown in the *National Fuel Gas Code*. To account for fittings, the concept of *equivalent length* was developed. Table 11-7 shows equivalent length values for threaded fittings, welding fittings, and certain valve types in pipe sizes from 3/8" to 8". The equivalent lengths are in terms of straight sections of Schedule 40 pipe.

Table 11-7 Equivalent Resistance of Bends, Fittings, and Valves. Length Expressed in Terms of Straight Pipe in Feet.

| Nominal Pipe or Tube Size (in.) | Screwed, Welded, Flanged, Flared, and Brazed Connections | | | | | | Copper | | | |
| | Elbows | | | Tees | | | Ells | | Tee | |
	90° Std	90° Long Rad	45° Std	Flow through branch	Straight-through Flow No reduction	Straight-through Flow Reduced 1/2	90°	45°	Side branch	Straight run
3/8	1.4	0.9	0.7	2.7	0.9	1.4	0.5	0.3	0.75	0.15
1/2	1.6	1.0	0.8	3.0	1.0	1.6	1	0.6	1.5	0.3
3/4	2.0	1.4	0.9	4.0	1.4	2.0	1.25	0.75	2	0.4
1	2.6	1.7	1.3	5.0	1.7	2.6	1.5	1.0	2.50	0.45
1-1/4	3.3	2.3	1.7	7.0	2.3	3.3	2	1.2	3	0.6
1-1/2	4.0	2.6	2.1	8.0	2.6	4.0	2.5	1.5	3.5	0.8
2	5.0	3.3	2.6	10	3.3	5.0	3.5	2	5	1
2-1/2	6.0	4.1	3.2	12	4.1	6.0	4	2.5	6	1.3
3	7.5	5.0	4.0	15	5.0	7.5	5	3	7.5	1.5
3-1/2	9.0	5.9	4.7	18	5.9	9.0	6	3.5	9	1.8
4	10	6.7	5.2	21	6.7	10	7	4	10.5	2
5	13	8.2	6.5	25	8.2	13	9	5	13	2.5
6	16	10	7.9	30	10	16	10	6	15	3
8	20	13	10	40	13	20				

NOTE: Gate valves are the same as pipe.
Globe valves: see catalog for valves selected.

Source: Rounded values from Crocker, S., and King, Reno Ceyton, ed. *Piping Handbook*. New York: McGraw-Hill, 1968.

For example, a 1" 90° elbow produces a certain pressure drop that is equal to the pressure drop of a 2.6' length of pipe. Likewise, a 1" 45° elbow has an equivalent length of 1.3' of pipe as does a tee, etc. Thus, if the fittings for a layout are tabulated and equivalent length information is available, the sum of equivalent lengths of all the fittings can be calculated. In order to arrive at the effective length of the pipe installation, the actual pipe length is increased by the equivalent length of the fittings. This equivalent length is the value selected from gas pipe capacity tables.

Example 6

What are the equivalent lengths of the following 8" items?

Table 11-8A   Example 6

Quantity	Fitting
1	8" tee (flow through branch)
6	8" 45° elbow
4	8" 90° elbow

Answer

Table 11-8B shows the values for the above list:

Table 11-8B   Answer to Example 6

Quantity	Fitting	Equivalent Length of Pipe	Total
1	8" Tee (flow through branch)	40'	40'
6	8" 45° elbow	10'	60'
4	8" 90° elbow	20'	80'

The usual procedure to make the initial design is to add 50% of the piping length as a fitting allowance, calculate the pipe sizes, and then check the equivalent lengths of the fittings for the pipe sizes actually selected. If the original 50% allowance checks out within 10% or 20%, no adjustment needs to be made. If the original allowance is in substantial error, then the piping should be resized based on the new equivalent length.

In residential work, this refinement of equivalent length is almost never worth the bother, but it should always be checked in commercial or industrial designs.

Example 7

Figure 11-3 shows a piping system with four connected loads and with sections identified by letters A through G. Size the system for natural gas, 1000 BTU/ft^3, 0.60

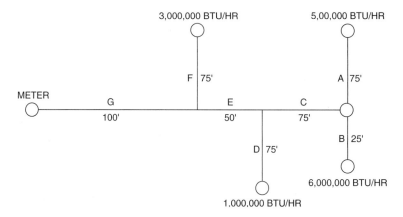

Figure 11-3 Commercial layout using standard natural gas.

specific gravity. The fittings used are welded type. The total pressure drop allowed is 0.5" water column and the gas pressure is less than 2 psig.

Answer

Fill in the table in the diagram. First, determine the gas volume required by dividing the heating load by the heating valuc of the gas. Next, calculate the length of pipe to the farthest appliance.

$$100' + 50' + 75' + 75' = 300'$$

Now, calculate an estimate of the equivalent length by increasing this amount by 50% to allow for fittings.

$$300' + 50\% \ (300') = 450'$$

Now, each section of pipe can be sized using Table 11-9. The answers are tabulated in Table 11-10.

Table 11-9 Pipe Sizes for Figure 11-3 Based on Longest Run Method

Pipe Section	Gas Volume Demand (cfh)	Pipe Size (in.)
A	5,000	5"
B	6,000	6"
C	5,000 + 6,000 = 11,000	8"
D	1,000	3"
E	11,000 + 1,000 = 12,000	8"
F	3,000	5"
G	12,000 + 3,000 = 15,000	8"

Table 11-10 Equivalent Length of Fittings

Fitting	Equivalent Length	Quantity	Total
8" ell	20'	2	40'
8" tee straight through	13'	2	26'
8" tee branch	40'	1	40'
5" ell	13'	2	26'
Total			132'

Now that the size of the fittings is estimated, a more accurate equivalent length can be calculated. Table 11-10 shows the new fitting allowances. Assume that the following fittings are used in the pipe run to the farthest appliance (some are added just as extra examples).

The actual fitting equivalent of 132' compares with the initial estimate of 150'. Note that the actual equivalent length did not change the sizing row that was used in Table 11-6, so the exercise is complete.

CORRUGATED STAINLESS STEEL TUBING (CSST)

Corrugated stainless steel tubing (CSST) is a product that has been on the market for several years now and is gaining in popularity. It consists of semi-flexible tubing made of 300 series stainless steel that has a corrugated shape (see Figure 11-4).

Figure 11-4 OmegaFlex corrogated stainless steel pipe (Trac Pipe). Notice the protective jacket is removed for installation of fitting. Always use the same manufacturer's fittings with its pipe.

The tubing is covered in a yellow polyethylene jacket that contains the following:

- The maximum working pressure
- The pipe size
- The testing standards and approval agencies

Each manufacturer makes its own special fitting that can be used to join its pipe to any standard pipe threaded fixture. CAUTION: Never mix and match different manufacturer fittings and pipe. This type of pipe gives the advantage of being flexible and lightweight so it can be installed in long continuous runs, without the need of couplings or elbows, which greatly reduces the installation time. Each manufacturer requires that the installer pass a test on its product before he's allowed to purchase the product. This does not substitute as a state or local license.

The same methods used to size gas piping systems can be used with one exception. CSST is sized not by its internal diameter, but by its *effective hydraulic diameter (EHD)*. Effective hydraulic diameter is a relative measure of flow capacity. This number is used to compare individual sizes between different manufacturers. The higher the EHD number, the greater flow capacity of the piping.

Example 8

Size the gas piping system for Figure 11-5 using Table 11-11, Table N-2A from OmegaFlex TracPipe, and the following gas conditions: natural gas with a specific gravity of 0.60 and 1000 BTU/ft^3; inlet gas pressure 7" W.C.; and a pressure drop of 1.0" W.C.

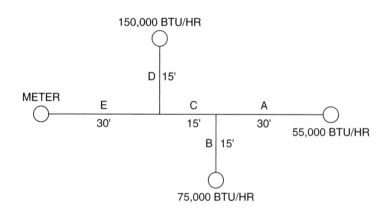

Figure 11-5 Gas pipe layout with CSST.

Table 11-11   Example 8 CSST Piping

Pipe Section	Gas Demand (cfh)	Pipe Size (EHD)
A	55	25
B	75	25
C	55 + 75 = 130	31
D	150	31
E	130 + 150 = 280	37

Answer

The longest run method dictates that 75' distance be used, but since there is no 75' column, 80' must be used.

TracPipe Corrugated Stainless Steel Tubing

Natural Gas With Pressure Drop Measured in Inches of Water Column:

Maximum Capacity of CSST in Cubic Feet per Hour of Gas
Gas Pressure: 0.5 psi 6-7 inches water column
Pressure Drop: 1 in w.c.
(Based on a 0.6 Specific Gravity Gas)

Tube Size (EHD)	Tubing Length (feet)																
	5	10	15	20	25	30	40	50	60	70	80	90	100	150	200	250	300
3/8" (15)	82	58	48	42	37	34	30	27	24	23	21	20	19	15	13	12	11
1/2" (19)	182	131	108	94	85	78	68	61	56	52	48	46	44	36	31	28	26
3/4" (25)	403	288	237	206	185	169	147	132	121	112	105	99	94	77	67	60	55
1" (31)	734	518	423	366	327	299	259	231	211	195	183	172	163	133	115	103	94
1 1/4" (37)	1324	901	720	614	542	490	418	369	334	306	284	266	251	201	171	151	137
1 1/2" (46)	2541	1790	1458	1261	1126	1027	888	793	723	669	625	589	559	455	393	351	320
2" (62)	5848	4142	3386	2934	2626	2398	2078	1860	1698	1573	1472	1388	1317	1076	933	835	762

REVIEW QUESTIONS

1 Using Table 11-1, what size pipe would be required to supply 120 cfh to a heater if the equivalent length of pipe were 48'?

2 Using Table 11-1, what size pipe would be required to supply 281 cfh to a heater if the equivalent length of pipe were 200'?

3 Using Table 11-6, what is the equivalent length of a 2" tee with flow through the branch outlet?

4 Using Table 11-6, what is the equivalent length of a 1" 90° elbow?

5 Using the OmegaFlex table, what size pipe would be required to supply 120 cfh to a heater if the equivalent length of pipe were 48'?

CHAPTER

12

Fuel Gas Piping, Fittings, and Connections

LEARNING OBJECTIVES

The student will:

- Summarize proper installation techniques.
- Describe pipe support spacing based on type and size of pipe.
- Discuss some of the more common prohibited fuel gas piping practices.

FUEL GAS PIPING INSTALLATION TECHNIQUES

The *National Fuel Gas Code* (NFGC) describes considerations, rules, and regulations concerning the installation or modification of fuel gas piping. General precautions and actions detailed by the code when extending or working on active fuel gas piping include the following items:

1. If the gas has to be turned off, notify all affected users and turn off all pilot burners. Relight all pilots later when gas pressure is restored.
2. The gas must be turned off.
3. If possible, turn off only the portion of the system that is to be worked on.
4. If job conditions require two or more workers, never work alone.
5. If a gas leak is suspected, use only approved flashlights or safety lamps for illumination.
6. If an electric switch is on, leave it on; if the switch is off, leave it off. Any spark may ignite the gas. Do not operate an electric switch.
7. Use only non-corrosive leak-detection fluid solutions or leak-detection instruments to search for gas leaks. Never use matches or open flames.
8. Do not smoke when working around an active gas line.
9. Do not leave the job in an unsafe condition. Even if you will be gone only briefly, plug or cap any openings, even valved lines.
10. Test the system by approved methods once the work is finished.
11. Relight pilots and be sure all appliances are returned to service.

The code requires that a plan or sketch of the work to be performed be developed before starting. This plan should include a plan view; show pipe sizes, pipe material, and meter location. It is advisable to include a schedule of equipment and list heating input and gas requirements in cubic feet per hour. If the system is to supply gas to more than one user or if an alternate fuel (stand-by fuel) will be interconnected to any part of the system, it must be clearly identified on the plans.

Once the plan is approved, code usually requires the following specific items:

1. Install valves, cocks, meters, unions, and sediment traps to be accessible.
2. If multiple users are involved, they should be supplied with valved connections from a header near the meter (or point-of-gas delivery).
3. Use care working with the pipe or tubing to ensure the highest quality workmanship. Keep burrs, chips, and joint compound out of the piping. Replace any defective material.
4. Piping must be supported to conform to the requirements of Table 18-1.
5. Carefully check any used or reused material to be sure it is satisfactory for the application proposed.
6. The pipe joint compound must be suitable and approved for the fuel gas to be conveyed.
7. If the gas is not dry, all piping must be protected from freezing.
8. Any turns in the piping must be made with fittings, factory bends, or field bends.

Table 12-1 shows the support required for piping. The table is divided: the left-hand side deals with rigid pipe; the right-hand side with tubing. CSST should be installed in accordance with the CSST manufacturer's instructions, which in most cases indicates closer spacing of supports.

FUEL GAS FITTINGS AND CONNECTIONS

Prohibited Materials and Methods

To stress the point, the prohibition will often be emphasized by special typeface or capital letters. For example, Section 8.1.2 of the 2006 *National Fuel Gas Code*

Table 12-1 **Support of Piping**

Nominal Size Steel Pipe (inches)	Maximum Spacing of Supports (feet)	Nominal Size of Tubing (inches, O.D.)	Maximum Spacing of Supports (feet)
HORIZONTAL			
1/2	6	1/2	4
3/4 or 1	8	5/8 to 3/4	6
1-1/4 or larger	10	7/8 or 1	8
VERTICAL			
All	Every floor level	All	Every floor level

discusses methods of pressure testing gas piping systems. Prominent in the text is the following statement:

"OXYGEN SHALL NEVER BE USED"

Although the experienced mechanic realizes that pure oxygen and almost any fuel can form an explosive mixture, the prohibition is stressed by using capital letters.

Other prohibited materials are piping substances, which are subject to chemical attack by the contained gas or the environment where the pipe is to be installed, subject to excessive physical abuse, or where the nature of the material makes it unsuitable under some circumstances.

Here are examples of each of the above problems:

- Copper tubing with gas that contains hydrogen sulfide
- Steel pipe without coatings buried outdoors
- Plastic pipe installed indoors or cast-iron pipe
- Cast-iron fittings that are brittle and subject to cracking or may contain casting sand holes. Cast-iron pipe and cast-iron fittings are permitted in some sizes.

Prohibited joints and connections include any types that are not as strong, leak-tight, or as suitable for gas as the piping itself.

1. Cast-iron fittings 4" and larger cannot be used indoors; and fittings 6" and larger cannot be used at all unless approved by the authority having jurisdiction.
2. Concealed joints cannot be made up of unions, tubing fittings (except brazed), running threads, or right and left couplings.
3. Reducing couplings are preferred to bushings, as bushings form an abrupt reduction of cross-section which could collect debris, and therefore, shall not be permitted by code.
4. Flexible gas appliance connectors are restricted to be six feet or less and cannot extend from one room to another, pass through any walls, ceilings or floors, or be installed after a shut-off valve. CSST cannot be used to direct connect to any movable gas appliance.

Common piping errors that come under a blanket prohibition include improper pipe end preparation, improper fabrication of flares or threads, poor welds, failure to ream the pipe end, or putting thread sealant in the female thread (excess sealant is pushed into the pipe).

Other mistakes include:

1. Improper or inadequate supports
2. Failure to slope the pipe to condensate traps (for gases other than dry)
3. Failure to protect buried pipe from aggressive soils or from damage during backfilling operations
4. Improper placing of buried piping in uncompacted soils so that settlement could damage the line
5. Failure to bury the pipe deep enough to clear garden and lawn tools and maintenance activities

Acceptable Joining Methods

Indoors

The most frequently used pipefittings are malleable iron, threaded, and 125 psi rated. These fittings are permitted for all sizes. For natural gas, however, they are commonly used up to and including the 4" size. Many installers use welding fittings over 2" pipe size.

LP gases and special-purpose gases use copper tubing with flare fittings in sizes up to and including 3/4". Above 3/4", steel pipe and malleable fittings are most often used.

Several other materials and joints are listed in the code, but you will find that they are generally used only for special purposes such as pilot burner supply lines.

Outdoors

Since a wider variety of materials are permitted outdoors, there is a wider variety of joining methods available for outdoor installations. The essential requirement is that the joining method selected must be appropriate for the piping material.

General Rules

Be sure that you blow out newly installed lines, provide sediment traps at low points, adjust all burners, and install all required auxiliary equipment. Also, always work to keep the jobsite safe during the installation process.

REVIEW QUESTIONS

1. Who must be notified that the gas will be turned off?
2. TRUE/FALSE   It is required by all gas suppliers that a drawing be submitted of the proposed gas system.
3. TRUE/FALSE   One half-inch black iron pipe can be supported every 32 inches horizontally.
4. TRUE/FALSE   All gas valves must be lever handled.
5. Why can't CSST be used to direct connect to a moveable appliance?

CHAPTER

13

Gas Appliances, Regulators, Meters, and Appliance Controls

LEARNING OBJECTIVES

The student will:

- Contrast the different types of gas burners.
- Summarize the importance and function of a gas meter and pressure regulator.
- Describe the function of different types of temperature controls.

GAS APPLIANCES

A gas appliance is any device that utilizes gas as a fuel or raw material to produce light, heat, or power. Gas appliances are available in many different styles, types, and sizes to perform a large variety of useful functions. They include fixed-in-place equipment such as furnaces and boilers for space heating, process heat for manufacturing, domestic water heaters, food-cooking devices of many sorts, decorative lamps, clothes dryers, refrigerators, air conditioners, incinerators, and stationary engines. Portable devices include certain cooking equipment, industrial heating apparatus, flame-cutting and flame-welding devices, and many sizes of engines.

The performance capacities and details vary from one manufacturer to another, but an important consideration for most appliances is the efficiency of operation. *High efficiency* means that most of the purchased energy supplied to the appliance is delivered as useful work by the appliance. Efficiency is expressed as the ratio of energy output to energy input. The efficiency of an appliance will never be greater than 100%.

Table 13-1 shows some of the items listed in Table 5.4.2.1 of the 2006 *National Fuel Gas Code*. These are typical input ratings for frequently encountered appliances, but many other models are used with different inputs.

Table 13-1   Approximate Gas Input for Typical Appliance

Appliance	Input (BTU/hr)
Warm air furnace, single family	100,000
Hydronic boiler, single family	100,000
Water heater, automatic storage 30–40 gallon tank	35,000
Water heater, instantaneous capacity at 4 gal/min	285,000
Free-standing range, domestic	65,000
Clothes dryer	35,000
Gas logs	80,000
Gas light	2,500

Source: Reprinted with permission from NFPA 54, National Fuel Gas Code, Copyright © 2006, National Fire Protection Association, Quincy, MA 02169. The reprinted material is not the complete and official position of the National Fire Protection Association on the referenced subject which is represented only by the standard in its entirely.

Some appliances are operated manually, but most are turned on and off by automatic controls. Such automatic units must have means to ensure safe operation. Different methods of control are discussed later in this chapter.

Gas Burners

Injection

Injection is the most often-used burner style for small-fired appliances. The *injection burner* is also called a *Bunsen burner* after the scientist who invented it. Figure 13-1 shows the essential components. This type of burner relies on natural draft and a

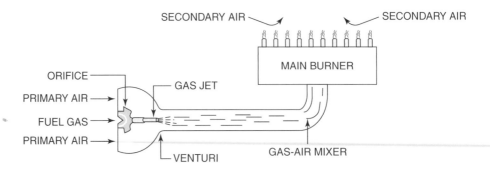

Figure 13-1 Cross-sectional view of injection gas burner.

venturi to draw air into the combustion chamber. This is commonly found in residential standing pilot water heaters.

Gas under slight pressure (usually 3" to 4" water column) passes through the orifice (sometimes referred to as *spud*). This jet of gas is released into the minimum diameter region of the venturi tube. The venturi produces a low-pressure region that draws air into the venturi tube so it can mix with the gas. This is known as *primary air*. The resulting gas-air mixture flows to the burner head and passes into the fire. The flame that results is a blue flame, very nearly invisible in a bright light. Air that mixes with gas in the venturi tube is the primary air required for combustion. The remainder of the needed oxygen comes from the surrounding air (secondary air) to allow complete combustion to take place.

Some burners are equipped with an adjusting shutter on the primary air intake, so the precise character of the flame can be modified. Reducing the primary air intake area will reduce the amount of gas being burned and result in a smaller, soft, and lazy flame. This would be referred to as *under firing* the appliance. Excessive amounts of air will increase the firing rate, which would produce a hard, noisy flame that sometimes lifts off the burner. If primary air is increased too much, the fire will become hard, short, and noisy.

With extreme excess primary air quantity, the flame may even propagate through the burner top and proceed to the orifice in the venturi. This flame is dirty, noisy, and must be corrected quickly.

Power

The second type is known as a *power burner*. This burner uses a motor-driven blower to deliver the air, both primary and secondary, to the combustion zone. The primary air and gas are premixed, as in the injection burner, and delivered to the combustion zone. Secondary air is also brought to the flame area through the combustion chamber and used to help complete combustion.

The blower can be arranged to blow air into the combustion chamber, which is a *forced draft*. It can also be arranged to pull air through the combustion chamber, which is an *induced draft*. On some large commercial equipment, there may be a combination of both forced and induced draft. These configurations are superior for larger combustion rates and provide better control of total air amount and fuel-air mixing. This improved air control results in higher operating efficiency. Figure 13-2 shows an illustration of both the forced and induced draft systems.

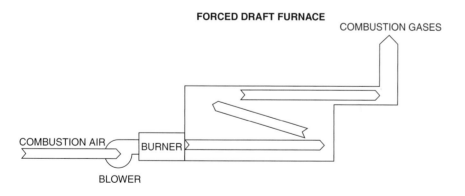

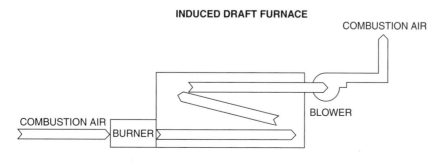

Figure 13-2 Notice how the position of the blower assembly changes depending on whether the air is being blown into (forced) or drawn through (induced).

Luminous

The third burner type is the *luminous burner*. This burner does not use any primary air for combustion—all air is secondary air. The result is a flame that undergoes incomplete combustion, and the glowing, unburned carbon is the source of light characteristic of this flame. The luminous flame will deposit soot on any surface placed in or above the flame and would be very inefficient as a heating flame. These are commonly used for decorative lighting purposes or for fireplace lighting.

Refrigeration and Air Conditioning

Some small refrigeration or air conditioning appliances such as those found in recreational vehicles use a gas burner to supply the energy for the refrigeration cycle. Commonly known as *absorption units,* the unit uses a mixture of ammonia and water as a refrigerant. Special training and tools are required for this type of work.

FUEL GAS REGULATORS AND METERS

In order to meet the growing demands of the public, natural gas is transported by cross-country pipeline at very high pressures (several hundred pounds per square inch). Local utility systems distribute natural gas at main pressures from 20 to 150 psi. Since these pressures are considered hazardous levels if used indoors, the gas industry developed pressure-reducing equipment and procedures to achieve appropriate natural gas pressures for distribution within buildings.

Pressure Regulator

Building Pressure Regulator

The principal control device used to achieve lower pressures in gas supply systems is called a *pressure regulator*. Figure 13-3 shows a diagram of a building pressure regulator.

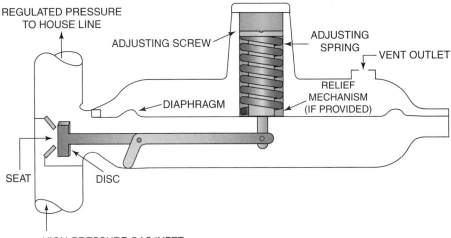

Figure 13-3 Cross-sectional view of building pressure regulator. Do not leave the vent outlet pointing up when installed outside.

IN THE FIELD

If the gas regulator is installed outside, always install it so that the vent opening is facing down; this will prevent rainwater from entering the assembly.

The area of the diaphragm, below the spring, is subjected to the low-pressure downstream gas that enters the low-pressure portion of the system. The valve disc is connected to the diaphragm, so an increase in downstream pressure causes the diaphragm to move the disc closer to the seat, thus reducing flow. With a reduction in downstream pressure, the spring moves the disc away from the seat, thus increasing the flow. The vent opening in the upper portion of the regulator ensures that the upper half of the diaphragm chamber is always at atmospheric pressure.

Smaller regulators, like the one shown in Figure 13-4, are intended for use in individual appliances equipped with an orifice in the vent opening. If the diaphragm should break, only a small amount of gas could escape into the room, so they do not have to be vented outdoors. This is sometimes referred to as a *vent limiter*.

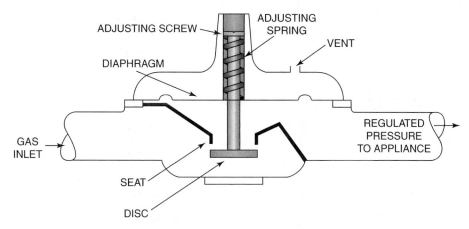

Figure 13-4 Cross-sectional view of appliance pressure regulator. Do not leave the vent outlet pointing up when installed outside.

Larger regulator vent openings are piped to the outdoors to permit gas to escape in the event of diaphragm rupture or relief valve opening (see discussion below). The outdoor end of this pipe is provided with fittings so that the terminal faces downward with an opening covered by a screen. Small insects have been known to build nests inside these vent pipes and prevent them from working properly.

These devices will provide nearly constant downstream pressure, up to the flow capacity of the seat-disc geometry, and down to zero flow if the seat-disc match is perfect. If there is any leakage past the disc, then the downstream pressure will gradually increase if none of the downstream appliances are burning gas. As this condition could result in overfiring the first appliance that turns on, overpressure must be guarded against. In addition, the regulator could fail in an open position that produces overpressure, so the protection must also be adequate for this condition.

One form of overpressure protection is to add a separate relief valve at the regulator station. The valve is set to open at a pressure slightly higher than the regulator control pressure. The discharge of this relief mechanism must be vented outside.

Building regulators are frequently equipped with a built-in internal relief valve mechanism; appliance regulators are not. Except for the relief valve feature, the two regulator types have similar construction.

One method formerly used for protecting the downstream pressure was to use a standpipe filled with mercury to the proper depth for the pressure desired. If the pressure exceeded the desired value, the mercury was blown out and the excess pressure vented. Instead of resealing automatically when the correct pressure was restored (after the overpressure was corrected), more mercury had to be used to reseal the relief system. Given that mercury is expensive and poisonous, this method is no longer used.

Another method is to use a spring-loaded valve disc and seat in the center of the diaphragm. If the relief mechanism operates, the excess pressure is vented into the upper regulator chamber and conveyed outdoors by the vent pipe. Once the overpressure is corrected, the valve will re-close. With either of the above methods, there is financial loss and the possible hazard of large volumes of gas escaping.

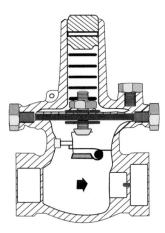

Figure 13-5 Cross-sectional view of security gas supply valve.

Security Valve

One preferred safety device is to use a *security valve* that trips closed when the line pressure exceeds a set value. When the overpressure is corrected, the security valve must be reset manually and the system is restored to service. Security valves are available that close in the event of fire or earthquake, with abnormally low pressure, high pressure, or excessive flow. Security valves are small, inexpensive, relatively maintenance-free, and some gas suppliers are offering to install them in new systems, especially residential. Figure 13-5 shows a security valve that trips on overpressure.

GAS METERS

Meter installations are necessary in order to charge users for taking gas from the system. Figure 13-6 shows a typical installation of high-pressure service, shutoff valve, pressure regulator, security valve, and meter. Note that the regulator is vented outdoors. (If vandalism is not a problem, all this equipment may be installed outside the building.) The two most common meter types currently in use are displacement and rotary, described below.

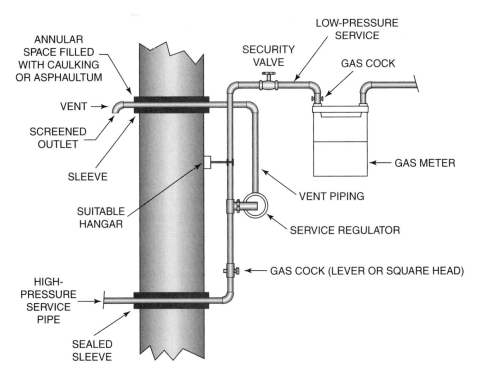

Figure 13-6 Typical layout for gas meter and pressure regulator piping. In colder climates or where vandalism is an issue, meters and regulators are sometimes installed in the building.

Displacement

Displacement meters (Figure 13-7) are used for smaller loads and lower pressures (up to 1000 cfh and 5 psig).

The meter consists of two flexible diaphragm compartments and two fixed compartments: (A) surrounding the diaphragms, a series of alternating valves and (B) directs gas into and then out of the various chambers, causing the diaphragms to alternately contract and expand. The alternations are linked to a counter to indicate the total flow through the meter since the volume change of the diaphragms is known.

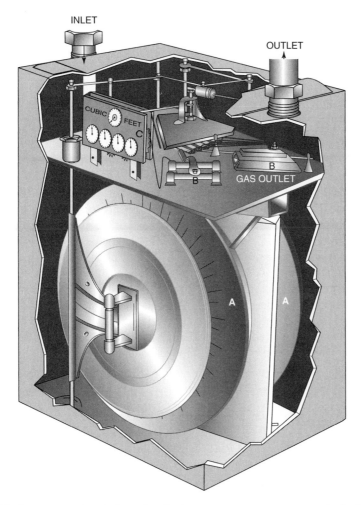

Figure 13-7 Cross-sectional view of a displacement type gas meter typical of most residential installations.

Rotary

Rotary meters are used for large volumes and are suitable for any pressure. Figure 13-8 shows a typical rotary meter. The unit contains impellers that turn as gas flows through the meter. This turning motion is coupled to a counter that records the volume.

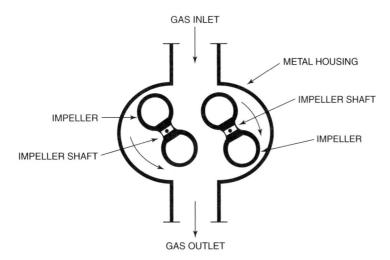

Figure 13-8 Conceptional illustration of a rotary type gas meter typical of large volume commercial installations.

All gas meters incorporate a test dial that makes a complete revolution for a very small amount of gas. For residential applications, the test dial shows a complete revolution for 1/2 ft of gas.

Some meters have dial-type registers and others have digital readout registers. Either way, the meters are capable of accurate recording of gas flow. Most utilities have regular programs of meter replacement and calibration to ensure the accuracy of their billings.

FUEL GAS APPLIANCE CONTROLS

Gas appliances operate either manually or by automatic controls. Some manual appliances utilize a standing pilot for the convenience of the user, but since the starting or stopping of the burner is in the control of a person, these units are still considered to be manual.

The essential consideration of automatic control is that it is unattended. Therefore, safety precautions must be provided to ensure that no hazardous condition can occur. The characteristics of each product type or appliance system are considered when safety devices are selected. As nearly all appliances produced by a manufacturer come as complete products, the installer rarely decides the selection of safety devices. The safety devices might require adjustment, repair, or replacement, so the basic operation of these products should be discussed.

Pressure

Pneumatic

A fluid is placed in a sealed system with a bulb (reservoir) in the controlled medium. The other end is attached to a bellows mechanism that operates a switch or valve plunger. When the bulb end is warmed, the fluid expands and causes the bellows to expand. As the bellows expands, a force is exerted on the other end that causes a switch or valve plunger to move (see Figure 13-9).

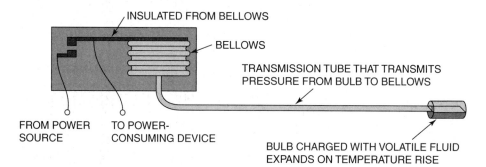

Figure 13-9 Bellows with sensing bulb. As bellows expand (because of temperature increase) the electrical contacts will break (open). Always be careful not to kink the transmission tube or the switch will be ruined.

Some similar devices, such as the thermostatic steam trap, have only a small bellows that expands and contracts to move the valve seat. By having a smaller bellows that acts directly on the valve seat, only a small amount of condensate is needed to cool the bellows and thereby open the valve. As soon as the condensate is discharged, steam entering the valve causes it to close.

Differential Expansion

Rod and Tube

The control consists of a rod within a tube. One end of the tube is attached to a stationary frame, while the other end is exposed to the medium being controlled.

Inside the tube is a rod that has one end attached to the end of the tube and the other end connected to a switch or valve. The rod material has a very low coefficient of expansion and the tube has a high coefficient of expansion. Thus, when the assembly is heated, the tube will lengthen more than the rod. This will move the switch (or valve) end of the rod farther into the tube. Figure 13-10 shows how this type of arrangement could be used to control the flow of gas into a gas-fired water heater.

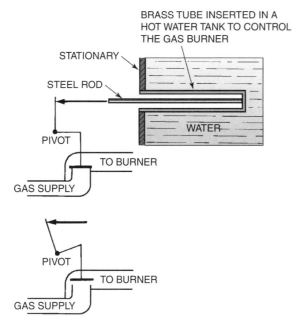

Figure 13-10 The rod and tube-type controller relies on the fact that the tube will expand and contract more than the rod. This causes the rod to move in and out of the tube.

As the water heats up, the brass tube expands more than the steel rod, so the result is that the end of the rod travels to the right and closes off the gas supply.

Bimetal

Two different strips of metal are laminated together by riveting, soldering, or brazing. One end of the combination is attached to a rigid support and the other end is free to move. The end of the combined strip bends as the temperature changes. This is probably the most common power element for room thermostats and it is frequently used for temperature limit switches (see Figure 13-11).

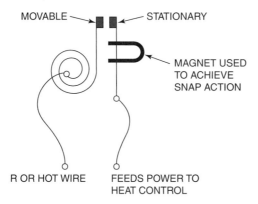

Figure 13-11 The bimetal temperature sensor is wrapped in a spiral with the center point fixed. As it heats up it will try to unwind and cause the contacts to close. This is similar to the common mechanical thermostat except a mercury bulb is used instead.

Another highly popular temperature-sensing device is the *thermal disk*, sometimes called a *snap disk* or *click on*. It can be used as the high-temperature limit for an electrical water heater or for the small, hermetically sealed refrigeration compressors found in small appliances (see Figure 13-12).

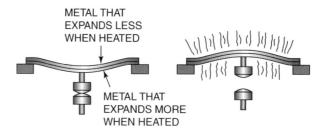

Figure 13-12 The bimetal disk will snap to open the contacts. Sometimes this type of sensor may have to be manually reset.

It consists of two dissimilar metals that are bonded together in the shape of a curved disk. When heat is applied, one side will expand more than the other, causing it to bend. This will cause the normally closed electrical contacts to open.

Thermoelectricity

Two different metals are joined together at one end. As that end is heated, an electrical potential appears between the other ends of the two metals. Commercial *thermocouples* are made with one metal in tube form and the other in wire form. The wire is placed inside the tube and insulated from it. A special end is installed that can be solidly attached to safety valves or safety switches (see Figure 13-13).

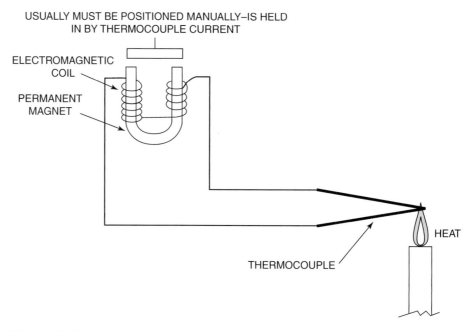

Figure 13-13 Conceptional illustration of a thermocouple. Make sure enough of the thermocouple tip is in the flame or the voltage output will be low.

The usual voltage level of the thermocouple output is approximately 10 to 25 millivolts (0.010–0.025 volts). This voltage can develop sufficient current flow to hold a small latched electromagnet that is connected to a switch or safety valve mechanism. In order for it to be held by the thermocouple, the safety device must be manually positioned to the "engaged" position. If the thermocouple cools down, the output is reduced and the safety device is released.

Electrically Activated

Because of the growing need for fuel economy and efficiency, many of the standing pilot appliances are being replaced with *intermittent-use devices*. This means that when a demand for heat is recognized (by the thermostat), an *integrated furnace controller (IFC)* will send power to the gas valve. These newer gas valves have two electrical coils in them. The first is a small coil that, when energized, opens the gas passage for the pilot flame. If a flame sensor such as a flame rod or photocell detects a flame, the IFC will then energize the second coil and the main gas burner will fire.

The IFC is a control module that monitors the entire system. It receives the temperature signal from the thermostat, starts the firing of the pilot light, checks to see if the pilot did light, and then turns on the main burner. It can also control other devices, such as blowers and draft control motors. Automatically controlled units must sense conditions that could be hazardous and stop the input of gas before the hazard develops. The principal hazards are overheating, overpressure, and gas flow without flame being present.

Flame Sensing

In addition to using thermocouples, flame sensor systems use flame rods, photocells, ultraviolet sensors, or lead sulfide cells to observe the flame. The principles used with these other devices will be described in later lessons.

> **IN THE FIELD**
>
> Review the safety devices provided with the appliances on which you are working. Be sure you understand the sequence of events and the problems to guard against in their control systems. Study the operating principles of the common controls used on the equipment you install and service and NEVER BYPASS A SAFETY DEVICE.

REVIEW QUESTIONS

1. Name three types of burners.
2. Which air (primary or secondary) if increased will increase the amount of gas being burned and increase the firing rate of the appliance?
3. Which temperature control device uses a small voltage generated by dissimilar metals to energize a coil?
4. What is an IFC?
5. Which type of gas meter is better suited for large volumes of gas?

Fuel Gas Piping, Corrosion, and Corrosion Protection

The student will:

- Explain some common terms associated with corrosion protection.
- Summarize the theory related to cathode protection for piping.

FUEL GAS PIPING, CORROSION, AND CORROSION PROTECTION

Corrosion of buried piping is frequently encountered in many types of systems. Corrosion of any material involves a chemical change. In many cases, this change is started or accelerated by an electrical change in the material. Piping systems are typically protected by either a pipe coating, cathode protection, or in some cases both. Both have their advantages and disadvantages. The piping could be purchased *precoated* from the manufacturer, which would minimize the field installation time; but this coating must be protected throughout the installation process because corrosion can occur with any failure of the pipe coating.

Cathode systems require a good bit of knowledge and training to be properly sized, so they can be costly for large systems. It has been proven that an electrical system, when properly applied, can prevent corrosion in buried piping by causing a sacrificial material to corrode instead. In order to understand how piping is to be protected, some common terminology must first be introduced. The following are some of the more common terms associated with pipe corrosion protection:

Definitions

Active

Active relates to the ability of a metal to produce current.

Anaerobic

Anaerobic describes a process that takes place in an oxygen-free environment.

Anion

An *anion* is a negatively charged ion or an atom or group of atoms carrying a negative electric charge. It is the negatively charged particle that migrates toward the anode with current flow.

Anode

The *anode* is a positive terminal of an electrical device or in this case, the area of metal from which current flows, i.e., it is the sacrificial metal that will be corroded.

Caldwell/Thermoweld

Caldwell and Thermoweld are exothermic welding processes. Both use powdered metals, aluminum, copper, or iron oxide, to create a superheated bonding material. Fusing temperatures have been known to exceed 2000°F. These processes allow for alloys such as copper or aluminum to bond to steel or iron, thus creating a super strong joint that is corrosion resistant and capable of handling a very high current.

Cathode

Cathode is the negative terminal of an electrical device or, in this case, the area of metal to which current flows (the protected metal).

Cathodic Protection

Cathodic protection is a method used to control corrosion by using a sacrificial anode system.

Circuit

A *circuit* refers to the complete path for current flow.

Direct Current

Direct current (DC) is current that flows in only one direction.

Electrolyte

An *electrolyte* is any solution or compound that allows current to flow.

Galvanic Anode

A *galvanic anode* is an electrode that will cause current to travel in one direction to protect a secondary metal from decomposition.

Ground Bed

The *ground bed* refers to the area in which the installation of the sacrificial anode takes place.

Insulating Point

An *insulating point* is a non-metallic joint to prevent current flow from one piece of pipe to another.

Ohm

An *ohm* is the unit of measure for electrical resistance to current flow.

Soil Resistance

Soil resistance is the degree to which water, soil, or the electrolyte resists the flow of electric current. A high-resistance soil is preferred to stop stray currents.

Stray Current

A *stray current* is any unpredicted current(s) that flows from one structure to another due to unrecognized and unintended connections between the structures.

Corrosion

Corrosion can occur in many different forms and is more prone to show up in certain areas. For example, *preferential corrosion* is corrosion that takes place in or around curves, threads, or welds. These are areas where the metal has been worked. Similarly, *stress corrosion* occurs in areas where the material has external or internal forces pushing or pulling against it. Areas where a weld was cooled too fast or places where pipes were forced into alignment will corrode while pieces of the same material under similar conditions show no signs of corrosion. The technician should pay special attention to these locations and make certain that they are well coated.

In any of these cases, the corrosion may show up in the form of *pitting, rusting,* or *slabbing*. Pitting is when the metal is removed in the shape of a cone. The base of the cone is at the surface of the metal and the point of the cone is below the surface. If the conical void is deep enough, the cone will pass through the pipe wall, forming a leak. Rusting occurs when iron is exposed to oxygen and moisture and reverts to iron oxide, a flaky scale that is easily dislodged. If it isn't inhibited or prevented, rusting can eventually destroy the iron (or steel) component. Slabbing is a combination of pitting and rusting. This condition rapidly deteriorates steel pipe.

As stated at the beginning of this chapter, corrosion can be greatly affected by electricity. This electricity can come from a number of sources: either from stray currents from obvious electric devices or from chemical reactions between different materials. Stray currents are commonly caused by pieces of equipment that have frayed wires or poor connections or from static electricity on pieces that are not properly grounded. They can usually be found in the vicinity of substantial electrical activity (plating works, electrical installations, welding processes, or electric railways). These currents remove metal from the place where the current leaves the original structure (i.e., the pipe main) and proceed along the stray path.

The ability to obtain electricity from combinations of different materials has been around for thousands of years. It is known as a *battery* or *battery action*. When a substance that will conduct electricity or allows electrons to flow connects two dissimilar metals, a process known as *galvanic action* occurs. Galvanic effects are named after the nineteenth-century physicist Galvani. He discovered and formulated the theory that different metals, when placed in an electrolyte, develop an electrical potential. He began the classification system that bears his name, the *Galvanic Series of Metals* (Table 14-1), also called the *Electromotive Series*. The farther apart two materials are on this scale, the greater the electric potential developed when placed in an electrolyte. The material higher in the table is the one that is corroded when the two interact in an electrolyte. Thus, the steel (i.e., iron) pipe is corroded when combined with copper.

Table 14-1 Standard Electrode Potential of the Elements at 25°C (77°F)

Element	Reference Ion	Potential in Volts
Lithium	Li^+	+2.9595
Potassium	K^+	+2.9241
Calcium	Ca^{++}	+2.763
Sodium	Na^+	+2.7145
Magnesium	Mg^+	+2.34
Aluminum	Al^{***}	1.70
Zinc	Zn^{**}	+0.7618
Chromium	Cr^{++}	+0.557
Chromium	Cr^{+++}	+0.505
Iron	Fe^{++}	+0.441
Cadmium	Cd^{++}	+0.402
Nickel	Ni^{++}	+0.231
Tin	Sn^{++}	+0.136
Lead	Pb^{++}	+0.122
Iron	Fe^{+++}	+0.045
Hydrogen	H^+	0.0000
Copper	Cu^{++}	−0.344
Copper	Cu^+	−0.522
Iodine	I^-	−0.5346
Silver	Ag^{++}	−0.7978
Mercury	Hg^{++}	−0.7986
Bromine	Br^-	−1.0645
Chlorine	Cl^-	−1.3588
Gold	Au^{++}	−1.36

Figure 14-1 shows the basic idea of how battery action can occur between a copper water line and a cast-iron sewer pipe.

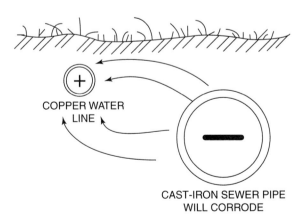

Figure 14-1 Different pipes in the same type of soil can react and cause a small current to flow. This can lead to corrosion.

In some cases, the corrosion can occur between two similar metals because of differences in levels of electrolytes in the soils. The electrolyte is changed in local regions by atmospheric effects, chemicals, biological materials, or soil differences.

The electrolyte will attempt to stabilize itself and corrosion may occur. Figure 14-2 shows an example of how a pipe entering a house would travel from one type of soil to another. This causes an imbalance in electrolytes.

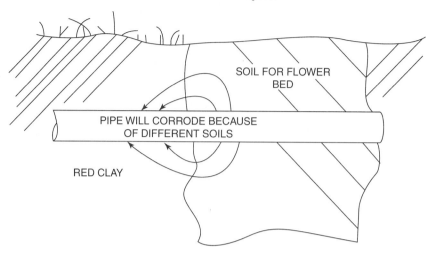

Figure 14-2 Different soils will react differently to the metal pipe and in some cases can cause corrosion.

PREVENTION METHODS

Pipe Covering

No matter what the precise mechanism of corrosion, certain protective measures are effective in minimizing or preventing it. The most important measure is to isolate the material to be protected from exposure to electric currents or chemicals. The pipe can be wrapped with a moisture proof, inert material, and any field-made joints can also be wrapped so the final installation is completely insulated from external events or materials. Thus, the pipe is chemically and electrically separate from any activity going on around it.

One serious problem with this method is the difficulty of keeping the pipe covering intact during the handling and installation phase of the project. To determine the integrity of the pipe covering, and even to repair it if necessary, various test methods and equipment have been developed.

Pipe-covering materials used in this century include the following substances:

- Asphalts   Used from 1900 to present
- Coal tar   Used from 1925 to present
- Waxes and greases   Used from 1932 to present
- Rubber mastics   Used from 1939 to present
- Plastics   Used from 1945 to present
- Epoxies   Used from 1954 to present
- Extruded plastics   Used from 1954 to present

Natural substances such as asphalt, waxes, and tar are becoming less common as new products such as polyethylene and polyvinyl chloride tapes are readily available. In many cases, shrink-wrap sleeves can be applied to cover joints or flaws in the protective coating. A *shrink-wrap sleeve* is a piece of plastic material that will shrink in size and bond to the adjoining pipe coatings when heat is applied.

Cathode Protection

Cathode protection takes advantage of the fact that galvanic action can occur and the results are easy to predict. It was well known that between the cathode and the anode, the anode would corrode while the cathode stayed pretty much intact. Thus, cathode protection systems are designed to create a direct current flow from the sacrificial material to the pipe. Depending on the condition of the soil, one of two methods is usually applied: either the *sacrificial anode* or the *impressed current system*.

Sacrificial Anodes

When sacrificial anodes are used, they typically consist of a magnesium or aluminum rod buried in a pit and are backfilled with a mix of specially prepared material that will aid in current flow. Figure 14-3 shows how the rod is connected to the pipe with a heavy gauge insulated wire.

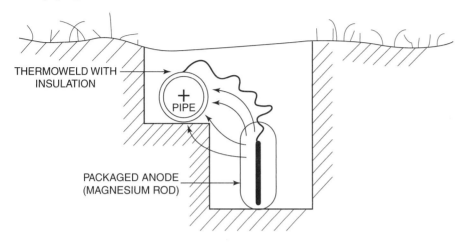

Figure 14-3 **The bottom of the anode should be installed at least 60" below grade. The top of the anode should always be below the bottom of the pipe.**

The wire is usually thermowelded to the pipe and the anode. The welded joint is then coated with a protective coating to prevent localized corrosion. Once the system is in place, the anode will gradually corrode while the pipe remains intact. Depending on the soil conditions and the size of the pipe and anode, the system may last for many years; but eventually the anode will have to be replaced.

Impressed Current

In cases where large amounts of current are needed to protect the pipe, an impressed current system should be installed when the original piping system is installed. Figure 14-4 shows an impressed current where an external source of direct current is being applied to the anode and cathode.

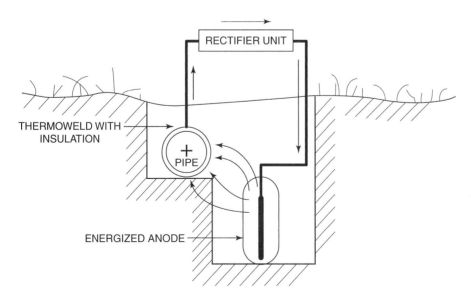

Figure 14-4 **The impressed current from the rectifier creates the force and path for the protective current flow.**

In this case, a rectifier is used to convert *alternating current (AC)* into a direct current source. This system helps drive more electron flow into the pipe and thereby increases the amount of protection. This type of system typically uses a graphite or high-silicon iron material for the anode. Graphite and silicon iron materials are usually consumed at a rate of about 1 pound of material for every ampere year. The benefits of the wide range of voltages and currents that the impressed current systems offer must be considered against the need for an external power source and the higher up-front costs of the installation.

Remarks

As to corrosion protection, it should be stressed that present industry thought is to use piping materials inherently resistant to the chemicals and corrosion factors that may exist at a given location. Thus, gas utilities and progressive contractors are installing piping that does not require labor-intensive coatings and test methods or any of the other preventive methods described. The techniques described in this chapter may be considered as historical information or as measures to be taken to protect the very small sections of susceptible buried gas or water piping that may be encountered in mechanical systems.

REVIEW QUESTIONS

❶ If a cathode protection system were made of copper and lead, which material would corrode?

❷ At what parts of a piping system is corrosion more likely to occur?

❸ What is a disadvantage of using precoated piping?

❹ What is battery action?

CHAPTER

15

Vents for Category 1 Appliances

The student will:

- Summarize the different types of drafts when working with fuel gas appliance vents.
- Contrast the differences in vent capacity based on appliance type and vent type.
- Discuss considerations for vent terminals and installation of vent piping systems.

VENTING MATERIALS

Venting of gas appliances is the process of removing the products of combustion from an appliance usually located within an enclosed living space and conveying the products to the outdoors. The flue pipe should be constructed to maintain a leak-free condition and flue gases entering the flue pipe should exit only through the opening to the outdoors. Since the flue gases for a Category 1 appliance must have a temperature high enough to avoid excessive condensation, arrangements must be made to keep surrounding building materials safe.

The principal products of combustion of any hydrocarbon are *water vapor* and *carbon dioxide*. If combustion is incomplete, other products that occur are *carbon monoxide* and *free carbon*. Because water vapor and carbon dioxide are the principal products present in the vent, some people ask why a vent system is necessary at all since water vapor and carbon dioxide are harmless. The answer is that without discharging the products outside, the combustion process would soon use up the oxygen in the room. Given that, the combustion process would soon deliver the dangerous product carbon monoxide and the noxious product free carbon. Some appliances (principally cooking equipment), however, are not vented outdoors because their cycle of use is very brief, the BTU firing rate is small, and/or the space where they are located is well ventilated.

Vent Definitions

Chimney

A *chimney* is a vertical (or nearly so) passageway that delivers products of combustion and dilution air to the outdoors (see Figure 15-1).

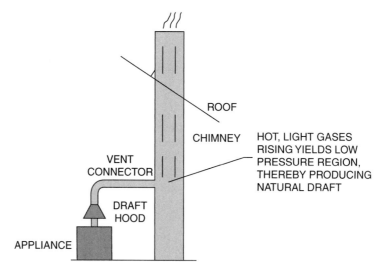

Figure 15-1 Natural draft through a chimney.

Dilution Air

Dilution air is room air that mixes with flue gas at the draft diverter. Dilution air reduces the temperature of the sum of gases in the vent connector and also acts to reduce draft variations at the appliance outlet.

Draft

The motive force that moves the combustion products through the vent pipe is called draft. *Draft* is defined in the *National Fuel Gas Code* as a pressure difference that causes gases or air to flow through a chimney, vent, flue, or appliance.

Natural Draft

Natural draft is produced by differences in density between the gases within the vent conduit and the surrounding atmosphere. The hot gases of combustion have a low density, which means they are lighter than the surrounding ambient air. Because they are lighter, they will tend to rise and be replaced with cooler air (see Figure 15-1).

The amount of natural draft is limited by the height of the chimney and the differences in gas densities; this, in turn, will limit the size and efficiency of the appliance suitable for this type of system.

Mechanical Draft

Mechanical draft is a draft that is developed by a fan or blower installed in the vent path. The two most common types of mechanical draft are *induced* and *forced*. An induced draft has the blower installed in such a way that the gases of combustion are drawn through the combustion chamber with a negative pressure (sucked through). A forced draft has the blower installed in such as way that the gases of combustion are forced through the combustion chamber with a positive pressure (blown through) (see Figure 15-2).

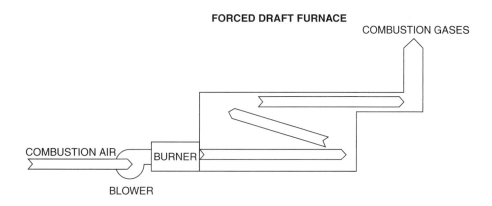

FORCED DRAFT FURNACE

COMBUSTION GASES

COMBUSTION AIR

BURNER

BLOWER

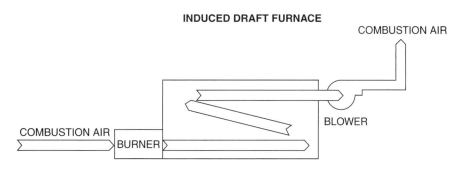

INDUCED DRAFT FURNACE

COMBUSTION AIR

BLOWER

COMBUSTION AIR

BURNER

Figure 15-2 Forced and induced draft. Notice the position of the draft blower.

Back (Reverse) Draft

An occasional problem with chimneys is *back drafts* or *reverse drafts*. This is when the flue gases are pushed back into the combustion chamber. In extreme cases, this movement can cause the flame to trip a safety limit or damage the appliance. Conditions that produce reverse chimney drafts include the following:

1. A gust of wind strikes the chimney top and produces a flow down the chimney, expelling combustion products from the draft hood.
2. Building pressure becomes less than chimney pressure, usually from the action of exhaust fans.
3. Building (or furnace room) is sealed so tightly that combustion products cannot go up the chimney because room pressure would drop.

4. At appliance startup (after a shutdown period), the chimney will temporarily chill the products of combustion so that the draft does not exist. After the chimney is warmed, normal draft will be restored. This condition is made less severe by using lightweight chimney materials and by having most of the chimney inside the building. The problem is made worse if the chimney is massive, oversized, and mostly exposed to outdoor temperatures. CAUTION: Newer technicians will see this same condensate coming down from the inside of the combustion chamber flue of a gas-fired water heater and assume that the tank is leaking.

Figure 15-3 Draft hood for natural gas water heater. Notice how it allows for air to enter the vent connector along with the flue gases. From this point on they are known as vent gases.

Draft Hood

The *draft hood* is a special fitting that receives the products of combustion from the appliance and admits dilution air to the stream. The draft hood acts to reduce the variation in draft seen by the appliance outlet (see Figure 15-3).

Flue Gas

Flue gas includes the products of combustion contained within the appliance flue as they travel to the draft hood. In order to slow the velocity of the gases and produce the turbulence necessary to extract the maximum heat from the flue gas stream, the appliance flue frequently contains a baffle.

Vent Connector

The *vent connector* is the conduit or passageway that connects a fuel-gas burning appliance to a vent or chimney. It is sometimes referred to as *breaching* (see Figure 15-1).

Vent Gas

Vent gas is the mixture that is produced when the room air mixes with the flue gas at the draft diverter. The temperature is in the 350° F–600° F range.

Vent Materials

Materials for vent components can be listed in two groups: (1) *metallic* and (2) *non-metallic* (see Figure 15-4).

Metallic

Single Wall Vent Connectors

Single wall vent connectors are usually made of galvanized steel not less than 0.0304" in thickness (23 gauge or heavier wall). Considerable clearance from combustible materials must be maintained, since the surface temperature of the single wall pipe is unpredictable. However, the temperature will be very high.

Single Wall Chimneys

Single wall chimneys may be used in some local codes, but these chimneys tend to cool the gases excessively. They must have approximately 6" of clearance from combustible materials and are known for corroding easily. Single wall chimneys are subject to special installation restrictions.

Double Wall, Type B

Type B vents are a factory-made combination of aluminum inner wall and steel outer wall with about a 3/4" space between the two walls. The ends of each length are fabricated with a spacer piece with ventilation openings, which makes the annular space continuously connected and ventilated in any complete assembly of Type B parts (see Figure 15-5).

The component parts made by different manufacturers are not interchangeable, so you must use pipe and fittings of the same brand for a proper job. *Type B vents* are used in lieu of chimneys for the following: appliances listed by the product standard for this vent material; appliances with draft hoods; and for vent connectors

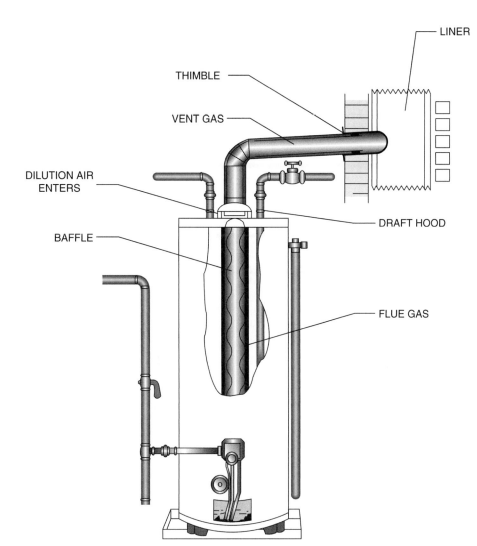

Figure 15-4 Components of the venting system.

Figure 15-5 Double wall Type B vent pipe.

when conditions call for minimum temperature loss in the vent connector. The most important limitations are 1" clearance from combustible materials and 400° F maximum flue gas temperature.

Double Wall, Type B-W

Type B-W material is similar to Type B, but its intended use is for wall furnaces. It is oval-shaped and designed to fit within a stud wall.

Double Wall, Type L

Double wall Type L is a special design vent piping for high temperature vent gases, but it is suitable and approved for gas-fired appliances.

Triple wall assemblies are used where extremely high temperatures are encountered. Stainless steel is typically used for the inner material.

Non-Metallic

Masonry Chimneys

These vents are usually in the form of *masonry chimneys* with a clay tile or Portland cement to contain the hot gases until they exit out the top. Approved listed metal liners are also available to line existing masonry chimneys.

Openings into these chimneys should always be made with a thimble or sleeve. This will ensure that the gases are conveyed into the liner area and guard against the pipe extending too far into the liner, where it could partially or totally block the internal area of the liner.

Asbestos-Cement

Asbestos-cement pipe may also be used as a non-metallic chimney. Although this material is no longer available, existing asbestos-cement chimneys are satisfactory for use.

Plastic

Plastic piping may be used for special venting systems, but typically not Category 1 systems. These will be discussed in a later chapter.

VENT SIZING FOR CATEGORY I THROUGH IV APPLIANCES

Within the past few years, many variations on typical gas-burning appliances have become available to the public. One way to identify categories of appliances is by the nature of the venting products each appliance produces.

The gas appliance industry has devised the following four categories to identify four venting conditions (check the code for exact wording).

Category I

Category I are non-positive vent static pressure appliances with a vent gas temperature that avoids excessive condensate production in the vent.

Category II

Category II is the same as Category I except it operates with a temperature that may cause excessive condensate production in the vent.

Category III

Category III appliances are positive vent static pressure with a vent gas temperature that avoids excessive condensate production in the vent. These will be either a forced or induced draft system.

Category IV

Category IV is the same as Category III except it operates with a temperature that may cause excessive condensate production in the vent. These will be either a forced or induced draft system.

The code only requires Category II, III, and IV appliances be vented with materials and sizes as recommended by the appliance manufacturer.

NOTE: The only practical appliances are either Category I or Category IV. However, installing a heat recovery unit in the flue gas travel could unintentionally convert a Category I appliance to a Category II by reducing the flue gas temperatures. In this case, rapid deterioration of the vent is probable as well as possible damage to the appliance itself.

Category I appliances have recently been divided into two sub-sets: (1) natural vented and (2) fan-assisted. Note that the fan in this latter group only assists the flue gas travel. It's not capable of developing forced draft pressures.

The tables in this chapter show headings for fan-assisted and for natural draft venting. This idea was first introduced with the 1992 *National Fuel Gas Code* as a minimum rating for a vent system in the fan-assisted columns. Laboratory and field testing show that a certain minimum energy is required to move products of combustion through the venting system. Thus, because oversized chimneys require additional heat to start the draft, oversized chimneys may fail to vent appliances, especially in mild weather or with intermittent firing.

In recent experiences with modern appliances, masonry chimneys have proven to be less suitable for gas vents than materials developed expressly for the purpose of serving as gas vents.

Under many probable size and firing patterns, such heavy chimneys may result in unsuccessful venting and hazardous conditions. Also, over time, the effect of condensate formed at firing startup can lead to deterioration and fume leakage into the building from chimney failures. In this connection, it is absolutely forbidden to use unlined masonry chimneys.

For a venting system to operate satisfactorily, it must be of proper size for the appliance load (or loads) connected. The manufacturer sizes the appliance flue, which is internal to the appliance, and the draft hood. The technician on the job frequently must select the size (or verify the adequacy) of the chimney and the size of the pipe from the appliance to the chimney. This latter pipe is called the vent connector.

Tables are introduced later in this chapter to provide the necessary information to size the vent connector and chimney. Or, the tables can be used to verify that existing vent connectors and/or chimneys are adequate for the connected appliance(s).

Appliances That Require Venting

The *National Fuel Gas Code* requires that the following appliance types be vented:

1. Boilers
2. Furnaces
3. Unit heaters
4. Duct furnaces
5. Incinerators
6. Water heaters
7. Listed built-in cooking units where the terms of the listing require venting
8. Room heaters listed to be vented
9. Type 1 and 2 clothes dryers
10. Appliances with conversion burners
11. Any appliance listed for venting

Some cooking ranges, domestic cooking units, or hot plates may not need to be vented. Domestic dryers should be vented outside, to avoid the buildup of humidity within the dwelling unit. Screws or any other means that extend into the duct which could catch lint and reduce the efficiency of the exhaust system should not be fastened to the dryer duct. Non-metallic flexible dryer vents are not permitted under some codes.

Table 15-1 (Table 13-1(a) of NFGC) shows the capacity of a Type B vent pipe and vent connector serving a single appliance. Figure 15-6 illustrates a typical installation of a single appliance with a double wall vent system.

IN THE FIELD
Dryer lint is extremely flammable and caution should be used when working with an open flame around areas where lint has been allowed to accumulate.

Example 1

A water heater with 50,000 BTU/hr input is to be vented with a Type B vent. What size is required if it is 25' high and the lateral offset is 2'?

Answer

Note that Table 15-1 does not have an entry for 25', so you must decide whether to use the 20' entry or the 30' entry. Since the entry for the higher chimney shows a greater capacity, a capacity that our lesser height could not develop, you should use the values given for 20' for the conservative choice.

Table 15-1   Capacity of Type B Vent Pipe and Vent Connector Serving a Single Appliance

Number of Appliances: Single
Appliance Type: Category I
Appliance Vent Connection: Connected Directly to Vent

Height H (ft)	Lateral L (ft)	Vent Diameter — D (in.)																				
		3			4			5			6			7			8			9		
		FAN		NAT	FAN		NAT	FAN		NAT	FAN		NAT	FAN		NAT	FAN		NAT	FAN		NAT
		Min	Max	Max	Min	Max	Max	Min	Max	Max	Min	Max	Max	Min	Max	Max	Min	Max	Max	Min	Max	Max
		Appliance Input Rating in Thousands of BTU per Hour																				
6	0	0	78	46	0	152	86	0	251	141	0	375	205	0	524	285	0	698	370	0	897	470
	2	13	51	36	18	97	67	27	157	105	32	232	157	44	321	217	53	425	285	63	543	370
	4	21	49	34	30	94	64	39	153	103	50	227	153	66	316	211	79	419	279	93	536	362
	6	25	46	32	36	91	61	47	149	100	59	223	149	78	310	205	93	413	273	110	530	354
8	0	0	84	50	0	165	94	0	276	155	0	415	235	0	583	320	0	780	415	0	1006	537
	2	12	57	40	16	109	75	25	178	120	28	263	180	42	365	247	50	483	322	60	619	418
	5	23	53	38	32	103	71	42	171	115	53	255	173	70	356	237	83	473	313	99	607	407
	8	28	49	35	39	98	66	51	164	109	64	247	165	84	347	227	99	463	303	117	596	396
10	0	0	88	53	0	175	100	0	295	166	0	447	255	0	631	345	0	847	450	0	1096	585
	2	12	61	42	17	118	81	23	194	129	26	289	195	40	402	273	48	533	355	57	684	457
	5	23	57	40	32	113	77	41	187	124	52	280	188	68	392	263	81	522	346	95	671	446
	10	30	51	36	41	104	70	54	176	115	67	267	175	88	376	245	104	504	330	122	651	427
15	0	0	94	58	0	191	112	0	327	187	0	502	285	0	716	390	0	970	525	0	1263	682
	2	11	69	48	15	136	93	20	226	150	22	339	225	38	475	316	45	633	414	53	815	544
	5	22	65	45	30	130	87	39	219	142	49	330	217	64	463	300	76	620	403	90	800	529
	10	29	59	41	40	121	82	51	206	135	64	315	208	84	445	288	99	600	386	116	777	507
	15	35	53	37	48	112	76	61	195	128	76	301	198	98	429	275	115	580	373	134	755	491

20	0	0	97	61	0	202	119	0	349	202	0	540	307	0	776	430	0	1057	575	0	1384	752
	2	10	75	51	14	149	100	18	250	166	20	377	249	33	531	346	41	711	470	50	917	612
	5	21	71	48	29	143	96	38	242	160	47	367	241	62	519	337	73	697	460	86	902	599
	10	28	64	44	38	133	89	50	229	150	62	351	228	81	499	321	95	675	443	112	877	576
	15	34	58	40	46	124	84	59	217	142	73	337	217	94	481	308	111	654	427	129	853	557
	20	48	52	35	55	116	78	69	206	134	84	322	206	107	464	295	125	634	410	145	830	537
30	0	0	100	64	0	213	128	0	374	220	0	587	336	0	853	475	0	1173	650	0	1548	855
	2	9	81	56	13	166	112	14	283	185	18	432	280	27	613	394	33	826	535	42	1072	700
	5	21	77	54	28	160	108	36	275	176	45	421	273	58	600	385	69	811	524	82	1055	688
	10	27	70	50	37	150	102	48	262	171	59	405	261	77	580	371	91	788	507	107	1028	668
	15	33	64	NA	44	141	96	57	249	163	70	389	249	90	560	357	105	765	490	124	1002	648
	20	56	58	NA	53	132	90	66	237	154	80	374	237	102	542	343	119	743	473	139	977	628
	30	NA	NA	NA	73	113	NA	88	214	NA	104	346	219	131	507	321	149	702	444	171	929	594
50	0	0	101	67	0	216	134	0	397	232	0	633	363	0	932	518	0	1297	708	0	1730	952
	2	8	86	61	11	183	122	14	320	206	15	497	314	22	715	445	26	975	615	33	1276	813
	5	20	82	NA	27	177	119	35	312	200	43	487	308	55	702	438	65	960	605	77	1259	798
	10	26	76	NA	35	168	114	45	299	190	56	471	298	73	681	426	86	935	589	101	1230	773
	15	59	70	NA	42	158	NA	54	287	180	66	455	288	85	662	413	100	911	572	117	1203	747
	20	NA	NA	NA	50	149	NA	63	275	169	76	440	278	97	642	401	113	888	556	131	1176	722
	30	NA	NA	NA	69	131	NA	84	250	NA	99	410	259	123	605	376	141	844	522	161	1125	670
100	0	NA	NA	NA	0	218	NA	0	407	NA	0	665	400	0	997	560	0	1411	770	0	1908	1040
	2	NA	NA	NA	10	194	NA	12	354	NA	13	566	375	18	831	510	21	1155	700	25	1536	935
	5	NA	NA	NA	26	189	NA	33	347	NA	40	557	369	52	820	504	60	1141	692	71	1519	926
	10	NA	NA	NA	33	182	NA	43	335	NA	53	542	361	68	801	493	80	1118	679	94	1492	910
	15	NA	NA	NA	40	174	NA	50	321	NA	62	528	353	80	782	482	93	1095	666	109	1465	895
	20	NA	NA	NA	47	166	NA	59	311	NA	71	513	344	90	763	471	105	1073	653	122	1438	880
	30	NA	NA	NA	NA	NA	NA	78	290	NA	92	483	NA	115	726	449	131	1029	627	149	1387	849
	50	NA	NA	NA	NA	NA	NA	NA	NA	NA	147	428	NA	180	651	405	197	944	575	217	1288	787

(continued)

Table 15-1 (Continued)

		Number of Appliances:																					Single			
		Appliance Type:																					Category I			
		Appliance Vent Connection:																					Connected Directly to Vent			
		Vent Diameter—D (in.)																								
		10			12			14			16			18			20			22			24		
		\multicolumn{24}{c}{Appliance Input Rating in Thousands of BTU per Hour}																						
Height H (ft)	Lateral L (ft)	FAN		NAT	FAN		NAT	FAN		NAT	FAN		NAT	FAN		NAT	FAN		NAT	FAN		NAT	FAN		NAT
		Min	Max	Max	Min	Max	Max	Min	Max	Max	Min	Max	Max	Min	Max	Max	Min	Max	Max	Min	Max	Max	Min	Max	Max
6	0	0	1121	570	0	1645	850	0	2267	1170	0	2983	1530	0	3802	1960	0	4721	2430	0	5737	2950	0	6853	3520
	2	75	675	455	103	982	650	138	1346	890	178	1769	1170	225	2250	1480	296	2782	1850	360	3377	2220	426	4030	2670
	4	110	668	445	147	975	640	191	1338	880	242	1761	1160	300	2242	1475	390	2774	1835	469	3370	2215	555	4023	2660
	6	128	661	435	171	967	630	219	1330	870	276	1753	1150	341	2235	1470	437	2767	1820	523	3363	2210	618	4017	2650
8	0	0	1261	660	0	1858	970	0	2571	1320	0	3399	1740	0	4333	2220	0	5387	2750	0	6555	3360	0	7838	4010
	2	71	770	515	98	1124	745	130	1543	1020	168	2030	1340	212	2584	1700	278	3196	2110	336	3882	2560	401	4634	3050
	5	115	758	503	154	1110	733	199	1528	1010	251	2013	1330	311	2563	1685	398	3180	2090	476	3863	2545	562	4612	3040
	8	137	746	490	180	1097	720	231	1514	1000	289	2000	1320	354	2552	1670	450	3163	2070	537	3850	2530	630	4602	3030
10	0	0	1377	720	0	2036	1060	0	2825	1450	0	3742	1925	0	4782	2450	0	5955	3050	0	7254	3710	0	8682	4450
	2	68	852	560	93	1244	850	124	1713	1130	161	2256	1480	202	2868	1890	264	3556	2340	319	4322	2840	378	5153	3390
	5	112	839	547	149	1229	829	192	1696	1105	243	2238	1461	300	2849	1871	382	3536	2318	458	4301	2818	540	5132	3371
	10	142	817	525	187	1204	795	238	1669	1080	298	2209	1430	364	2818	1840	459	3504	2280	546	4268	2780	641	5099	3340
15	0	0	1596	840	0	2380	1240	0	3323	1720	0	4423	2270	0	5678	2900	0	7099	3620	0	8665	4410	0	10,393	5300
	2	63	1019	675	86	1495	985	114	2062	1350	147	2719	1770	186	3467	2260	239	4304	2800	290	5232	3410	346	6251	4080
	5	105	1003	660	140	1476	967	182	2041	1327	229	2696	1748	283	3442	2235	355	4278	2777	426	5204	3385	501	6222	4057
	10	135	977	635	177	1446	936	227	2009	1289	283	2659	1712	346	3402	2193	432	4234	2739	510	5159	3343	599	6175	4019
	15	155	953	610	202	1418	905	257	1976	1250	318	2623	1675	385	3363	2150	479	4192	2700	564	5115	3300	665	6129	3980

Height (ft)	Lateral (ft)	Min	FAN Max	NAT Max	Min	FAN Max	NAT Max	Min	FAN Max	NAT Max	Min	FAN Max	NAT Max	Min	FAN Max	NAT Max	Min	FAN Max	NAT Max	Min	FAN Max	NAT Max	Min	FAN Max	NAT Max
20	0	0	1756	930	0	2637	1350	0	3701	1900	0	4948	2520	0	6376	3250	0	7988	4060	0	9785	4980	0	11,753	6000
	2	59	1150	755	81	1694	1100	107	2343	1520	139	3097	2000	175	3955	2570	220	4916	3200	269	5983	3910	321	7154	4700
	5	101	1133	738	135	1674	1079	174	2320	1498	219	3071	1978	270	3926	2544	337	4885	3174	403	5950	3880	475	7119	4662
	10	130	1105	710	172	1641	1045	220	2282	1460	273	3029	1940	334	3880	2500	413	4835	3130	489	5896	3830	573	7063	4600
	15	150	1078	688	195	1609	1018	248	2245	1425	306	2988	1910	372	3835	2465	459	4786	3090	541	5844	3795	631	7007	4575
	20	167	1052	665	217	1578	990	273	2210	1390	335	2948	1880	404	3791	2430	495	4737	3050	585	5792	3760	689	6953	4550
30	0	0	1977	1060	0	3004	1550	0	4252	2170	0	5725	2920	0	7420	3770	0	9341	4750	0	11,483	5850	0	13,848	7060
	2	54	1351	865	74	2004	1310	98	2786	1800	127	3696	2380	159	4734	3050	199	5900	3810	241	7194	4650	285	8617	5600
	5	96	1332	851	127	1981	1289	164	2759	1775	206	3666	2350	252	4701	3020	312	5863	3783	373	7155	4622	439	8574	5552
	10	125	1301	829	164	1944	1254	209	2716	1733	259	3617	2300	316	4647	2970	386	5803	3739	456	7090	4574	535	8505	5471
	15	143	1272	807	187	1908	1220	237	2674	1692	292	3570	2250	354	4594	2920	431	5744	3695	507	7026	4527	590	8437	5391
	20	160	1243	784	207	1873	1185	260	2633	1650	319	3523	2200	384	4542	2870	467	5686	3650	548	6964	4480	639	8370	5310
	30	195	1189	745	246	1807	1130	305	2555	1585	369	3433	2130	440	4442	2785	540	5574	3565	635	6842	4375	739	8239	5225
50	0	0	2231	1195	0	3441	1825	0	4934	2550	0	6711	3440	0	8774	4460	0	11,129	5635	0	13,767	6940	0	16,694	8430
	2	41	1620	1010	66	2431	1513	86	3409	2125	113	4554	2840	141	5864	3670	171	7339	4630	209	8980	5695	251	10,788	6860
	5	90	1600	996	118	2406	1495	151	3380	2102	191	4520	2813	234	5826	3639	283	7295	4597	336	8933	5654	394	10,737	6818
	10	118	1567	972	154	2366	1466	196	3332	2064	243	4464	2767	295	5763	3585	355	7224	4542	419	8855	5585	491	10,652	6749
	15	136	1536	948	177	2327	1437	222	3285	2026	274	4409	2721	330	5701	3534	396	7155	4511	465	8779	5546	542	10,570	6710
	20	151	1505	924	195	2288	1408	244	3239	1987	300	4356	2675	361	5641	3481	433	7086	4479	506	8704	5506	586	10,488	6670
	30	183	1446	876	232	2214	1349	287	3150	1910	347	4253	2631	412	5523	3431	494	6953	4421	577	8557	5444	672	10,328	6603
100	0	0	2491	1310	0	3925	2050	0	5729	2950	0	7914	4050	0	10,485	5300	0	13,454	6700	0	16,817	8600	0	20,578	10,300
	2	30	1975	1170	44	3027	1820	72	4313	2550	95	5834	3500	120	7591	4600	138	9577	5800	169	11,803	7200	204	14,264	8800
	5	82	1955	1159	107	3002	1803	136	4282	2531	172	5797	3475	208	7548	4566	245	9528	5769	293	11,748	7162	341	14,204	8756
	10	108	1923	1142	142	2961	1775	180	4231	2500	223	5737	3434	268	7478	4509	318	9447	5717	374	11,658	7100	436	14,105	8683
	15	126	1892	1124	163	2920	1747	206	4182	2469	252	5678	3392	304	7409	4451	358	9367	5665	418	11,569	7037	487	14,007	8610
	20	141	1861	1107	181	2880	1719	226	4133	2438	277	5619	3351	330	7341	4394	387	9289	5613	452	11,482	6975	523	13,910	8537
	30	170	1802	1071	215	2803	1663	265	4037	2375	319	5505	3267	378	7209	4279	446	9136	5509	514	11,310	6850	592	13,720	8391
	50	241	1688	1000	292	2657	1550	350	3856	2250	415	5289	3100	486	6956	4050	572	8841	5300	659	10,979	6600	752	13,354	8100

For SI units, 1 in. = 25.4 mm, l ft = 0.305 m, 1000 BTU/hr = 0.293 kW, 1 in.² = 645 mm².

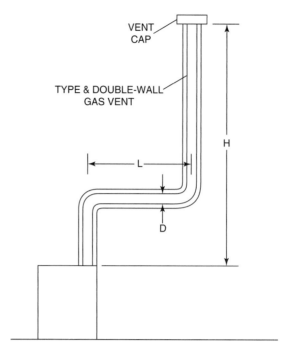

Figure 15-6 Single appliance with a double wall vent system.

For a 2' offset, you see that a 3" vent has a capacity of 51,000 BTU/hour, which is greater than the load of the heater. The notes, which apply to Table 15-1, indicate that you should not connect a 4" draft hood to a 3" vent, so the 3" vent selection is satisfactory only if the heater is furnished with a 3" draft hood. Note that if this water heater were fan-assisted, the 3" vent is satisfactory for a range of 10,000 to 75,000 Btu/hour.

When using the vent sizing tables in the *National Fuel Gas Code,* the following definitions apply:

Table 15-2   Description of Common Abbreviations Used in Vent Tables

Appliance Categorized Vent Area/Diameter	The minimum vent area/diameter permissible for Category I appliances to maintain a non-positive vent static pressure when tested in accordance with nationally recognized standards.
Fan-Assisted Combustion System	An appliance equipped with an integral mechanical means either to draw or force products of combustion through the combustion chamber or heat exchanger.
FAN min	The minimum appliance input rating of a Category I appliance with a fan-assisted combustion system that could be attached to the vent.
FAN max	The maximum appliance input rating of a Category I appliance with a fan-assisted combustion system that could be attached to the vent.
NAT max	The maximum appliance input rating of a Category I appliance equipped with a draft hood that could be attached to the vent. There is no minimum appliance input rating for the draft hood equipped appliances.
FAN + FAN	The maximum combined appliance input rating of two or more Category I fan-assisted appliances attached to the common vent.
FAN + NAT	The maximum combined appliance input rating of two or more Category I fan-assisted appliances and one or more Category I draft hood-equipped appliances attached to the common vent.
NAT + NAT	The maximum combined appliance input rating of two or more Category I draft hood-equipped appliances attached to the common vent.
NR	The vent configuration is not recommended due to potential for condensate formation or pressurization of the venting system.
NA	The vent configuration is not applicable due to physical or geometric constraints.

Notes for Single Appliance Vents (when using the following tables)

1. For single wall metal connector pipe, use Table 15-3.
2. If the vent size determined from the tables is smaller than the appliance draft hood outlet or flue collar, the smaller size shall be permitted to be used provided:
   (a) The total vent height "H" is at least 10 feet.
   (b) Vents for appliance draft hood outlets or flue collars 12" in diameter or smaller are not reduced more than one table size (12" to 10" is a one-size reduction).
   (c) Vents for appliance draft hood outlets or flue collars larger than 12" in diameter are not reduced more than two table sizes (24" to 20" is a two-size reduction).
   (d) The maximum capacity listed in the tables for a fan-assisted appliance is reduced by 10% (0.90 × maximum table capacity).
   (e) The draft hood outlet is greater than 4" in diameter. Do not connect a 3" diameter vent to a 4" diameter draft hood outlet. This provision does not apply to fan-assisted appliances.
3. Single appliance venting configurations with zero *(0)* lateral lengths must have no elbows in the venting system. For vent configurations with lateral lengths, the venting tables include allowance for two 90-degree turns. For each additional 90-degree turn or equivalent, the maximum capacity listed in the venting tables shall be reduced by 10% (0.90 × maximum table capacity). NOTE: Two 45-degree turns are equivalent to one 90-degree turn.
4. Zero (0) lateral "L" shall apply only to a straight vertical vent attached to a top outlet draft hood or flue collar.
5. Sea level input ratings shall be used when determining maximum capacity for high-altitude installation. Actual input (derated for altitude) shall be used for determining minimum capacity for high-altitude installation.
6. Numbers followed by an asterisk (*) in Tables 15-7 and 15-8 indicate the possibility of continuous condensation depending on locality. Consult appliance manufacturer, local serving gas supplier, or authority having jurisdiction.
7. For appliances with more than one input rate, the minimum vent capacity (FAN Min) determined from the tables shall be less than the lowest appliance input rating, and the maximum vent capacity (FAN Max/NAT Max) determined from the tables shall be greater than the highest appliance rating input (see Table 15-2).
8. Listed corrugated metallic chimney liner systems in masonry chimneys must be sized by using Table 15-1 for Type B vents, with the maximum capacity reduced by 20% (0.80 × maximum capacity) and the minimum capacity as shown in Table 15-1. Corrugated metallic liner systems installed with bends or offsets shall have their maximum capacity additionally reduced (see Note 3 above).
9. If the vertical vent has a larger diameter than the vent connector, the vertical vent diameter shall be used to determine the minimum vent capacity and the connector diameter shall be used to determine the maximum vent capacity. The flow area of the vertical vent shall not exceed seven times the flow area of the listed appliance categorized vent area, flue collar area, or draft hood outlet area unless designated in accordance with approved engineering methods.
10. The tables included should be used for chimneys and vents, not to the outdoors below the roof line. Chimneys or vents exposed to the outdoors below the roofline may experience continuous condensation depending on locality. Consult the appliance manufacturer, the local serving gas supplier, or the authority having jurisdiction. A Type B vent or listed chimney lining system

passing through an unused masonry chimney flue are not considered to be exposed to the outdoors.

11. Vent connectors must not be upsized more than two sizes greater than the listed appliance categorized vent diameter, flue collar diameter, or draft hood outlet diameter.

12. In a single run of vent or vent connector, more than one diameter and types are permitted to be used provided that all the sizes and types are permitted by the tables.

13. Interpolation is permitted in calculating capacities for vent dimensions that fall between table entries.

14. Extrapolation beyond the table entries is not permitted.

Example 2

In Figure 15-7, 210,000 BTU/hr heater is connected to a masonry chimney. If the height is 9' and the lateral offset is 5', what is the diameter of the vent connector and the minimum required area of the flue liner?

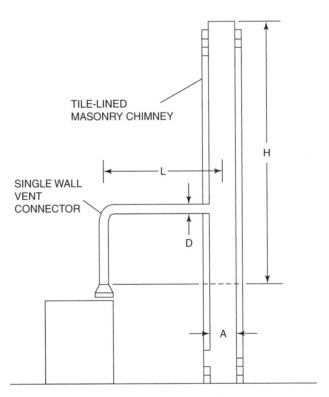

Figure 15-7 Single wall vent connector into masonry chimney.

Answer

Table 15-3, which covers masonry chimneys with single wall connectors, must be used with the 8' height column. Select the row for 5' lateral. An 8" vent connector is required with a chimney liner minimum size of 63 sq in (read from second to last row). Table 15-4 shows that an 8" × 12" liner provides 63 sq in.

Figure 15-8 and Table 15-5 pertain to combination systems connected to Type B double wall vents. You must first size the connectors for the individual appliances and then size the chimney. Select the vent connector size from the first part, selecting on the basis of whether the appliance is fan-assisted or natural draft. The combined vent size is then selected and the choices are fan plus fan, fan plus natural, and natural plus natural draft.

Table 15-3 Masonry Chimneys with Single Wall Connectors

Number of Appliances:	Single
Appliance Type:	Category I
Appliance Vent Connection:	Single-Wall Metal Connector

Single-Wall Metal Connector Diameter — *D* (in.)

To be used with chimney areas within the size limits at bottom

Appliance Input Rating in Thousands of Btu per Hour

Height *H* (ft)	Lateral *L* (ft)	3 FAN Min	3 FAN Max	3 NAT Max	4 FAN Min	4 FAN Max	4 NAT Max	5 FAN Min	5 FAN Max	5 NAT Max	6 FAN Min	6 FAN Max	6 NAT Max	7 FAN Min	7 FAN Max	7 NAT Max	8 FAN Min	8 FAN Max	8 NAT Max	9 FAN Min	9 FAN Max	9 NAT Max	10 FAN Min	10 FAN Max	10 NAT Max	12 FAN Min	12 FAN Max	12 NAT Max
6	2	NA	NA	28	NA	NA	52	NA	NA	86	NA	NA	130	NA	NA	180	NA	NA	247	NA	NA	319	NA	NA	400	NA	NA	580
	5	NA	NA	25	NA	NA	48	NA	NA	81	NA	NA	116	NA	NA	164	NA	NA	230	NA	NA	297	NA	NA	375	NA	NA	560
8	2	NA	NA	29	NA	NA	55	NA	NA	93	NA	NA	145	NA	NA	197	NA	NA	265	NA	NA	349	382	725	445	549	1021	650
	5	NA	NA	26	NA	NA	51	NA	NA	87	NA	NA	133	NA	NA	182	NA	NA	246	NA	NA	327	NA	NA	422	673	1003	638
	8	NA	NA	23	NA	NA	47	NA	NA	82	NA	NA	126	NA	NA	174	NA	NA	237	NA	NA	317	NA	NA	408	747	985	621
10	2	NA	NA	31	NA	NA	61	NA	NA	102	NA	NA	161	NA	NA	220	216	518	297	271	654	387	373	808	490	536	1142	722
	5	NA	NA	28	NA	NA	56	NA	NA	95	NA	NA	147	NA	NA	203	NA	NA	276	334	635	364	459	789	465	657	1121	710
	10	NA	NA	24	NA	NA	49	NA	NA	86	NA	NA	137	NA	NA	189	NA	NA	261	NA	NA	345	547	758	441	771	1088	665
15	2	NA	NA	35	NA	NA	67	NA	NA	113	NA	NA	178	166	473	249	211	611	335	264	776	440	362	965	560	520	1373	840
	5	NA	NA	32	NA	NA	61	NA	NA	106	NA	NA	163	NA	NA	230	261	591	312	325	755	414	444	942	531	637	1348	825
	10	NA	NA	27	NA	NA	54	NA	NA	96	NA	NA	151	NA	NA	214	NA	NA	294	392	722	392	531	907	504	749	1309	774
	15	NA	NA	NA	NA	NA	46	NA	NA	87	NA	NA	138	NA	NA	198	NA	NA	278	452	692	372	606	873	481	841	1272	738
20	2	NA	NA	38	NA	NA	73	NA	NA	123	NA	NA	200	163	520	273	206	675	374	258	864	490	252	1079	625	508	1544	950
	5	NA	NA	35	NA	NA	67	NA	NA	115	NA	NA	183	NA	NA	252	255	655	348	317	842	461	433	1055	594	623	1518	930
	10	NA	NA	NA	NA	NA	59	NA	NA	105	NA	NA	170	NA	NA	235	312	622	330	382	806	437	517	1016	562	733	1475	875
	15	NA	NA	NA	NA	NA	NA	NA	NA	95	NA	NA	156	NA	NA	217	NA	NA	311	442	773	414	591	979	539	823	1434	835
	20	NA	NA	NA	NA	NA	NA	NA	NA	80	NA	NA	144	NA	NA	202	NA	NA	292	NA	NA	392	663	944	510	911	1394	800

(continued)

Table 15-3   (Continued)

Number of Appliances: Single
Appliance Type: Category I
Appliance Vent Connection: Single-Wall Metal Connector

Single-Wall Metal Connector Diameter — D (in.)
To be used with chimney areas within the size limits at bottom

Appliance Input Rating in Thousands of Btu per Hour

Height H (ft)	Lateral L (ft)	3 FAN Min	3 FAN Max	3 NAT Max	4 FAN Min	4 FAN Max	4 NAT Max	5 FAN Min	5 FAN Max	5 NAT Max	6 FAN Min	6 FAN Max	6 NAT Max	7 FAN Min	7 FAN Max	7 NAT Max	8 FAN Min	8 FAN Max	8 NAT Max	9 FAN Min	9 FAN Max	9 NAT Max	10 FAN Min	10 FAN Max	10 NAT Max	12 FAN Min	12 FAN Max	12 NAT Max
30	2	NA	NA	41	NA	NA	81	NA	NA	136	NA	NA	215	158	578	302	200	759	420	249	982	556	340	1237	715	489	1789	1110
	5	NA	NA	NA	NA	NA	75	NA	NA	127	NA	NA	196	NA	NA	279	245	737	391	306	958	524	417	1210	680	600	1760	1090
	10	NA	NA	NA	NA	NA	66	NA	NA	113	NA	NA	182	NA	NA	260	300	703	370	370	920	496	500	1168	644	708	1713	1020
	15	NA	NA	NA	NA	NA	NA	NA	NA	105	NA	NA	168	NA	NA	240	NA	NA	349	428	884	471	572	1128	615	798	1668	975
	20	NA	NA	NA	NA	NA	NA	NA	NA	88	NA	NA	155	NA	NA	223	NA	NA	327	NA	NA	445	643	1089	585	883	1624	932
	30	NA	NA	NA	NA	NA	NA	NA	NA	NA	NA	NA	NA	NA	NA	182	NA	NA	281	NA	NA	408	NA	NA	544	1055	1539	865
50	2	NA	NA	NA	NA	NA	91	NA	NA	160	NA	NA	250	NA	NA	350	191	837	475	238	1103	631	323	1408	810	463	2076	1240
	5	NA	NA	NA	NA	NA	NA	NA	NA	149	NA	NA	228	NA	NA	321	NA	NA	442	293	1078	593	398	1381	770	571	2044	1220
	10	NA	NA	NA	NA	NA	NA	NA	NA	136	NA	NA	212	NA	NA	301	NA	NA	420	355	1038	562	447	1337	728	674	1994	1140
	15	NA	NA	NA	NA	NA	NA	NA	NA	124	NA	NA	195	NA	NA	278	NA	NA	395	NA	NA	533	546	1294	695	761	1945	1090
	20	NA	NA	NA	NA	NA	NA	NA	NA	NA	NA	NA	180	NA	NA	258	NA	NA	370	NA	NA	504	616	1251	660	844	1898	1040
	30	NA	NA	NA	NA	NA	NA	NA	NA	NA	NA	NA	NA	NA	NA	NA	NA	NA	318	NA	NA	458	NA	NA	610	1009	1805	970
Minimum internal area of chimney (in.²)				12			19			28			38			50			63			78			95			132
Maximum internal area of chimney (in.²)				49			88			137			198			269			352			445			550			792

For SI units, 1 in. = 25.4 mm, 1 ft = 0.305 m, 1000 BTU/hr = 0.293 kW, 1 in.^{2} = 645 mm^2.

Table 15-4   Masonry Chimney Liner Dimensions with Circular Equivalents

Nominal Liner Size (in.)	Inside Dimensions of Liner (in.)	Inside Diameter or Equivalent Diameter (in.)	Equivalent Area (in.²)
4 × 8	2 1/2 × 6 1/2	4.0	12.2
		5.0	19.6
		6.0	28.3
		7.0	38.3
8 × 8	6 3/4 × 6 3/4	7.4	42.7
		8.0	50.3
8 × 12	6 1/2 × 10 1/2	9.0	63.6
		10.0	78.5
12 × 12	9 3/4 × 9 3/4	10.4	83.3
		11.0	95.0
12 × 16	9 1/2 × 13 1/2	11.8	107.5
		12.0	113.0
		14.0	153.9
16 × 16	13 1/4 × 13 1/4	14.5	162.9
		15.0	176.7
16 × 20	13 × 17	16.2	206.1
		18.0	254.4
20 × 20	16 1/2 × 16 3/4	18.2	260.2
		20.0	314.1
20 × 24	16 1/2 × 20 1/2	20.1	314.2
		22.0	380.1
24 × 24	20 1/4 × 20 1/4	22.1	380.1
		24.0	452.3
24 × 28	20 1/4 × 24 1/4	24.1	456.2
28 × 28	24 1/4 × 24 1/4	26.4	543.3
		27.0	572.5
30 × 30	25 1/2 × 25 1/2	27.9	607.0
		30.0	706.8
30 × 36	25 1/2 × 31 1/2	30.9	749.9
		33.0	855.3
36 × 36	31 1/2 × 31 1/2	34.4	929.4
		36.0	1017.9

For SI units, 1 in. = 25.4 mm, 1 in.² = 645 mm².
Note: When liner sizes differ dimensionally from those shown in this table, equivalent diameters can be determined from published tables for square and rectangular ducts of equivalent carrying capacity or by other engineering methods.

Source: Reprinted with permission from NFPA 54, *National Fuel Gas Code*, Copyright © 2006, National Fire Protection Association, Quincy MA 02169. This reprinted material is not the complete and official position of the National Fire Protection Association on the referenced subject which is represented only by the standard in its entirety.

Note that these tables are complicated! The tables must be used along with the many notes that modify and limit their applications.

Example 3

Using Figure 15-8, a 50,000 BTU/hr water heater and an 180,000 BTU/hr fan-assisted boiler are connected with Type B vent connectors to a double-wall Type B vent. The water heater rise is 4" and the rise for the boiler is 2'. The total chimney height is 15'. What are the required sizes?

Answer

The water heater connector (Table 15-5) is selected from the 15' height, 1' rise, rating 53,000 BTU/hour for 4". Thus, the water heater is connected with 4" Type B material.

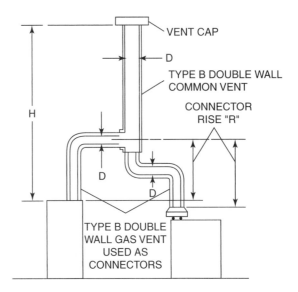

Figure 15-8 Double wall Type B vent pipe serving two or more appliances.

The boiler connector (Table 15-5) is selected as 6" from the 15', 2' rise row, which shows a capacity rating of 66,000 BTU/hour minimum to 235,000 BTU/hour maximum.

The common vent is sized from Table 15-5 after combining the inputs of the individual appliances. In this case, the total is:

50,000 + 180,000 = 230,000 BTU/hour

The 15' height column (in the common vent portion) for fan plus natural from Table 15-5 shows 7" to have a capacity of 352,000 BTU/hour. Therefore, 7" is the minimum size of the common vent serving these appliances.

Note that special remarks follow these tables that limit or aid in the use of the tables.

Example 4

For the same appliances and chimney dimensions as given in Example 3, what are the required sizes for a masonry chimney and single wall vent connectors?

Answer

From Table 15-6, the water heater vent connector must be 5" and the boiler 5" with 3' rise. Note that for 2' rise and 6" the minimum is 207,000 BTU/hour. From Table 15-6, the chimney must have a minimum area of 28 sq in. The smallest suitable standard tile size is 4" × 8" (from Table 15-4). Experience shows that most installers would choose an 8" × 8" flue liner.

Note that the water heater connector could be 4" if a 2' rise is possible.

ADDITIONAL INFORMATION CONCERNING MULTIPLE APPLIANCE VENTING

1. These vent tables should not be used where obstructions are installed in the venting system.
2. The maximum vent connector horizontal length is 1-1/2' (18") for each inch of connector diameter as follows:

Table 15-5 Type B Double Wall Vent

Number of Appliances:	Two or More
Appliance Type:	Category I
Appliance Vent Connection:	Type B Double-Wall Connector

Vent Connector Capacity

Type B Double-Wall Vent and Connector Diameter — D (in.)

Appliance Input Rating Limits in Thousands of BTU per Hour

Vent Height H (ft)	Connector Rise R (ft)	3 FAN Min	3 FAN Max	3 NAT Max	4 FAN Min	4 FAN Max	4 NAT Max	5 FAN Min	5 FAN Max	5 NAT Max	6 FAN Min	6 FAN Max	6 NAT Max	7 FAN Min	7 FAN Max	7 NAT Max	8 FAN Min	8 FAN Max	8 NAT Max	9 FAN Min	9 FAN Max	9 NAT Max	10 FAN Min	10 FAN Max	10 NAT Max
6	1	22	37	26	35	66	46	46	106	72	58	164	104	77	225	142	92	296	185	109	376	237	128	466	289
6	2	23	41	31	37	75	55	48	121	86	60	183	124	79	253	168	95	333	220	112	424	282	131	526	345
6	3	24	44	35	38	81	62	49	132	96	62	199	139	82	275	189	97	363	248	114	463	317	134	575	386
8	1	22	40	27	35	72	48	49	114	76	64	176	109	84	243	148	100	320	194	118	408	248	138	507	303
8	2	23	44	32	36	80	57	51	128	90	66	195	129	86	269	175	103	356	230	121	454	294	141	564	358
8	3	24	47	36	37	87	64	53	139	101	67	210	145	88	290	198	105	384	258	123	492	330	143	612	402
10	1	22	43	28	34	78	50	49	123	78	65	189	113	89	257	154	106	341	200	125	436	257	146	542	314
10	2	23	47	33	36	86	59	51	136	93	67	206	134	91	282	182	109	374	238	128	479	305	149	596	372
10	3	24	50	37	37	92	67	52	146	104	69	220	150	94	303	205	111	402	268	131	515	342	152	642	417
15	1	21	50	30	33	89	53	47	142	83	64	220	120	88	298	163	110	389	214	134	493	273	162	609	333
15	2	22	53	35	35	96	63	49	153	99	66	235	142	91	320	193	112	419	253	137	532	323	165	658	394
15	3	24	55	40	36	102	71	51	163	111	68	248	160	93	339	218	115	445	286	140	565	365	167	700	444
20	1	21	54	31	33	99	56	46	157	87	62	246	125	86	334	171	107	436	224	131	552	285	158	681	347
20	2	22	57	37	34	105	66	48	167	104	64	259	149	89	354	202	110	463	265	134	587	339	161	725	414
20	3	23	60	42	35	110	74	50	176	116	66	271	168	91	371	228	113	486	300	137	618	383	164	764	466
30	1	20	62	33	31	113	59	45	181	93	60	288	134	83	391	182	103	512	238	125	649	305	151	802	372
30	2	21	64	39	33	118	70	47	190	110	62	299	158	85	408	215	105	535	282	129	679	360	155	840	439
30	3	22	66	44	34	123	79	48	198	124	64	309	178	88	423	242	108	555	317	132	706	405	158	874	494
50	1	19	71	36	30	133	64	43	216	101	57	349	145	78	477	197	97	627	257	120	797	330	144	984	403
50	2	21	73	43	32	137	76	45	223	119	59	358	172	81	490	234	100	645	306	123	820	392	148	1014	478
50	3	22	75	48	33	141	86	46	229	134	61	366	194	83	502	263	103	661	343	126	842	441	151	1043	538
100	1	18	82	37	28	158	66	40	262	104	53	442	150	73	611	204	91	810	266	112	1038	341	135	1285	417
100	2	19	83	44	30	161	79	42	267	123	55	447	178	75	619	242	94	822	316	115	1054	405	139	1306	494
100	3	20	84	50	31	163	89	44	272	138	57	452	200	78	627	272	97	834	355	118	1069	455	142	1327	555

(continued)

Table 15-5 (Continued)

Common Vent Capacity

Type B Double-Wall Common Vent Diameter — D (in.)

Combined Appliance Input Rating in Thousands of BTU per Hour

Vent Height H (ft)	4 FAN +FAN	4 FAN +NAT	4 NAT +NAT	5 FAN +FAN	5 FAN +NAT	5 NAT +NAT	6 FAN +FAN	6 FAN +NAT	6 NAT +NAT	7 FAN +FAN	7 FAN +NAT	7 NAT +NAT	8 FAN +FAN	8 FAN +NAT	8 NAT +NAT	9 FAN +FAN	9 FAN +NAT	9 NAT +NAT	10 FAN +FAN	10 FAN +NAT	10 NAT +NAT
6	92	81	65	140	116	103	204	161	147	309	248	200	404	314	260	547	434	335	672	520	410
8	101	90	73	155	129	114	224	178	163	339	275	223	444	348	290	602	480	378	740	577	465
10	110	97	79	169	141	124	243	194	178	367	299	242	477	377	315	649	522	405	800	627	495
15	125	112	91	195	164	144	283	228	206	427	352	280	556	444	365	753	612	465	924	733	565
20	136	123	102	215	183	160	314	255	229	475	394	310	621	499	405	842	688	523	1035	826	640
30	152	138	118	244	210	185	361	297	266	547	459	360	720	585	470	979	808	605	1209	975	740
50	167	153	134	279	244	214	421	353	310	641	547	423	854	706	550	1164	977	705	1451	1188	860
100	175	163	NA	311	277	NA	489	421	NA	751	658	479	1025	873	625	1408	1215	800	1784	1502	975

Number of Appliances: Two or More
Appliance Type: Category I
Appliance Vent Connection: Type B Double-Wall Connector

Type B Double-Wall Vent and Connector Diameter — D (in.)

Appliance Input Rating Limits in Thousands of BTU per Hour

Vent Height H (ft)	Connector Rise R (ft)	12 FAN Min	12 FAN Max	12 NAT Max	14 FAN Min	14 FAN Max	14 NAT Max	16 FAN Min	16 FAN Max	16 NAT Max	18 FAN Min	18 FAN Max	18 NAT Max	20 FAN Min	20 FAN Max	20 NAT Max	22 FAN Min	22 FAN Max	22 NAT Max	24 FAN Min	24 FAN Max	24 NAT Max
6	2	174	764	496	223	1046	653	281	1371	853	346	1772	1080	NA	NA	NA	NA	NA	NA	NA	NA	NA
6	4	180	897	616	230	1231	827	287	1617	1081	352	2069	1370	NA	NA	NA	NA	NA	NA	NA	NA	NA
6	6	NA	NA	NA	NA	NA	NA	NA	NA	NA	NA	NA	NA	NA	NA	NA	NA	NA	NA	NA	NA	NA
8	2	186	822	516	238	1126	696	298	1478	910	365	1920	1150	NA	NA	NA	NA	NA	NA	NA	NA	NA
8	4	192	952	644	244	1307	884	305	1719	1150	372	2211	1460	471	2737	1800	560	3319	2180	662	3957	2590
8	6	198	1050	772	252	1445	1072	313	1902	1390	380	2434	1770	478	3018	2180	568	3665	2640	669	4373	3130
10	2	196	870	536	249	1195	730	311	1570	955	379	2049	1205	NA	NA	NA	NA	NA	NA	NA	NA	NA
10	4	201	997	664	256	1371	924	318	1804	1205	387	2332	1535	486	2887	1890	581	3502	2280	686	4175	2710
10	6	207	1095	792	263	1509	1118	325	1989	1455	395	2556	1865	494	3169	2290	589	3849	2760	694	4593	3270
15	2	214	967	568	272	1334	790	336	1760	1030	408	2317	1305	NA	NA	NA	NA	NA	NA	NA	NA	NA
15	4	221	1085	712	279	1499	1006	344	1978	1320	416	2579	1665	523	3197	2060	624	3881	2490	734	4631	2960
15	6	228	1181	856	286	1632	1222	351	2157	1610	424	2796	2025	533	3470	2510	634	4216	3030	743	5035	3600
20	2	223	1051	596	291	1443	840	357	1911	1095	430	2533	1385	NA	NA	NA	NA	NA	NA	NA	NA	NA
20	4	230	1162	748	298	1597	1064	365	2116	1395	438	2778	1765	554	3447	2180	661	4190	2630	772	5005	3130
20	6	237	1253	900	307	1726	1288	373	2287	1695	450	2984	2145	567	3708	2650	671	4511	3190	785	5392	3790
30	2	216	1217	632	286	1664	910	367	2183	1190	461	2891	1540	NA	NA	NA	NA	NA	NA	NA	NA	NA
30	4	223	1316	792	294	1802	1160	376	2366	1510	474	3110	1920	619	3840	2365	728	4861	2860	847	5606	3410
30	6	231	1400	952	303	1920	1410	384	2524	1830	485	3299	2340	632	4080	2875	741	4976	3480	860	5961	4150

H (ft)	Rise (ft)	12 FAN Min	12 FAN Max	12 NAT Max	14 FAN Min	14 FAN Max	14 NAT Max	16 FAN Min	16 FAN Max	16 NAT Max	18 FAN Min	18 FAN Max	18 NAT Max	20 FAN Min	20 FAN Max	20 NAT Max	22 FAN Min	22 FAN Max	22 NAT Max	24 FAN Min	24 FAN Max	24 NAT Max
50	2	206	1479	689	273	2023	1007	350	2659	1315	435	3548	1665	NA	NA	NA	NA	NA	NA	NA	NA	NA
	4	213	1561	860	281	2139	1291	359	2814	1685	447	3730	2135	580	4601	2633	709	5569	3185	851	6633	3790
	6	221	1631	1031	290	2242	1575	369	2951	2055	461	3893	2605	594	4808	3208	724	5826	3885	867	6943	4620
100	2	192	1923	712	254	2644	1050	326	3490	1370	402	4707	1740	NA	NA	NA	NA	NA	NA	NA	NA	NA
	4	200	1984	888	263	2731	1346	336	3606	1760	414	4842	2220	523	5982	2750	639	7254	3330	769	8650	3950
	6	208	2035	1064	272	2811	1642	346	3714	2150	426	4968	2700	539	6143	3350	654	7453	4070	786	8892	4810

Common Vent Capacity

Type B Double-Wall Common Vent Diameter — D (in.)

Combined Appliance Input Rating in Thousands of BTU per Hour

Vent Height H (ft)	12 FAN+FAN	12 FAN+NAT	12 NAT+NAT	14 FAN+FAN	14 FAN+NAT	14 NAT+NAT	16 FAN+FAN	16 FAN+NAT	16 NAT+NAT	18 FAN+FAN	18 FAN+NAT	18 NAT+NAT	20 FAN+FAN	20 FAN+NAT	20 NAT+NAT	22 FAN+FAN	22 FAN+NAT	22 NAT+NAT	24 FAN+FAN	24 FAN+NAT	24 NAT+NAT
6	900	696	588	1284	990	815	1735	1336	1065	2253	1732	1345	2838	2180	1660	3488	2677	1970	4206	3226	2390
8	994	773	652	1423	1103	912	1927	1491	1190	2507	1936	1510	3162	2439	1860	3890	2998	2200	4695	3616	2680
10	1076	841	712	1542	1200	995	2093	1625	1300	2727	2113	1645	3444	2665	2030	4241	3278	2400	5123	3957	2920
15	1247	986	825	1794	1410	1158	2440	1910	1510	3184	2484	1910	4026	3133	2360	4971	3862	2790	6016	4670	3400
20	1405	1116	916	2006	1588	1290	2722	2147	1690	3561	2798	2140	4548	3552	2640	5573	4352	3120	6749	5261	3800
30	1658	1327	1025	2373	1892	1525	3220	2558	1990	4197	3326	2520	5303	4193	3110	6539	5157	3680	7940	6247	4480
50	2024	1640	1280	2911	2347	1863	3964	3183	2430	5184	4149	3075	6567	5240	3800	8116	6458	4500	9837	7813	5475
100	2569	2131	1670	3732	3076	2450	5125	4202	3200	6749	5509	4050	8597	6986	5000	10,681	8648	5920	13,004	10,499	7200

For SI units, 1 in. = 25.4 mm, 1 in.² = 645 mm², 1 ft = 0.305 m, 1000 Btu per hr = 0.293 kW.

Source: Reprinted with permission from NFPA 54, *National Fuel Gas Code*, Copyright © 2006, National Fire Protection Association, Quincy, MA 02169. This reprinted material is not the complete and official position of the National Fire Protection Association on the referenced subject which is represented only by the standard in its entirety.

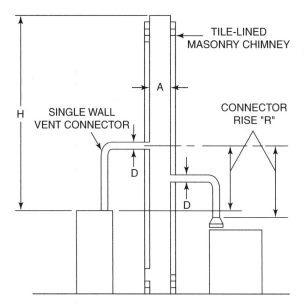

Figure 15-9 Typical piping layout for single wall connector and a masonry chimney when dual appliances are being used.

3. The vent connector must be routed to the vent utilizing the shortest possible of route. Connectors with longer than those listed above are permitted under the following conditions (see Table 15-7).

   (a) The maximum capacity (FAN Max or NAT Max) of the vent connector must be reduced 10% for each additional multiple of the length listed above. For example, the maximum length listed above for a 4" connector is 6'. With a connector greater than 6' but not exceeding 12', the maximum capacity must be reduced by 10% (0.90 × maximum vent connector capacity). With a connector length greater than 12' but not exceeding 18', the maximum must be reduced by 20% (0.80 × maximum vent capacity).

   (b) The minimum capacity (FAN Min) shall be determined by referring to the corresponding single-appliance table. In this case, for each appliance, the entire vent connector and common vent from the appliance to the vent termination is treated as a single-appliance vent as if the other appliances were not present.

4. If the vent connectors are combined prior to entering the common vent, the maximum common vent capacity listed in the common venting tables is reduced by 10% (0.90 × maximum common vent capacity). The length of the common vent connector manifold (L) will not exceed 1-1/2' (18") for each inch of common vent connector manifold diameter (D).

5. If the common vertical vent is offset as shown, the maximum common vent capacity listed in the common venting tables is reduced by 20% (0.80 × maximum common vent capacity), the equivalent of two 90-degree turns. The horizontal length of the common vent offset (L) does not exceed 1-1/2' for each inch of common vent diameter (D).

6. Excluding elbows counted in Note 5, for each additional 90-degree turn in excess of two, the maximum capacity of that portion of the venting system is reduced by 10% (0.90 × maximum common vent capacity).

   NOTE: Two 45-degree turns are equivalent to one 90-degree turn.

7. The common vent diameter is at least as large as the largest vent connector diameter.

8. Interconnection fittings are the same size as the common vent.

9. Sea level input ratings are used when determining maximum capacity for high-attitude installation.

Table 15-6 Masonry Chimney

Number of Appliances:	Two or More
Appliance Type:	Category I
Appliance Vent Connection:	Single-Wall Metal Connector

Vent Connector Capacity

Single-Wall Metal Vent Connector Diameter — D (in.)

Appliance Input Rating Limits in Thousands of BTU per Hour

Vent Height H (ft)	Connector Rise R (ft)	3 FAN Min	3 FAN Max	3 NAT Max	4 FAN Min	4 FAN Max	4 NAT Max	5 FAN Min	5 FAN Max	5 NAT Max	6 FAN Min	6 FAN Max	6 NAT Max	7 FAN Min	7 FAN Max	7 NAT Max	8 FAN Min	8 FAN Max	8 NAT Max	9 FAN Min	9 FAN Max	9 NAT Max	10 FAN Min	10 FAN Max	10 NAT Max
6	1	NA	NA	21	NA	NA	39	NA	NA	66	179	191	100	231	271	140	292	366	200	362	474	252	499	594	316
	2	NA	NA	28	NA	NA	52	NA	NA	84	186	227	123	239	321	172	301	432	231	373	557	299	509	696	376
	3	NA	NA	34	NA	NA	61	134	153	97	193	258	142	247	365	202	309	491	269	381	634	348	519	793	437
8	1	NA	NA	21	NA	NA	40	NA	NA	68	195	208	103	250	298	146	313	407	207	387	530	263	529	672	331
	2	NA	NA	28	NA	NA	52	137	139	85	202	240	125	258	343	177	323	465	238	397	607	309	540	766	391
	3	NA	NA	34	NA	NA	62	143	156	98	210	264	145	266	376	205	332	509	274	407	663	356	551	838	450
10	1	NA	NA	22	NA	NA	41	130	151	70	202	225	106	267	316	151	333	434	213	410	571	273	558	727	343
	2	NA	NA	29	NA	NA	53	136	150	86	210	255	128	276	358	181	343	489	244	420	640	317	569	813	403
	3	NA	NA	34	97	102	62	143	166	99	217	277	147	284	389	207	352	530	279	430	694	363	580	880	459
15	1	NA	NA	23	NA	NA	43	129	151	73	199	271	112	268	376	161	349	502	225	445	646	291	623	808	366
	2	NA	NA	30	92	103	54	135	170	88	207	295	132	277	411	189	359	548	256	456	706	334	634	884	424
	3	NA	NA	34	96	112	63	141	185	101	215	315	151	286	439	213	368	586	289	466	755	378	646	945	479
20	1	NA	NA	23	87	99	45	128	167	76	197	303	117	265	425	169	345	569	235	439	734	306	614	921	387
	2	NA	NA	30	91	111	55	134	185	90	205	325	136	274	455	195	355	610	266	450	787	348	627	986	443
	3	NA	NA	35	96	119	64	140	199	103	213	343	154	282	481	219	365	644	298	461	831	391	639	1042	496
30	1	NA	NA	24	86	108	47	126	187	80	193	347	124	259	492	183	338	665	250	430	864	330	600	1089	421
	2	NA	NA	31	91	119	57	132	203	93	201	366	142	269	518	205	348	699	282	442	908	372	613	1145	473
	3	NA	NA	35	95	127	65	138	216	105	209	381	160	277	540	229	358	729	312	452	946	412	626	1193	524
50	1	NA	NA	24	85	113	50	124	204	87	188	392	139	252	567	208	328	778	287	417	1022	383	582	1302	492
	2	NA	NA	31	89	123	60	130	218	100	196	408	158	262	588	230	339	806	320	429	1058	425	596	1346	545
	3	NA	NA	35	94	131	68	136	231	112	205	422	176	271	607	255	349	831	351	440	1090	466	610	1386	597
100	1	NA	NA	23	84	104	49	122	200	89	182	410	151	243	617	232	315	875	328	402	1181	444	560	1537	580
	2	NA	NA	30	88	115	59	127	215	102	190	425	169	253	636	254	326	899	361	415	1210	488	575	1570	634
	3	NA	NA	34	93	124	67	133	228	115	199	438	188	262	654	279	337	921	392	427	1238	529	589	1604	687

(continued)

Table 15-6 (Continued)

Common Vent Capacity

	12			19			28			38			50			63			78			113		
Vent Height H (ft)	FAN +FAN	FAN +NAT	NAT +NAT	FAN +FAN	FAN +NAT	NAT +NAT	FAN +FAN	FAN +NAT	NAT +NAT	FAN +FAN	FAN +NAT	NAT +NAT	FAN +FAN	FAN +NAT	NAT +NAT	FAN +FAN	FAN +NAT	NAT +NAT	FAN +FAN	FAN +NAT	NAT +NAT	FAN +FAN	FAN +NAT	NAT +NAT
	Combined Appliance Input Rating in Thousands of BTU per Hour																							
6	NA	NA	25	NA	118	45	NA	176	71	NA	255	102	NA	348	142	NA	455	187	NA	579	245	NA	846	NA
8	NA	NA	28	NA	128	52	NA	190	81	NA	276	118	NA	380	162	NA	497	217	NA	633	277	1136	928	405
10	NA	NA	31	NA	136	56	NA	205	89	NA	295	129	NA	405	175	NA	532	234	771	680	300	1216	1000	450
15	NA	NA	36	NA	NA	66	NA	230	105	NA	335	150	NA	400	210	677	602	280	866	772	360	1359	1139	540
20	NA	NA	NA	NA	NA	74	NA	247	120	NA	362	170	NA	503	240	765	661	321	947	849	415	1495	1264	640
30	NA	NA	NA	NA	NA	NA	NA	NA	135	NA	398	195	NA	558	275	808	739	377	1052	957	490	1682	1447	740
50	NA	NA	NA	NA	NA	NA	NA	NA	NA	NA	NA	NA	NA	612	325	NA	821	456	1152	1076	600	1879	1672	910
100	NA	NA	NA	NA	NA	NA	NA	NA	NA	NA	NA	NA	NA	NA	NA	NA	NA	494	NA	NA	663	2006	1885	1046

Minimum Internal Area of Masonry Chimney Flue (in.²)

For SI units, 1 in. = 25.4 mm, 1 in.² = 645 mm², 1 ft = 0.305 m, 1000 Btu per hr = 0.293 kW.

Source: Reprinted with permission from NFPA 54, *National Fuel Gas Code*, Copyright © 2006, National Fire Protection Association, Quincy, MA 02169. This reprinted material is not the complete and official position of the National Fire Protection Association on the referenced subject which is represented only by the standard in its entirety.

Table 15-7 Connector Diameter to Horizontal Length Ratios

Connector Diameter Maximum (inches)	Connector Horizontal Length (feet)
3	4.5
4	6
5	7.5
6	9
7	10.5
8	12
9	13.5
10	15
12	18
14	21
16	24
18	27
20	30
22	33
24	36

10. For multiple units of gas utilization equipment all located on one floor, available total height (H) is measured from the highest draft hood outlet or flue collar up to the level of the cap or terminal. Connector rise (R) is measured from the draft hood outlet or flue collar to the level where the vent gas streams come together (not applicable to multistory).

11. For multistory installations, available total height (H) for each segment of the system is the vertical distance between the highest draft hood outlet or flue collar entering that segment and the centerline of the next higher interconnection tee.

12. The size of the lowest connector and the size of the vertical vent leading to the lowest interconnection of a multistory system, must be in accordance with Table 15-1 for available total height (H) up to the lowest interconnection.

13. Where used in multistory systems, vertical common vents must be Type B double wall and have no offsets.

14. Where two or more appliances are connected to a vertical vent or chimney, the flow area of the largest section of vertical vent or chimney must not exceed seven times the smallest listed appliance categorized vent areas, flue collar area, or draft hood outlet area, unless designed in accordance with approved engineering methods.

15. For appliances with more than one input rate, the minimum vent connector capacity (FAN Min), determined from the tables, must not be less than the lowest appliance input rating and the maximum vent connector capacity (FAN Max or NAT Max), determined from the tables, must be greater than the highest appliance input rating.

16. Listed, corrugated metallic chimney liner systems in masonry chimneys are sized by using Table 15-5 for Type B vents, with the maximum capacity reduced by 20% (0.80 × maximum capacity) and the minimum capacity as shown in Table 15-5. Corrugated metal vent systems installed with bends or offsets require additional reduction of the vent maximum capacity (see Note 6).

17. The tables included must be used for chimneys and vents not exposed to the outdoors below the roof line. Chimney or vents exposed to the outdoors below the roof line may experience continuous condensation, depending on locality. Consult the appliance manufacturer, the local serving gas supplier, or the authority having jurisdiction. The Type B vent or listed chimney lining system, passing through an unused masonry chimney flue, must not be considered to be exposed to the outdoors.

18. Vent connectors will not be upsized more than two sizes greater than the listed appliance categorized vent diameter, flue collar diameter, or draft hood outlet diameter. Vent connectors must not be smaller than the listed appliance categorized vent diameter, flue collar diameter, or draft hood outlet diameter.

19. All combinations of pipe size, single wall, and double wall metal pipe shall be allowed within any connector run(s) or within the common vent provided ALL of the appropriate tables permit ALL of the desired sizes and types of pipe as if they were used for the entire length of the subject connector or vent. If single wall and Type B double wall metal pipes are used for vent connectors, the common vent must be sized using Table 15-6.

20. Where a table permits more than one diameter of pipe to be used for a connector or vent, all the permitted sizes are permitted to be used. NOTE: In general, it is preferable to use the smallest diameter permitted to minimize heat loss.

21. Interpolation is permitted in calculating capacities for vent dimensions that fall between table entries.

22. Extrapolation beyond the table entries is not permitted.

NOTE: Part 12 of the *National Fuel Gas Code* contains more tables than are presented in this chapter. See the *National Fuel Gas Code* for these additional tables. In order to build skill and familiarity with the vent sizing tables, practice sizing various appliances and appliance combinations using these tables.

VENTS FOR CATEGORY I APPLIANCES

Draft Hood (Draft Diverter)

The first item in the venting system to be installed is the *draft hood,* also called the *draft diverter*. This unit is usually furnished by the appliance manufacturer and is installed on or near the appliance. Frequently, the draft diverter is built into the appliance. If the manufacturer provides the unit, separately or built-in, it must be installed exactly as received.

Figures 15-10 and 15-11 illustrate vertical and horizontal draft diverters. Vertical draft hoods should be installed plumb; horizontal units should be level or sloping slightly upward in the direction of discharge flow. Hangers and straps should support the horizontal diverter.

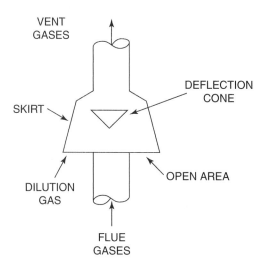

Figure 15-10 Vertical draft diverter.

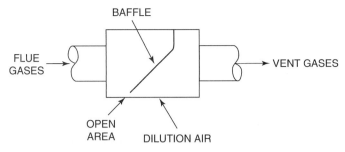

Figure 15-11 Horizontal draft diverter.

Vent Connector

The *vent connector* connects the draft hood to the chimney. The vent connector is usually single wall vent pipe, but in a difficult situation, double wall vent pipe may be used to solve problems. One complication is a long vent connector with very little upward slope available. Another difficulty occurs when a vent connector must extend through an unheated space. The earlier section of this chapter describes the effect on capacity of various vent configurations.

Whatever pipe material is selected, the assembly should be made as recommended by the manufacturer. Figure 15-12 shows the installation of Type B vents. The vent connector should slope upward at least 1/4" per foot in the direction of flow so that condensate (which could form during startup) will not accumulate in trapped sections of the pipe.

Since much water vapor is formed in the products of combustion of natural gas, condensation can form if any part of the venting conduit is cooler than the dew point of the gases exiting the appliance. If condensation can form in the piping, provision should be made to collect it and drain it away. Remember that products of combustion frequently contain trace amounts of corrosive chemicals, so accumulations of condensate should be avoided if possible.

When adding or subtracting an appliance to an existing venting system, determine the existing load and, using the tables in the earlier part of this chapter, verify that the venting system is adequate for the new total load.

General Rules for Vent Connectors

Installation procedures normally follow these guidelines:

1. Do not use reducers in the vent connector unless approved for a conversion burner installation.
2. Avoid unnecessary bends and use 45° bends instead of 90° where possible.
3. Use at least three sheet metal screws (on metallic vent type) per joint on single wall connectors. No screws are permitted on double wall connectors or vents that penetrate or distort the inner liner.
4. Make the vent connector as short and as straight as possible.
5. The length of a single wall vent connector should not exceed 75% of the height of the chimney.
6. Use a thimble or sleeve to enter a masonry chimney so the connector will be conveyed to the liner, but not into it.

Masonry Chimney

Figure 15-13 shows a masonry chimney with several possible defects that are described below. The clay tile liner should be sized properly and thimbles used to receive vent connectors. After cutting or drilling the hole through the chimney wall, use a mirror to look up the flue liner to be certain that the chimney is open.

Figure 15-12 Connected sections of Type B vent pipe.

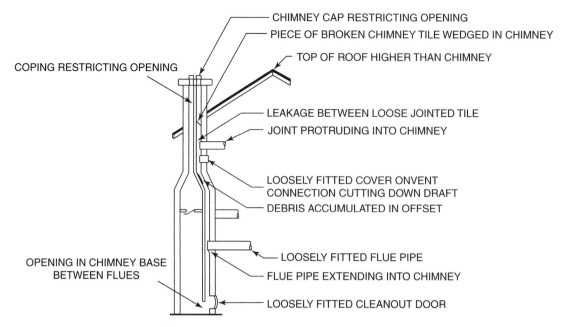

Figure 15-13 Masonry chimney and possible defects.

Chimney Problems

Chimney problems that must be checked include the following items:

1. Chimney leakage
   Failed or deteriorated mortar joints and cracked or open joints in the liner permit chimney gases to escape into the building.
2. Chimney obstruction
   Any object within the liner obstructs gas flow and produces a very dangerous situation.
3. Undersized or oversized
   Undersized or grossly oversized chimneys will produce poor and inadequate draft.
4. Improper connection
   Improper connection of other appliances to the chimney may lead to gas leaks or prevent other equipment from working properly.
5. Insufficient height
   Insufficient height of the chimney above the roof could lead to back draft or other problems.
6. Deflection of wind
   Deflection of wind into the chimney by trees or nearby buildings is problematic.

Other Chimneys

Type B, B-W, or masonry chimneys are the preferred types to use. Single wall types, unless large, usually cool the vent gases excessively. Many high-efficiency appliances—especially Category IV furnaces and boilers—use plastic pipe for vents. As in all cases, manufacturer's instructions must be followed.

Type B chimneys must be properly supported and they must maintain one-inch clearance from combustibles. Figure 15-14 shows a chimney that requires support at every floor.

A standard support item is a plate that holds the pipe and is nailed to joists or rafters. It also includes the required spacing from combustibles. Install these chimneys plumb by using a plumb bob. Joints are held together with the interlocking features in the pipe ends. Do not use sheet metal screws with Type B vents. Also, be sure to use a weather cap or other approved top fitting for the assembled chimney.

If it serves appliances burning liquid or solid fuels or incinerators, check the venting system carefully. Recent code changes permit the use of single chimneys for

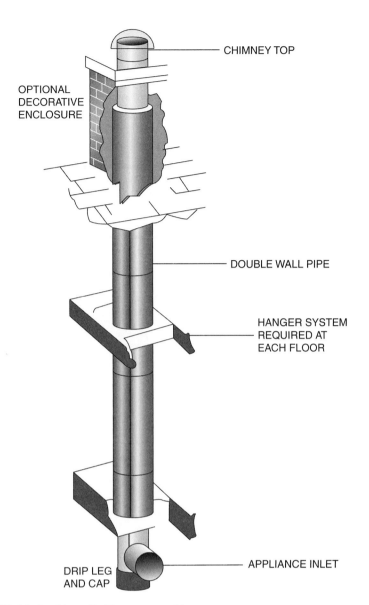

CHIMNEY TOP

OPTIONAL
DECORATIVE
ENCLOSURE

DOUBLE WALL PIPE

HANGER SYSTEM
REQUIRED AT
EACH FLOOR

DRIP LEG
AND CAP

APPLIANCE INLET

Figure 15-14 Double wall chimney assembly.

different fuels, provided certain precautions are observed—principally, the gas appliance draft hood must have a spill switch mounted on it to shut down the unit if flue gases should spill from the draft hood.

If liquid or solid fuel devices are used extensively in your area, any Type B vent should be labeled plainly; state that the vent should only be used for gas-fired equipment.

Single wall pipe should not be used in concealed spaces, attics, within wall structures, or through an unprotected floor. Do not install this pipe where cold drafts can affect it as this will produce heavy condensation within the pipe. When passing near combustible material, provide the proper clearances to maintain a safe installation.

VENT INSTALLATION DETAILS

All flue and vent piping operate at warm-to-hot temperatures. Even for those systems that are designed to operate at relatively cool temperatures, it is wise to consider the possible effects of partially blocked flues, inoperative fan motors, broken blower belts, dirty filters, and similar malfunctions on the vent gas temperatures.

Keep in mind that gas appliances are designed to deliver most of the heat from the combustion process to the heat transfer fluid—air, water, or steam—but if this transfer is prevented from taking place, the flue and vent gas temperatures will be abnormally high.

Because of the above concerns and to ensure safe operation under routine conditions, minimum clearance-to-combustible dimensions are given for each vent material type.

Usual combustible materials can be ignited at fairly low temperatures if exposed continuously to temperatures in the range of 250° F to 300° F. The high temperatures gradually drive the moisture from the material and when the moisture level is low enough, the material can spontaneously ignite. Years of industry experience have produced the minimum safe clearances-from-combustibles for vent connectors listed in Table 15-8.

Table 15-8   Vent Connector Clearance

| Listed Appliance | Minimum Distance From Combustible Material | | |
	Listed Type B Gas Vent Material	Listed Type L Material	Single Wall Metal Pipe
Listed appliances with draft hoods and appliances listed for use with Type B gas vents	As Listed	As Listed	6"
Boilers and furnaces with listed gas conversion burner and with draft hood	6"	6"	9"
Appliances listed for use with Type L venting systems	Not Permitted	As Listed	9"
Residential incinerators	Not Permitted	9"	18"
Unlisted residential appliances with draft hood	Not Permitted	6"	9"
Residential and low-heat appliances other than those above	Not Permitted	9"	18"

Example 5

A single wall vent connector serves a listed water heater with draft hood. What is the clearance space required from combustible material as shown in Table 15-8?

Answer

The first row of the table is for listed appliances with a draft hood and the connector is single wall metal pipe. Therefore the required clearance is 6".

Example 6

A residential incinerator vent must be fabricated. A single wall vent connector is used to a double wall chimney. What type chimney is needed and what clearances are required?

Answer

Notice that Type B cannot be used. The single wall vent connector to Type L vent must have 18" minimum clearance to combustible materials.

Clearance from combustibles is ensured and maintained by using manufactured spacers and supports to install the materials. Such supports are quicker and more reliable than most homemade hangers. Figures 15-15 and 15-16 show typical devices.

DWB' WALL BAND

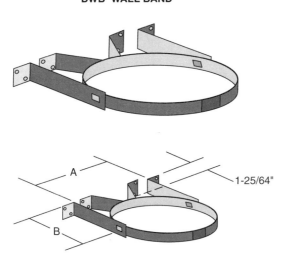

Figure 15-15 Wall support band used to support vertical flue pipes.

DFW* FIRE STOP SPACER. WALL

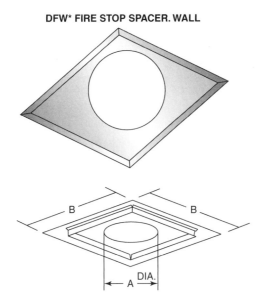

Figure 15-16 Fire stop spacers help maintain the minimum clearance from combustables and help stop the spread of flames between floors.

Vent Terminal

Vent terminals are an important part of the chimney. The location and type of terminal fitting have a major effect on the operation of the chimney.

Masonry Chimney

Masonry chimneys are usually open on top. They must be 3' above the highest point where it passes through the roof and at least 2' above any portion of the building within 10' as shown in Figure 15-17. In addition, the chimney should be at least 5' above the highest connected vent pipe.

Double Wall Unit

For double wall units (Types B, B-W, and L), the following conditions should be observed:

1. Use a listed vent cap or roof assembly to prevent rain, snow, or debris from entering the chimney.

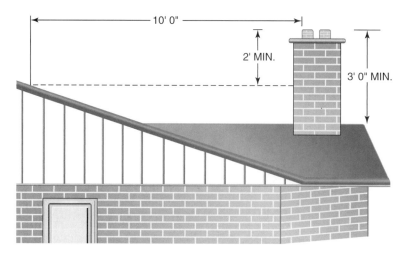

Figure 15-17 Always check for vent terminal clearance issues before cutting holes and installing the flue pipe.

2. The chimneys should be at least 3' higher than the highest point where they pass through the roof and 2' higher than any portion of the building within 10' unless the terms of the listing and manufacturer's instructions state otherwise.
3. Types B and L vents should be at least 5' above the highest draft hood inlet.
4. Type B-W should be at least 12' above the bottom of the wall furnace.
5. Vents that extend through an outside wall should not terminate adjacent to the wall or below eaves or parapets. This limitation does not apply to listed appliances and mechanical draft systems.

Outside Chimney

Outside chimneys are subjected to greater cooling effects than chimneys that are within the building. Type B and Type L chimneys may be permitted locally to be installed outside, but this should be avoided whenever possible. If this construction is necessary, the pipe must be substantially supported at the base, supported at 8' intervals, installed plumb and true, and be accessible for inspection and cleaning.

Design Example

As a design example, assume a house with an unlined masonry chimney is to be remodeled. A Type B chimney is to be installed to serve the new appliances.

Connectors must be sized first and then the chimney is sized to serve the total load. The total chimney height must be determined such that the chimney will be at least 5' above the highest vent connector. The chimney must meet the height-above-roof requirements as well.

If several appliances are to be vented, check to see the advantages of more than one Type B vent as compared to one large chimney. Several smaller sizes could cost less and be easier to install than a single large chimney. Figure 15-18 illustrates this idea.

Always check to see if offsets will be required and try to position the chimney to eliminate or minimize the need for offsets. Try to keep the chimney inside if possible; otherwise, careful consideration must be done in sizing an outside installation.

Order the materials needed and check the packing list carefully. Locate and cut the holes required, using a plumb bob to be sure the installed stack will be plumb. Be sure the holes are cut properly for the required clearance-from-combustibles. Also, place the support brackets as required for the chimney. Most codes require that fire stops be installed at each floor level.

Install the Type B piping in the brackets. If care was exercised in the preliminary work, the installed vent will be plumb and true. Install the roof assembly and clamp

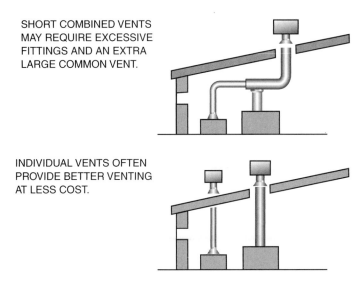

SHORT COMBINED VENTS MAY REQUIRE EXCESSIVE FITTINGS AND AN EXTRA LARGE COMMON VENT.

INDIVIDUAL VENTS OFTEN PROVIDE BETTER VENTING AT LESS COST.

Figure 15-18 In some cases it is more cost effective to vent appliances separately.

the piece firmly in the roof support. The top terminal must have a weather cap or approved weather shield.

Frequently, the portion of the vent above the roof is covered with an enclosure that simulates a brick chimney. Assemble this appearance enclosure according to the manufacturer's instructions. Be sure to have the lip of the high side of the enclosure under the roofing and have the lower part above the roof. Figure 15-19 shows a typical assembly.

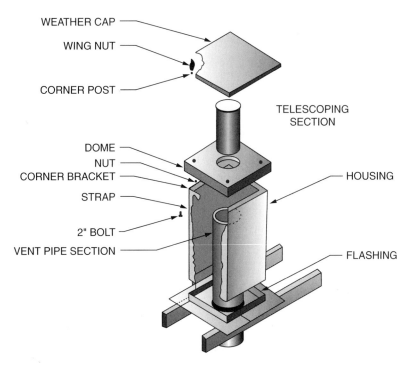

WEATHER CAP
WING NUT
CORNER POST
TELESCOPING SECTION
DOME
NUT
CORNER BRACKET
STRAP
2" BOLT
VENT PIPE SECTION
HOUSING
FLASHING

Figure 15-19 Vent housing assemblies are decorative features that should be coordinated with the roof installer.

If the vent pipe itself is to be used without a decorative cover, an appropriate flashing is used to seal the pipe at the roof line and the vent top is installed. Figure 15-20 shows a typical vent cap.

Remember to always check local code for any permitted variation from the methods described in this chapter.

Figure 15-20 The vent protection cap keeps rain, leaves, and some small animals from entering the vent system.

Acknowledgement

Material in this section relies extensively on material from the *National Fuel Gas Code*. This material has been reprinted with permission as follows:

Reprinted with permission from NFPA 54-2006, *National Fuel Gas Code,* Copyright 2006, National Fire Protection Association, Quincy, MA 02169. This reprinted material is not the complete and official position of the National Fire Protection Association on the referenced subject, which can only be represented by the standard in its entirety.

REVIEW QUESTIONS

1 What are the two types of mechanical drafts?

2 What is the purpose of the draft hood?

3 What is the difference between a Category I and a Category IV appliance?

4 When using the vent capacity tables, what does NAT Max refer to?

5 What is the maximum horizontal length for a 6" vent connector?

Gas Combustion and Controls

The student will:

- Explain the fuel gas combustion process.
- Describe the methods for providing ventilation for the combustion process.
- Describe controls and accessory items used in fuel gas burning appliances.

GAS COMBUSTION

For combustion to take place, oxygen, fuel, and a source of ignition (heat) must be present. Since the air being supplied to the burner contains approximately 21% oxygen, it takes about 10 cubic feet of air for every 1 cubic foot of natural gas burned. Note that LP gases require different quantities of air per unit volume of gas, but the quantity of air is constant for any given heating value. If the gas undergoes *perfect combustion*, then carbon dioxide (CO_2), water (H_2O), and heat will be produced. Perfect combustion occurs when all of the gas is burned and no excess air is left. This rarely happens; any changes in the air supply would make the fuel to air ratio wrong. If there is not enough oxygen, then *incomplete combustion* will occur. Incomplete combustion is wasteful and, because carbon monoxide (CO) is produced, it is dangerous. Incomplete combustion will also produce soot. Soot can build up and will eventually reduce the amount of heat transfer to the transfer media (water in a water heater, air in a furnace). If left long enough, the soot can reduce the amount of draft and cause the flame to *roll out* of the combustion chamber, which could cause serious damage to the equipment.

In order to prevent incomplete combustion, most burners are set up for complete combustion, even in the worst conditions. Complete combustion is when all of the fuel gas is burned and only a small amount of excess air is used. Introducing too much excess air could cause the flue gases to cool and the flame to become unstable and difficult to monitor. Atmospheric burners will use about 50% excess air and power burners should require about 25% excess air.

Air

As discussed in earlier chapters, primary air is mixed with the gas in the venturi of the burner. Secondary air combines with the gas at the point of combustion as shown in Figure 16-1

After the gases of combustion have passed through the heat exchanger and leave the appliance, additional air can be supplied through the draft hood or, on very large burners, from another blower. This additional air is call *dilution air*. Dilution

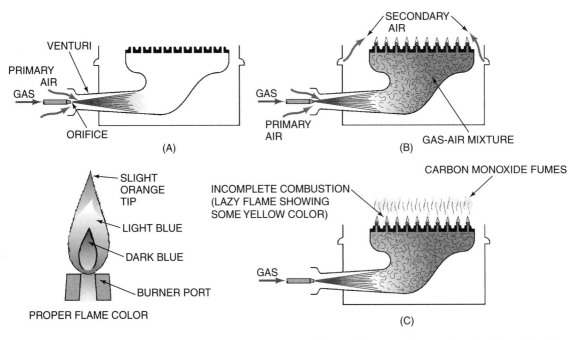

Figure 16-1 (a) The velocity of gas pressure in the reduced opening (venturi) causes primary air to be drawn in and mixed with the gas. (b) After the gas/air mixture has been ignited, secondary air is introduced to complete combustion. (c) Incomplete combustion will yield a yellow tipped flame.

air helps maintain a more constant draft pressure and helps lower the temperature of chimney gases.

Combustion Air Supplies

In order for an appliance to operate properly, all three–air types–primary, secondary, and dilution–must be in sufficient supply. In the past, buildings were built in such a way that sufficient fresh air came in through cracks around windows and doors and it was assumed that the appliance would draw air from within the building. In current times, the two most common practices are to supply combustion air through (1) air duct and grilles or (2) appliances that are direct vented.

Direct Vented Appliances

When an appliance is *direct vented,* fresh air is brought directly to the unit through a pipe that runs from outside the building to the combustion chamber of the appliance. This is sometimes referred to as *balanced flue.* The combustion chamber is sealed so all incoming air passes through the burner onto the heat exchanger and then out through the vent system. Figure 16-2, which is for a high-efficiency gas furnace, shows a dual pipe configuration of this process.

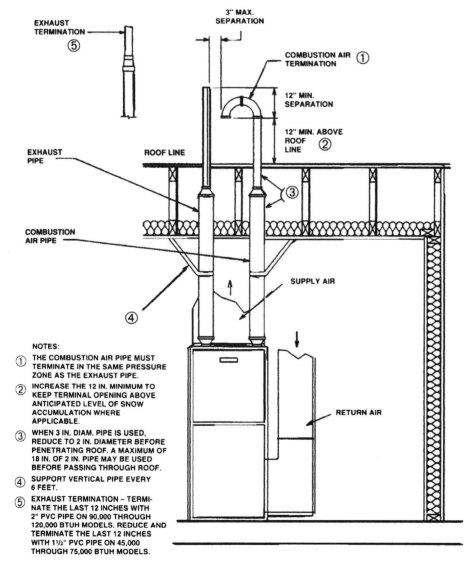

Figure 16-2 A direct vented high-efficiency gas furnace using two separate pipes. Courtesy of Rheem Manufacturing Company, Water Heater Division.

Another version of this same idea is for the exhaust flue gas pipe to be located inside the combustion air intake pipe. This should not be confused with Type B vent pipe.

The sizing of these systems is accomplished by strictly following the sizing charts supplied by the manufacturers. In some cases, high-efficiency models will lower the flue gases to such an extreme that they are vented with PVC pipe. The direct vent systems have a great advantage because other fuel burning appliances or exhaust fans do not affect the combustion air. Direct vent systems also reduce the amount of heat loss to the building.

Combustion Air Ducts and Grilles

When the appliance relies on combustion air from its surroundings, make sure a steady supply of fresh air is available. The gas code specifies the area of grilles or ducts required to ensure an adequate supply of air to the appliance space. Various situations are described below.

Enclosed Furnace Room Adjoins Space with Adequate Infiltration from Outside

Each opening must have a free area of not less than one square inch per 1000 BTU per hour of the total input rating of all appliances in the enclosure. Cut an opening of 12" minimum above the floor and a similar opening of 12" minimum below the ceiling, then install a grille or louver (see Figure 16-3). The minimum size of these openings is 100 in².

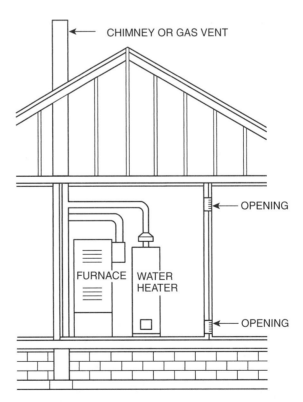

Figure 16-3 All combustion air from adjacent indoor spaces is through indoor combustion air openings. Reprinted with permission from NFPA 54, National Fuel Gas Code, Copyright © 2006, National Fire Protection Association, Quincy, MA 02169. This reprinted material is not the complete and official position of the National Fire Protection Association on the referenced subject, which is represented only by the standard in its entirety.

Example 1

An 80,000 BTU/hr furnace and 40,000 BTU/hr water heater are installed in a furnace room. What size metal grille must be installed in the combustion air openings (assume the metal grille has 75% free area)?

Answer

$$\frac{80{,}000 + 40{,}000}{1{,}000} = \frac{120{,}000}{1{,}000} = 120 \text{ sq in}$$

With an effective area of 75%, the grille area is

$$\frac{120}{.075} = 160 \text{ sq in}$$

A square grille of about 12-3/4" is therefore required. Or, 12" × 12" for each of two openings: one 12" above the floor and the other 12" below the ceiling.

Furnace Room Connected to a Freely Ventilated Attic and to a Freely Ventilated Crawl Space by Ducts

Figure 16-4 shows a situation in which all make up air is supplied through either the ventilated attic or crawl space. The inlet and outlet air openings must each have a free area of not less than one square inch per 4000 BTU per hour of the

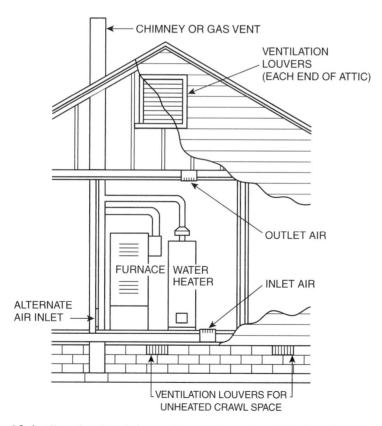

Figure 16-4 All combustion air from outdoors is through ventilated crawl space and outlet air to a ventilated attic. Reprinted with permission from NFPA 54, *National Fuel Gas Code*, Copyright © 2006, National Fire Protection Association, Quincy, MA 02169. This reprinted material is not the complete and official position of the National Fire Protection Association on the referenced subject which is represented only by the standard in its entirety.

total input rating of all appliances in the enclosure. (There is no minimum size for these openings.)

An alternate for this method is to use two ducts from the attic with one ending within 12" from the floor of the furnace room and the other extending from the ceiling into the attic. Figure 16-5 shows this arrangement.

This last arrangement is especially popular in slab built homes. Pay special attention so these vents are not covered by insulation.

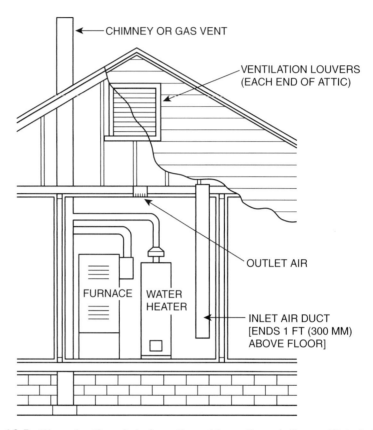

Figure 16-5 All combustion air is from the outdoors through the ventilated attic. Reprinted with permission from NFPA 54, *National Fuel Gas Code,* Copyright © 2006, National Fire Protection Association, Quincy MA 02169. This reprinted material is not the complete and official position of the National Fire Protection Association on the referenced subject which is represented only by the standard in its entirety.

Furnace Room Located a Distance from the Outside Wall

Figure 16-6 shows a furnace room located toward the center of the building—all outside walls are some distance away. Furnace room air may be supplied through two ducts that extend to the outside. The free openings are sized one square inch for each 2000 BTU/hr input. If the furnace room wall is an outside wall, the area of each of the free openings through the wall is sized one square inch per 4000 BTU/hr total input. The minimum dimension for any of these ducts or grilles is 3".

Other Considerations

All of the above discussions concern conventional appliances that take combustion air from the atmosphere adjacent to the appliance. Mechanical make up air systems can be used for larger installations or for spaces where some hazard may exist. The mechanical system can also be arranged and controlled to reduce infiltration and heat loss that occurs with the fixed openings described above. The basic requirement is to provide sufficient air for combustion and for draft diverter dilution.

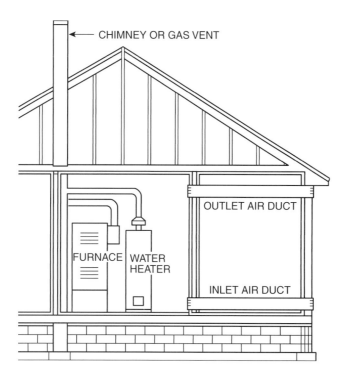

Figure 16-6 **All combustion air is from the outdoors through horizontal ducts. Reprinted with permission from NFPA 54, *National Fuel Gas Code,* Copyright © 2006, National Fire Protection Association, Quincy, MA 02169. This reprinted material is not the complete, and official position of the National Fire Protection Association on the referenced subject which is represented only by the standard in its entirety.**

If a building is equipped with significant exhaust fans, it is best practice to isolate the furnace room from the rest of the building and provide adequate air from outside by one of the above methods. If isolation is not possible, a mechanical make up system is all that will do the job.

Remember that inadequate ventilation will produce improper combustion in an appliance. Partial ventilation may permit satisfactory operation of some appliances in a furnace room, but when the major appliance is on, the inadequate ventilation will produce problems.

It is also important to remember that a fireplace with a roaring fire is exhausting a very large volume of air from the space. That factor combined with the tight construction used in modern buildings can produce back drafts in chimneys and combustion problems in appliances connected to the chimney. NOTE: When replacing an appliance with a unit with a greater firing rate, it is necessary to check all sizing to be sure the replacement unit will operate satisfactorily. Check fuel line sizing as well as vent connector and chimney sizes.

> **IN THE FIELD**
>
> In cases where a customer is complaining about the pilot light going out on a standing pilot appliance or when there are flame failures such as a roll-out switch being tripped on a furnace, look for large exhaust fans or other appliances that may cause changes in building pressures. A flickering flame could be a sign of combustion air problems or a cracked heat exchanger.

COMBUSTION CONTROLS

Automatic Vent Dampers

After the energy crisis of the early 1970s, a variety of vent dampers were developed (see Figure 16-7). The *vent damper* is intended to close as soon as the firing cycle is completed, thus minimizing the loss up the chimney of the residual heat in the appliance. Once the appliance has stopped firing, heat from the tank would warm the air in the flue pipe. Natural circulation will cause this warm air to rise and a draft will still exist. The flue damper would block the tendency for natural circulation.

One model is spring-loaded to open and motorized to close. When the electric gas valve is off, the damper motor drives the damper closed. On call for heat, the

TO
CHIMNEY

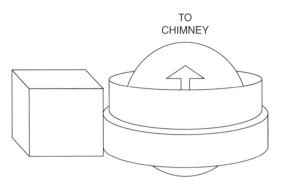

Figure 16-7 Automatic flue damper.

damper first opens and then an end-switch on the damper turns on the gas valve, ensuring a safe operation. It is estimated that these devices can save 10%–20% of the annual fuel cost by reducing standby losses when the heating system is off.

The first requirement for these devices is that they be fail-safe; that is, any malfunction of any component results in the damper opening to a safe position. Another requirement is that when the damper is open, the free area of the opening equals or exceeds the area of the vent pipe. An indicator showing damper position is also required. Finally, there needs to be a means of keeping the gas valve off until the damper is fully open, i.e., the end-switch referred to above.

Vent dampers should have the American Gas Association (AGA) seal, indicating that the design has been tested for all required operating characteristics.

A warning label is required on the device, which reads:

"This device is potentially dangerous if improperly installed. ANSI specifications require the installation to be performed by a qualified installing agency, in full compliance with all applicable building and/or safety codes."

Figure 16-8 shows another type of vent damper that operates with damper blades made of bimetal construction. These dampers open when heated and close when cooled without the aid of a motor. These dampers are not always permitted by code. Check your local code for clarification.

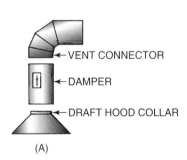

←VENT CONNECTOR

←DAMPER

←DRAFT HOOD COLLAR

(A)

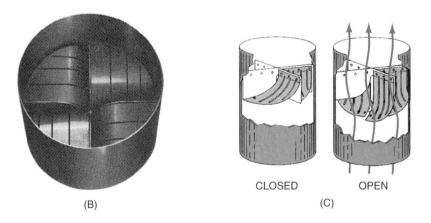

(B)

CLOSED   OPEN

(C)

Figure 16-8 An automatic vent damper that uses bimetal blades to open and close based on heat from the flue gases.

The following considerations should be noted about vent dampers:

1. Their function is to minimize standby losses when the appliance is not firing, but if any circumstance permits firing while the damper is closed, the situation is unsafe.
2. If the oxygen in a room is depleted and replaced with carbon dioxide and other products of combustion, anyone in the room is in danger of asphyxiation.
3. It is recommended that vent dampers be installed only by trained, certified technicians.

It should be noted that vent dampers have not lived up to their promise to make significant operating cost savings. Other means of saving energy must be used if real gains are to be realized.

Electric Pilot Igniter

One other energy-saving control is the *electric pilot igniter*. The pilot flame is extinguished during the off cycle and relit electrically upon each call for heat. These devices are more often found on new equipment, but existing devices can be retrofitted with intermittent pilots.

The device works by providing an electric valve on the pilot gas line and some means of developing a spark or heating a coil to glowing temperature at the pilot burner. On call for heat, the pilot gas valve opens and the igniting means is activated. Once the pilot flame is established and proved by a control sensor, the main gas valve is opened and full heat is provided. Figure 16-9 shows the pilot flame, spark electrode, flame sensor, and controller.

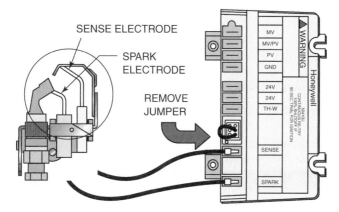

Figure 16-9 Flame controller with separate wiring for electronic spark ignition and flame sensor.

Compared with the standing pilot, this interrupted pilot system requires more control equipment, is more critical for adjustment, and an electric power source is required. This type of control system is found on higher-end appliances that have high-efficiency ratings. Chapters 30 and 31 deal with understanding and troubleshooting this type of control system; a thorough understanding is necessary in order for repair work to be performed correctly. In any case, be sure that only certified installers put these devices into operation.

Large Burners

Larger burners use extremely sensitive systems to prove the existence of flame, when the flame is supposed to be present. Depending on the fuel being burned, different methods of detection are used. For example, *photocells* are used to detect the light given off by the flame, which is fine for large coal or wood burning equipment; but since a natural gas flame is blue and gives off very little visible light, photocells would not be effective.

The *flame rod* works well with the gas-fired appliances because the flame is clean and conducts electricity. The flame rod is used in conjunction with a *flame rectification system* as shown in Figure 16-10.

It was found that a flame will conduct an electrical current, but has a very high resistance. Rectification refers to the process of converting alternating current (AC)

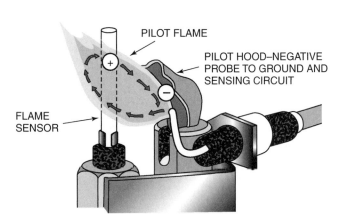

FLAME RECTIFICATION
PILOT & PROBE

PILOT FLAME

PILOT HOOD–NEGATIVE
PROBE TO GROUND AND
SENSING CIRCUIT

FLAME
SENSOR

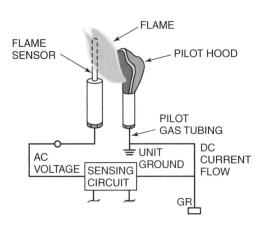

RECTIFICATION CIRCUIT

FLAME

FLAME
SENSOR

PILOT HOOD

PILOT
GAS TUBING

AC
VOLTAGE

SENSING
CIRCUIT

UNIT
GROUND

DC
CURRENT
FLOW

GR

Figure 16-10 Flame rectification using a grounded burner and the flame as part of the conducting circuit.
Courtesy of Johnson Controls.

into direct current (DC) and sending it through the flame. The amount of current (measured in micro amps) being conducted depends on how much of the flame sensor is exposed to the flame. As long as the controller receives approximately 6 micro amps, the unit will continue to run.

Since the flame rod relies on a completed electrical path, the problems in Table 16-1 can occur and cause the unit to fail.

These systems will be further discussed in detail in later chapters.

Table 16-1 Flame Rod Troubleshooting

Problem	Symptom
Too much gas pressure	Flame will not make contact with the burner, therefore the circuit will be broken
Too little gas pressure	Flame will be small or flicker; not making good contact with flame sensor
Equipment not grounded well	Rectification relies on a good ground; system will fail to run
Incoming power polarity reversed	In most newer equipment all the blowers will start at once, but there is no firing of the fuel system
Flame sensor dirty	Low current being produced; equipment will go out on flame failure, usually at start up

REVIEW QUESTIONS

❶ If a 50,000 BTU/hr water heater is being supplied with combustion air, as shown in Figure 16-5, how many square inches of free area grille is required?

❷ If the same 50,000 BTU/hr water heater were being supplied with combustion air, as shown in Figure 16-3, how many square inches of free area grille is required?

❸ What are the byproducts of the combustion of natural gas?

❹ What is the purpose of a vent damper?

❺ What is a symptom of reversed polarity in a gas-fired appliance?

CHAPTER

17

Study of Local Fuel Gas Codes

STUDY OF LOCAL CODE

This chapter is to be used to review the pertinent aspects of the fuel gas code in your area. While this book has consistently focused on the *National Fuel Gas Code,* there is also an *International Fuel Gas Code* that has been adopted by many states. Both the *National* and *International Fuel Gas Codes* are approved by the American Gas Association (AGA) and are very similar. It is important to remember that local authorities can make changes to the national codes (plumbing, electrical, mechanical, and fuel) as long as the changes are viewed as being more restrictive than the existing codes. These changes are usually driven from local conditions such as northern states requiring additional freeze protection or increased burial depths for piping while southern states do not. Or, in locations prone to earthquakes or hurricanes, additional structural supports or seismic bracing may be required. The following subjects are items that should be checked whenever starting work in a new area:

1. Acceptable materials for piping and venting
2. Joints and connections
3. Valve types and locations required
4. Mobile home and trailer park systems
5. Rooftop appliance procedures and requirements
6. Electrical requirements
7. Installation of specific appliances
8. Venting of specific appliances
9. Placing an appliance in operation

Whenever a new product comes to market, it may take time for it to be accepted by the local authorities. You should always check the manufacturer's requirements; for example, some manufacturers do not list or approve of their appliances being installed in mobile homes.

Inspection and Testing

Testing a gas piping installation starts with testing the pipe system. Disconnect any appliance(s) and/or cap any opening(s). Remember that some items such as gas pressure regulators have diaphragms and other sensitivity components that, if exposed to the high pressures of the test, could become damaged. The piping system should be filled with air or an inert gas and pressurized to 1-1/2 times the working pressure of the line (3 psig minimum). Never use oxygen to test the system!

The test measuring devices (manometer or pressure gauge) are suitable to indicate a pressure loss during the test period. Usually the range on any pressure gauge should be double the test pressure; this ensures that the test pressure will be at mid-range of the gauge. Bourdon tube gauges are most accurate in the middle range.

The test duration is one half hour for each 500 cu ft or less of pipe volume. For systems of 10 cu ft or less or for residential systems, the test duration is 10 minutes minimum. At the end of the test period, if the pressure gauge or manometer shows no change, there is no leak. If a leak is indicated, test all joints to find the leak. Repair it and test again.

After the piping test, connect the appliances. Again, test the appliance connections as follows:

1. Close all outlets (including pilot valves), turn on gas, and check for leaking joints at the appliance connections.
2. The leak test can be made with a soapy solution applied to all joints, observing the *test hand* on a gas meter or using an approved gas detector. Never test for gas leaks with an open flame!
3. Observing the test hand for 10–15 minutes performs the meter test. If the test hand does not move, it is assumed that there is no leak. To verify this

assumption, make sure the meter shows flow by lighting a pilot burner after completing a "successful" test.

4. If fuel gas is present in the atmosphere in the vicinity of a leaking joint, the gas detector shows a needle movement.

5. The non-corrosive leak detection fluid solution method is satisfactory for a few joints, but to use this method for a large system requires excessive time.

6. After piping test and appliance connection test, open the appliance valves and light standing pilot burners.

7. Adjust the burner fire as recommended by the manufacturer. If possible, verify that the firing rate is equal to or less than the rated value. Appliance input must be reduced below nameplate rating if your location is more than 2500 feet above sea level (see manufacturer's instructions). Adjust the primary air to produce a blue primary flame without luminous tips. Verify that all safety and operating controls perform properly and that automatic pilot ignition systems are adjusted correctly.

8. To check the vent systems, place a small flame at the diverter inlet. The flame should be drawn into the vent system. Note that if the chimney is cold, there may be very little draft!

9. Check the draft (after the chimney is at equilibrium temperature) to be sure it is sufficient for proper operation.

10. Notify the equipment user that the job is complete and operating satisfactorily.

11. Always demonstrate the operation of the equipment and its safety features.

REVIEW QUESTIONS

1 What does AGA stand for?

2 What does placing a lit match near the draft hood prove?

3 Who makes the final decision if there is a conflict between differing codes?

4 What is the minimum test duration for an 8 cubic foot gas system?

CHAPTER

18

Level, Transit, Elevations, and Grade

LEARNING OBJECTIVES

The student will:

- Identify elevation and grades using site plans common to the plumbing industry.
- Solve practical problems relating to elevation and grades.
- Demonstrate a working knowledge of the builders level and transit.

SITE PLANS

When building first commences, several different trades are called upon to complete the task. In order for the building system to work as a whole, each craftsman must know how his work will relate to the others. A set of construction plans (blueprints) should show where the building will set and at what height, how the land will lay around the building, and the property borders. A plumbing technician must be able to use these plans to identify major obstacles and use the elevations on those plans in order to perform the work. For this reason, the following three chapters will explain some common terminology, symbols, and tools used to determine and check elevation and grade.

A *site plan* is an overhead view of a property, drawn to scale, showing the location of buildings, streets, sidewalks, alleys, and easements for utilities. It will show all significant existing items on the site. Special attention should be paid to estimated traffic patterns. Items such as septic tanks should not be located under patios, decks, and driveways. Figure 18-1 shows one style of site plan that uses lines of constant elevation to describe the existing and finished ground surfaces.

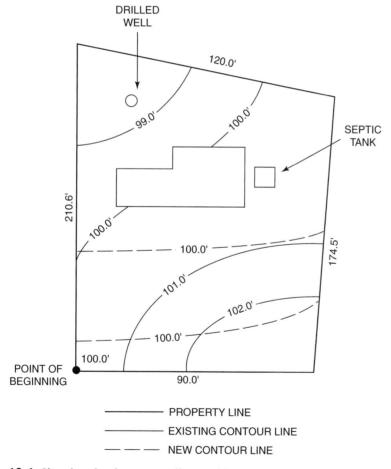

Figure 18-1 Site plan showing contour lines and boundaries. Notice how the point of beginning is located on the property line. This point also serves as the local benchmark.

These lines are known as *contour lines*. Common drafting practice uses continuous lines for existing elevations and dashed lines for proposed elevations. The contour lines that represent the elevation of the land when construction is complete are commonly referred to as the *grade line*. The grade line is the elevation at which the earth touches the foundation of the building. Elevation values, expressed in feet, are noted on the lines (see Figure 18-1). The numbers located on the lines are elevations or heights, above or below an expectable reference point commonly known as a *benchmark*.

A benchmark is any point used as the base elevation for job measurements. It is advisable to select a point that is unlikely to be changed or affected by the construction activity. When possible, select a point that has been established by the United States National Geodetic Survey (NGS), which can be found online at http://www.ngs.noaa.gov.

Benchmarks can be marked with brass plates indicating the elevation above mean sea level at that point. Such benchmarks can be located from a NGS topographic map, available at engineering supply stores or from the U.S. Government Printing Office. Over the past few years, many surveyors have started using *GPS or Global Position Systems*. GPS is a tool that is used for precise positioning of points. It operates through satellites that send out signals to a receiver that usually mounted on a tripod. The receiver then transmits signals to a data collector that stores the data. The user holds the data collector. Afterward, the surveyor downloads the data into a computer and the computer software resolves from the data the exact position of the points within a fraction of an inch or millimeter. Since very few plumbers are using this technology, the focus of this chapter will be on the traditional instruments currently in use.

A benchmark should be a firm and definite point such as a spike in a tree, a bolt on a fire hydrant, or a square chisel mark on a curb. The benchmark should be located sufficiently away from the construction site so that construction processes will not disturb it. On large sites, multiple benchmarks can be used. If the benchmark is located on one of the property corners, it is referred to as the *point of beginning (POB)*.

USING SITE ELEVATIONS TO LOCATE SYSTEM COMPONENTS

Once the plumber knows the route in which the piping systems will be laid, it will be helpful to know of all the elevations along that path. *Profile leveling* is the process of determining the elevations of the ground surface along the path of the pipe. Figure 18-2 shows a profile of all important points along the path; these points could be turns in the pipe or sudden changes in elevation.

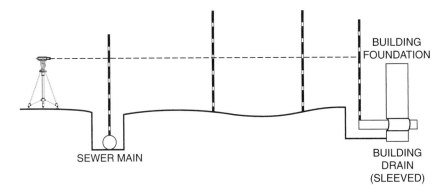

Figure 18-2 Profile leveling of land. Notice how intermediate elevations are taken at changes of elevation.

Now that the plumber knows the profile of the land, the depths of excavation can be determined. The pipe can be installed at the correct slope and below the minimum burial depths as described by code.

Slope is the term commonly used to describe the measurement of the steepness, incline, or grade of a straight line; meaning the higher the slope, the steeper the incline. It is usually expressed as the *rise over run*, which simply means how far the line drops or rises for every foot (or any other unit of length) of distance it travels. This term is sometimes called *pitch or grade*.

Example 1 demonstrates how a builders level can be used to measure elevations of the ground between the building foundation and the sewer main. The common surveying practice is to set up the transit in a position that can observe all points along the path. If one of the points is of known elevation, such as a benchmark, that point will be referred to as a *backsight*. Surveyors use this terminology because in many cases, they have to move the level and tripod and then refer back to the previous location to continue measuring.

Example 1

Determine the difference in elevation of the land between points A and B (see Figure 18-3).

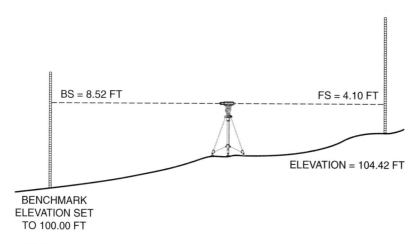

Figure 18-3 Differential leveling with an established benchmark of 100.

BS = Backsight = a reading on a rod on a point of known elevation
HI = Height of Instrument = the elevation of the line of sight of the telescope
FS = Fore Sight = a reading on a rod at any point to determine its elevation
HI = BS + 100
HI = 8.52' + 100'
HI = 108.52'
Elevation at point B = HI − FS
Elevation at point B = 108.52' − 4.10'
Elevation at point B = 104.42'

Answer

The difference in elevation from point A to point B is 4.42' (108.52' − 104.42').

SLOPE OF DRAINAGE LINES

Table 18-1 lists the flow rate and velocities for water flowing in sloping pipes. Note that lesser slopes are not suitable for small pipe sizes since the velocity of flow is inadequate for these conditions.

Table 18-1   Approximate Flow Rates and Velocities in Sloping Drains

Actual inside diameter of pipe-inches	1/16 in./ft. slope		1/8 in./ft. slope		1/4 in./ft. slope		1/2 in./ft. slope	
	Flow gpm	Vel. fps	Flow gpm	Vel. fps	Flow gpm	Vel. fps	Flow gpm	Vel. fps
1.250					2.40	1.25	3.40	2.77
1.375					3.10	1.35	4.38	1.91
1.500					3.90	1.41	5.52	1.99
1.625					4.83	1.50	6.83	2.12
2					8.41	1.73	11.9	2.44
3			17.5	1.60	24.8	2.26	35.0	3.20
4	26.7	1.36	37.7	1.92	53.4	2.72	75.5	3.84
5	48.4	1.58	68.4	2.23	96.8	3.36	137	4.47
6	78.7	1.79	111	2.53	157	3.57	223	5.05
8	169	2.17	240	3.06	339	4.33	479	6.13
10	307	2.51	435	3.55	615	5.02	869	7.10
12	500	2.84	707	4.01	999	5.67	1413	8.02
15	906	3.29	1281	4.66	2812	6.59	2563	9.32

Courtesy of Plumbing-Heating-Cooling Contractors–National Association.

Table 18-2 presents the information given in Table 18-1 in a form converted to drainage fixture units.

Table 18-2   Building Drains and Sewers

Maximum Number of Drainage Fixture Units (DFU) That May Be Connected to Any Portion of the Building Drain or the Building Sewer.

Pipe Size Inches	Slope Per Foot			
	1/16-Inch	1/8-Inch	1/4-Inch	1/2-Inch
2			21	26
3			42[2]	50[2]
4		180	216	250
5		390	480	575
6		700	840	1,000
8	1,400	1,600	1,920	2,300
10	2,500	2,900	3,500	4,200
12	3,900	4,600	5,600	6,700
15	7,000	8,300	10,000	12,000

Courtesy of Plumbing-Heating-Cooling Contractors–National Association.

[1] On-size sewers that serve more than one building may be sized according to the current standards and specifications of the authority having jurisdiction for the public sewers.
[2] See Sections 11.5.6.d, 11.5.6.e, and 11.5.6.f.

When establishing the slope of drainage lines, two possible conditions may be encountered:

1. The length of run is short enough and the difference in elevation great enough so that the desired slope can be achieved. In this case, the slope is selected to develop the required flow rate and the *fixture unit capacity* for the pipe size intended to be used.
2. The second possible condition occurs when the length is too great for the available difference in elevation for the desired slope. In this case, use Table 18-2 to select a larger pipe to obtain the desired *dfu capacity* at a lesser slope.

Example 2

If the length of the sewer is 250', the invert at the building 105.2', the connection at the sewer tap 99.0', and the sewer load 200 dfu, what size sewer pipe is required?
Determine the available elevation change:

$$105.2' - 99.0' = 6.2'$$

Next, calculate the required elevation change:
(From Table 18-2, 200 dfu requires 4" at 1/4"/ft.)

$$250' \times 1/4 \text{ inch per foot} \times 62.5"$$

Convert this value to feet (in decimal form):

$$\frac{62.5"}{12} = 5.2'$$

Since the available fall is greater than what is required for 1/4" per foot, the sewer can be installed as required.

Example 3

Solve Example 2, except that the sewer tap is 350' from one building.
Available fall for the building sewer:

$$105.2' - 99.0' = 6.2'$$

Check 4": 4" requires 1/4" per foot for 200 dfu
Required total drop for a 4" pipe is

$$350' \times (1/4") = 87.5" = 7' 3.5"$$

which is greater than the available fall.
Check 6": 6" requires 1/8" per ft (5" is probably not available in sewer pipe materials).
Required total drop for 6" pipe is

$$350' \times (1/8") = 43.75"$$

which is less than available fall. Therefore, use a 6" sewer (double-check to see if 5" is available and cost less than 6" pipe).

ELEVATIONS AND GRADES

The usual expression of pipe slope is given in inches per foot. The ratio of total fall to total length can be expressed as a percentage. A pipe that drops 12" or 1' over a length of 100' would have a slope of 1'/100' or 1% (the unit of measure has to be the same). An example will demonstrate how to use this form.

Example 4

If a pipeline with a slope of 2% is 548' long, what is the total fall and what is the slope expressed in inches per foot?

$$Total\ fall = 2\%(548\ ft)$$
$$Total\ fall = 0.02(548\ ft)$$
$$Total\ fall = 10.96\ ft$$
$$rounded\ to\ 11\ ft$$

Convert this to a slope in inches per foot:

$$\frac{11}{548} = 0.02 \ in/ft$$

$$0.02 \ ft\left(\frac{12 \ in}{ft}\right) = 0.24 \ inches/foot$$

or knowing 2% is 2 ft per 100 ft

$$\frac{24"}{100} = 0.24 \ inches/foot$$

or $\frac{1}{4}$ *inch/ft (approximately)*

Example 5

What is the total slope and percent slope of a pipe 200' long installed at 1/8" per foot?
Convert to percentage:

$$\frac{\frac{1}{8} \ inch}{1 \ ft} = \frac{\frac{1}{8} \ inch}{12 \ inch} = \frac{1}{8} \times \frac{1}{12} = \frac{1}{96} \ or \ about \ 1\%$$

$$Total \ fall = 200 \ ft \times 1\%$$

$$= 200 \ ft \times (.01)$$

$$Total \ fall = 2 \ ft$$

Using the percentage method is usually easier than working with fractions for sewers and other long runs. For calculating slopes within the building, however, it is usually easier to use the fractional form.

Example 6

A slope of 1/4" per foot is required for a 4" pipe. If the two ends of the line are at 89.5' and 96.2', how long is the line?

$$Total \ fall = 96.2 \ ft - 89.5 \ ft = 6.7 \ ft$$

$$Total \ length = \frac{6.7 \ ft}{\frac{1}{4} \ inch} = 6.7 \ ft\left(\frac{12 \ inch}{ft}\right) \times \frac{4}{1}$$

$$Total \ length = 6.7 \times 12 \times 4 = 321.6 \ ft$$

Example 7

A sewer line is 1085' long at 2% grade. The initial invert is 106.2'. What is the downstream elevation of the invert?

$$Total \ fall = 1085 \ ft \ (2\%)$$

$$= 1085 \ ft \ (0.02)$$

$$Total \ fall = 21.7 \ ft$$

$$Final \ elevation = 106.2 \ ft - 21.7 \ ft$$

$$= 84.5 \ ft \ at \ final \ terminal \ of \ sewer$$

Work additional problems of this type to develop your skills in this area.

Table 18-3 shows the approximate relationship between slope expressed in inches per foot and as a percentage.

Table 18-3   Equivalence of Slopes

Inches/foot	%	Inches/foot	%
1/16	0.5	1/4	2
1/8	1.0	1/2	4

LEVEL AND TRANSIT

To determine or measure the elevations discussed so far, an instrument that is accurate, portable, easy to use, and versatile is needed. The two most common instruments available are: (1) the engineers (or builders) level and (2) the engineers (or builders) transit.

Builders Level

The builders level (see Figure 18-4) is made up of the following elements:

* A bed plate that can be attached to a tripod
* A rotating plate marked with 360 fine lines for the 360 degrees in a circle
* A locking means to hold the rotating plate fixed if desired by the user
* An attachment hook and short chain at the center of the rotating system to permit use of a string and plumb bob for accurate locating over a point on the ground
* A telescope and level-vial combination that is attached to the rotating plate

Figure 18-4 Builders level.

The rotating system is attached to the bed plate by a large ball and socket. The ball-socket joint provides lateral and leveling adjustment. The ball and socket joint is made tight and fixed in position by four leveling screws spaced 90 degrees apart.

The telescope/level-vial combination is adjusted to be level in all directions with the four leveling screws. Once established, the line of sight through the telescope is level in all directions. The telescope is equipped with horizontal and vertical crosshairs so that precise readings and settings can be made.

Builders Transit

A *transit* (see Figure 18-5) is similar in construction to the level with an added construction feature: the telescope is mounted on a horizontal pivot so that it can be rotated vertically to measure vertical angles in addition to all the other level functions.

A variation in the construction described above is in the leveling system. A spherical joint is locked with a single nut when two leveling vials at right angles to each other are leveled. Someone who has experience working with this option can level these very quickly.

Figure 18-5 Builders transit has the same features as the builders level with the additional ability to pivot vertically.

Telescope

Telescopes are marked with magnifying power, e.g., 10× means that the view through the telescope is ten times that seen by the naked eye. Laser units with digital sensors are now used extensively.

The *transit telescope* is provided with two more horizontal crosshairs positioned a precise distance above and below the center horizontal crosshair. These additional lines are called *stadia lines*.

Stadia

The *transit stadia lines* are used to view stadia poles, marked with alternating red and white bands. Each band is a precise length. The distance from the instrument location to a stadia rod is a function of how many bands are seen between the stadia

IN THE FIELD

There are many new laser levels coming onto the market. The less expensive models typically have an accuracy of plus or minus 1/4" per 100' which is comparable to a typical builders transit. Some of the more expensive ones have an accuracy of plus or minus 1/8" per 100'. Several models only require one person to operate.

lines in the telescope. In this way, reasonably accurate distances can be measured with the transit without having to use a tape measure.

Tripod

A *tripod* is a necessary item when using either a level or transit. The most versatile tripod has adjustable legs to make rough terrain setups easier.

Leveling Rod

A six-foot or eight-foot ruler can be used for short distance sightings, but for general work a *leveling rod*, calibrated in feet and decimals, is used to make vertical measurements. These rods are made to telescope and can be obtained in various lengths (up to 20'). To aid accurate observations, beyond the range where the instrument person can actually read the rod, a large target (bull's eye) is slid up or down on the pole (the person at the telescope signals by arm motions) so that the rod person actually reads the elevation values when the telescope person signals that the target is at the correct point.

Accessory Items

Other accessory items include tape measures, measuring chains, markers, stakes, plumb bobs, and special poles called *stadia rods,* as referred to above.

Precautions

Precautions for the use of this equipment include the following:

- Only qualified and trained persons should use the instruments.
- When not in use, the equipment should be stored in protective cases.
- After a job is completed, carefully clean, pack, and store the instruments.
- Be sure the leveling screws are loosened before storing the instruments.
- The telescope lenses should be covered at all times except when actually using the equipment. Clean the lenses with a clean, soft cotton rag.
- Clean the tripod feet and base of the leveling rod.
- Record the serial numbers of your instruments and mark all the equipment with a company mark or logo.

Builders Level

LEARNING OBJECTIVES

The student will:

- Demonstrate the use of a builders level to lay out angles in a horizontal plane.
- Demonstrate the use of a builders level to determine elevations.

METHODS FOR USING A BUILDERS LEVEL

In Chapter 18, some practical uses for a builders level and transit were discussed. This chapter will focus on the methodology used. Although most builders levels are similar in construction, you should always follow the directions in the owner's manual for the specific one being used.

Set Up

Most builders levels can be set up using the following procedure:

1. Place the tripod on solid ground with the top approximately level. Remove the thread protector from the top of the tripod. (Not all tripods have thread protectors.)
2. Remove the level from its case and fasten it to the tripod by the threaded base of the level. Do not overtighten (see Figure 19-1).
3. Level the telescope by setting it across a pair of adjusting screws, bringing the vial to a level position, then turn it 90° and repeat the leveling process. Once again, rotate the level 90° in order to check the first adjustment. Repeat the process until the leveling vial on top of the telescope reads level in all directions (see Figure 19-2).
4. If the level is not going to be used for a period of time, loosen two adjacent adjusting screws to relieve the stress in the mounting system. Re-level the instrument when it is time to begin using it again.
5. When ready to begin, remove the telescope lens cap and eyepiece cover and clean the lenses if necessary.

Figure 19-1 Mounting system for builders level and transit. Caution: do not over tighten the plastic adjusting screws.

Laying Out Angles and Distances

A common task that can be performed with a builders level is the laying out of angles in a horizontal plane. This involves using a plumb bob, hung from the center of rotation of the instrument to locate the tripod precisely over a point on the ground. Measuring tapes or chains can then be used for accurate measurements to other points, which will make possible calculations of fall or slope for drainage lines. Angles can also be laid out with precision by using the 360° base plate. The next example demonstrates how to use a builders level to accurately lay out a 40' by 40' square (see Figure 19-3).

Figure 19-2 Mounting system uses a push/pull system to hold scope in place.

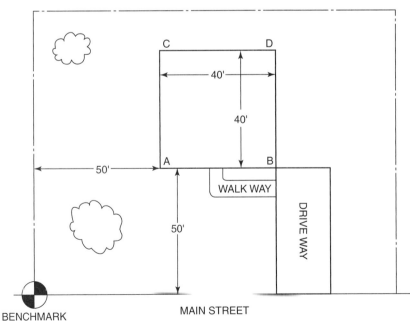

Figure 19-3 Plot showing a foundation to be laid out. Plan where materials can be stored and avoid natural obstacles. Also notice where the benchmark is located.

Example 1

Figure 19-3 shows a plot of land containing a 40' by 40' foundation. Once the corner point A is located and the direction of the line AB is determined, the instrument can be positioned precisely over point A by using a plumb bob. A steel pin, a 1/2" pipe, or a finishing nail driven into the top of a wooden stake would typically be used to locate point A. Next, measure 40' in the direction of point B and by using the vertical crosshair, set a stake at point B. Point B would normally be located so that the structure would be parallel to the road or some other feature on the plot; however, the architect or owner decides. At this time, the zero on the 360° base plate should be rotated until it lines up with the zero on the rotating portion of the level (see Figure 19-4).

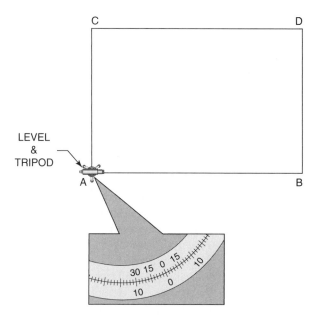

Figure 19-4 Use line AB to set the zero reference on the scale. Now point C will be 90 degrees from point B. (If all four sides of the building are equal and square then point D will be at 45 degrees for point B.)

The level can now be turned 90° to the right so that the zero on the builders level rotating base aligns with the 90° mark on the fixed base. Any object that lines up with the vertical crosshair of the level is now 90° from the original baseline AB. Using this vertical crosshair and the measuring tape, a stake can be located on point C. Relocate the instrument to point C and lay out point D by measuring a 90° angle from line AC and measuring 40' from point C (see Figure 19-5).

Now, relocate the instrument to point D and verify that point B is accurately located. It should be 40' from point D and 90° from line CD. Make a final check for squareness of the foundation by measuring the diagonals AD and BC. If all sides are parallel and the corners are 90°, the diagonals should be equal no matter what the length of the sides of a rectangle. Use the builders level and similar techniques to lay out other geometric shapes.

It should be pointed out that if the foundation is square, then all points could be laid out without moving the tripod. Point D will be at a 45° angle from line AB; its distance can be determined by using the Pythagorean theorem. The Pythagorean theorem states that the square of the diagonal of a right triangle will be equal to the sum of the squares of the sides, or simply $A^2 + B^2 = C^2$. Just remember that C must be the diagonal, while A or B can be either side.

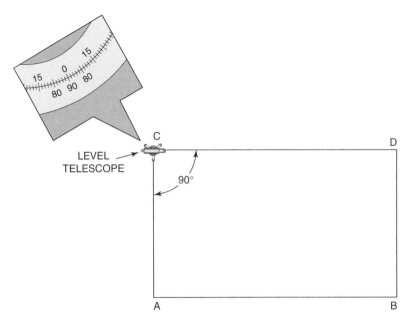

Figure 19-5 After rezeroing (AC = 0), swing 90 degrees to sight line CD and locate point D. Use the plumb bob to help center the level over the point.

Determining an Elevation Difference

Make a sketch and log sheet for your observations (see Table 19-1 for a typical example). Once the instrument is leveled, the leveling rod is placed on an object of known elevation. As stated in Chapter 18, the benchmark or point of beginning, which is found on the plot plan, may be assigned an elevation value. The reading at the telescope crosshair is known as the *Height of Instrument*. Record this reading. The elevation of any other point (compared to the known reference point) is then determined by reading the leveling rod at that location and subtracting that value from the Height of Instrument and then adding the known elevation of the original point. Example 2 demonstrates how to determine the elevation difference between two points.

Table 19-1   Height of Instrument = Benchmark Elevation + Reading to Benchmark

Location	Reading	Elevation	Height of Instrument
Benchmark	6.45	100	106.45
A	12.60	106.45 − 12.60 = 93.85	
B	14.75	106.45 − 14.75 = 91.70	
C	16.33	106.45 − 16.33 = 90.12	

Example 2

Determine elevation of points A, B, and C along a sewer line (see Figure 19-6). The level is set up at a convenient point from which the benchmark and points A, B, and C can all be seen and the readings are given in Table 19-1

Alternate Method

If a person did not want to assign the benchmark a value of 100 and was only interested in the elevation difference between points A, B, and C, then only those readings need to be recorded. Once the reading at A was known to be 12.60' and the reading at B was known to be 14.75', it could be determined that point B is 2.15' lower than point A (14.75 − 12.60). The same procedure could then be used to determine that point C is 1.58' lower than point B (16.33 − 14.75).

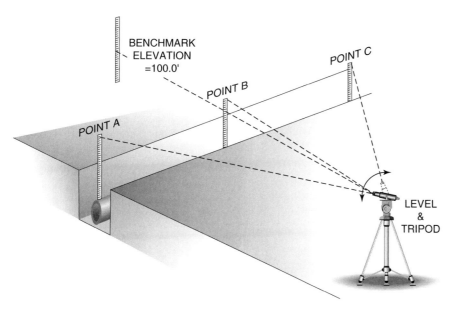

Figure 19-6 Tripod does not have to be in line with the elevation points.

Unfortunately with this method, once the tripod has been moved there is no basis for comparing these elevations to any other point on the job site.

In the next example, two points are at different elevations, but within a distance of 200' and in sight of one and another.

Example 3

Using the setup as shown in Figure 19-7, determine the elevation difference between points A and B.

1. Set up the instrument approximately midway between the points at any convenient spot (this need not be on or even near the line between A and B).
2. Level the instrument, then sight back to A to determine the Height of Instrument.

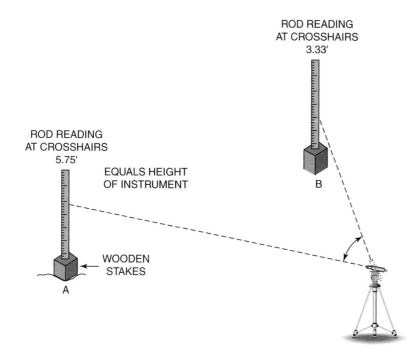

Figure 19-7 Differences in elevation. Remember to hold stadia rods vertical.

The elevation of point B is the Height of Instrument minus the reading of the leveling rod set on B.

The Height of Instrument is 5.75'. The leveling rod reading at B is 3.33'.

Therefore, the elevation of B is 5.75' − 3.33' = 2.42' higher than A.

Occasionally, the second elevation is above the level plane of the telescope. In this case, the reading of the leveling rod is added to the Height of Instrument. Figure 19-8 shows this arrangement.

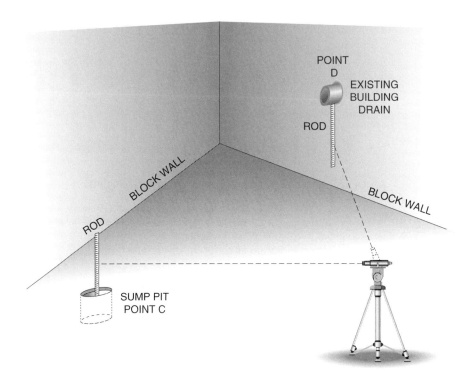

Figure 19-8 Differences in elevation when one point is higher than the height of the instrument. Remember the length measured at point D will need to be added to determine the difference in elevation.

Point C, the bottom of a sump pit, gives a rod reading (Height of Instrument) of 4.54'. The bottom of the pipe, which is above the level plane, is point D. When the rod is held upside down so that the zero is now against the pipe, it reads 7.78' above the Height of Instrument. Thus, the elevation of point D is 4.54' + 7.78' = 12.32' above the bottom of the pit. Practice exercises of this sort to develop skill for this type of problem.

Occasionally there are cases when the two points cannot be seen from one location. It may be because of a great distance, an obstruction, or a big difference in elevation. In any of these above instances, multiple readings will be required. This will take good teamwork and precise note taking. In the next example, there is a large difference in elevation between the two points of interest. The same steps described earlier will be followed, but the process will allow the elevation to be broken into manageable steps.

Example 4

Using the setup shown in Figure 19-9, determine the difference in elevation between points A and D. Notice that because of the steep slope, multiple readings will be taken. Point A is assigned the value of 100' so that all of the following readings will have a positive value. Once all readings and calculations are complete, any value less than 100 will be below point A while any value higher than 100 will be

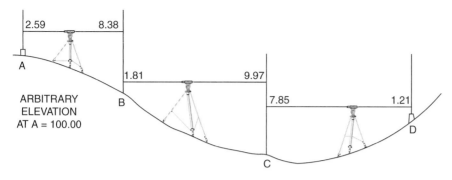

Figure 19-9 Step method for sighting at different elevations. Keep good notes and be precise.

above point A. The elevation difference between any two points can be found simply by subtracting one from the other.

After points A and D are determined, the distance will be broken up into steps where readings can be taken (see Figure 19-9). If the hill has a steep slope, the intermediate points will be close together. Set up at a convenient place between points A and B. The starting point (A) is usually a place of known elevation (or an arbitrary assigned elevation of 100'). The Height of Instrument above A is determined by the value read on the rod. The rod is then sighted at point B and the elevation of point B calculated by using the formula:

Elevation at B = Elevation at A + HI − Sighting at B

The instrument is then relocated to sighting position 2 and the Height of Instrument is determined above point; then point C can be calculated using the same procedure and formula that was just used to find elevation B at position 1. The sightings are thus chained across the hillside until the final point is determined. For clarity, the readings are show in both Figure 19-9 and in Table 19-2.

Table 19-2   Elevation Chart-Elevation at Point A set to 100.00 ft

Setup Position	Height of Instrument (HI)	Sighting at:	Elevation at:
1	2.59'	B: 8.38'	B: 94.21'
2	1.81'	C: 9.97'	C: 86.05'
3	7.85'	D: 1.21'	D: 92.69'

The equation to use: Elevation at A + HI − sighting at B = Elevation at B
Elevation at B + HI − sighting at C = Elevation at C

Example 5

Another common task for using the builders level is to determine the points of elevation along a sewer line. Consider a jobsite condition such as pictured in Figure 19-10.

On this particular jobsite, a building addition will require a sewer line to be run from the house addition to the existing building sewer. In order to work properly, the new line must have sufficient slope, so elevations and distances need to be determined. Assume that the proposed new line be a 4" diameter pipe at a 1/4" per foot slope. Following the route A, B, C, D, E, the line will have an approximate length of 110'. This means that point A will have to be 2.29' higher than point E. If possible, position the tripod at a point that will allow all of the points along the route to be seen. In this case, the tripod is positioned at point F.

If no excavation has been done at this point, all measurements can be from the existing grade. You can calculate digging depths later using the heights listed in

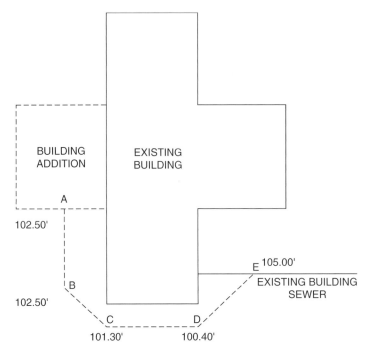

Figure 19-10 Laying out a sewer connection with a builders level. Remember to set the level at a position that allows the sighting of all points.

Table 19-3 and the actual depth of the sewer connection. The instrument height (HI) and the elevation of point E will be referenced from the point above the existing sewer. Using the same procedure as in Example 3, determine the elevations for points D, C, B, and A (see Table 19-3).

Table 19-3   Elevations Along Sewer Line (all measurements in feet)

Location	Reading	Elevation	Height of Instrument
Benchmark E	5.00	100.00	105.00
D	4.60	105.00 − 4.60 = 100.40	
C	3.70	105.00 − 3.70 = 101.30	
B	2.50	105.00 − 2.50 = 102.50	
A	2.50	105.00 − 3.70 = 102.50	

Since point A (elevation 102.50') is 2.5' higher than the sewer connection point E (elevation 100'), there will be plenty of slope for the proposed sewer line.

ESTABLISHING BATTER BOARDS FOR UNDERGROUND PIPING

Once the trench is dug for the sewer line in Example 4, the pipe must be installed. To ensure that the pipe is at a constant slope, batter boards can be erected at the same points (A, B, C, D, E) and a small vertical board can be attached over the centerline of the pipe at the correct elevation. For example, if the top of the small vertical board was located at a rod reading of 105.00' at point E, then another small vertical board could be installed at point D at 104.58'. This would give a 1/4" per

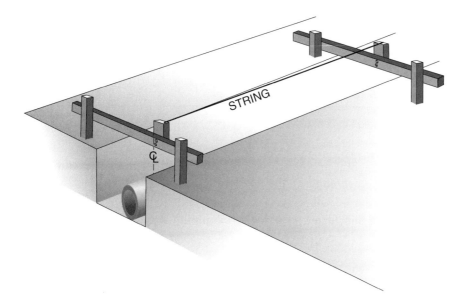

Figure 19-11 Trench with batter boards across it. Remember to practice good safety when working around trenches.

foot rise over the 20' length of pipe. The same procedure could be used to locate batter boards at the remaining points.

LASER APPLICATIONS FOR SEWERS

Today, main sewers are installed with laser beams to define the alignment and slope. A laser source is placed on the manhole base and the laser beam is aimed through the pipe opening in the manhole wall. The beam is aligned on the sewer centerline and slope. Then, as each length of pipe is laid, a target circle is placed in the hub and the pipe is adjusted so the center of the target is on the laser beam. In this way, the entire section of sewer is aligned horizontally and straight along the slope line.

REVIEW QUESTIONS

❶ Elevation of point A = 372.64'
Reading of level rod at A = 4.55'
Reading of level rod at B = 3.21'
Elevation of point B = _____?

❷ Elevation of A = 816.31'
Reading of rod at A = 3.26'
Reading of rod at B = −10.61'
Elevation of B = _____?

❸ Using Table 19-4, set up the record format if the following readings are observed:
Elevation of point A = 146.22'
Set up 1, HI   = 3.26'
Reading at B   = 9.47'
Set up 2, HI   = 2.69'
Reading at C   = 7.93'
Set up 3, HI   = 2.71'
Reading at D   = 8.68'

Table 19-4   Elevation Chart-Elevation at Point A Set to 100.00 ft

Setup	Reading	Elevation	Height of Instrument
1	B		
2	C		
3	D		

The equation to use: Elevation at A + HI − sighting at B = Elevation at B

4 If the horizontal angle reading at point A were 25°, what would be the included angle if the horizontal angle reading at point B were 120°?

5 Why is the benchmark or point of beginning assigned the value of 100?

20

Builders Transit

The student will:

- Describe methods of sloping walls when working in trenches.
- Describe how a trench atmosphere can be contaminated.
- Demonstrate the use of a builders transit to check objects and vertical lines for plumb.
- Demonstrate the use of the builders transit to set points in a straight line.
- Summarize the procedures for proper care of the builders level and transit.

TRENCH SAFETY

In the chapter on Sewers in Plumbing 201, the topic of trench safety was discussed and three simple facts were identified:

- A person does not have to be totally buried to suffocate.
- A person can live only minutes without breathing.
- No job is so important that safety can be ignored.

Since most accidents occur as a result of cave-ins, lack of oxygen from insufficient ventilation is common. In this section, sloping of trench walls to avoid cave-ins and the atmospheric conditions in the trench will be discussed.

The Occupational Safety and Health Administration (OSHA) defines a trench as " . . . a narrow excavation (in relation to its length) made below the surface of the ground. The depth is greater than the width, but the width of a trench (measured at the bottom) is not greater than 15 feet (4.6 m) . . ." (see Figure 20-1).

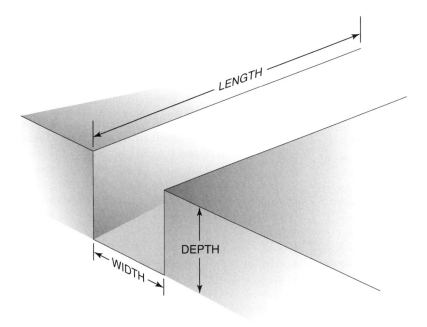

Figure 20-1 Dimensions of a trench as defined by OSHA. Notice that width is measured at the bottom of the trench.

Slopes and Grades

Grade is the fall or *slope* of a line in reference to a horizontal plane. This line usually coincides with the centerline of a pipe or the surface of the land. Plumbers usually refer to grade in terms of inches per foot such as 1/4" per foot. This simply means that for each foot of pipe there will be a 1/4" rise or fall in height (see Figure 20-2).

Slope can also be described as a percent grade based on a length of 100 feet. It is important to make sure all of the measurements are expressed in the same units when using a percentage. In Figure 20-2, the 1/4" must be converted into a portion of a foot before the percent can be calculated.

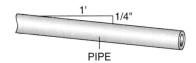

Figure 20-2 Pay attention to the units that are given. In this case the 1/4 is in inches while the 1 is in feet. This cannot be converted into a percent grade (%) without first changing both to the same units (either feet or inches).

Example 1

Convert 1/4" per foot into a percent grade.

$$1/4 \text{ inch} = \frac{.25 \text{ inches}}{12 \text{ inches}} = 0.0208 \text{ } feet$$

$$\frac{0.0208 \text{ } ft}{1 \text{ } ft} = 0.0208 \times 100\%$$

2.08%

Determining Slope of Trench Walls

OSHA safety guidelines must be followed unless:

- The trench is excavated in stable rock.
- The trench is less than 5 feet deep and a competent person has no reason to believe a cave-in is possible.

OSHA defines a "competent person" (in regard to excavations) as someone who:

- Has training, experience, and knowledge of:
  - Soil analysis
  - Use of protective systems
  - Requirements of OSHA 25 CFR Part 1926 Subpart P
- Has the ability to detect:
  - Conditions that could result in cave-ins
  - Failures in protective systems
  - Hazardous atmospheres
  - Other hazards including those associated with confined spaces
- Has authority to take prompt corrective measures to eliminate existing and predictable hazards and to stop work when required.

It is important to note that any excavation 20' or greater must be sloped or shored under the direction of a registered professional engineer. One of the biggest factors in considering cave-in possibilities is the type of soil being excavated. Generally, the *four basic types of soil* conditions are: stable rock, type A, type B, and type C. (OSHA soil classifications are based on Appendix A to Subpart P of Part 1926. See http://www.osha.gov.) While stable rock is usually easy to identify, there is some confusion about the differences between types A, B, and C. *Type A is the most stable, followed by type B, and then type C.*

Soil Tests

In order to determine the type of soil condition present, there are five popular test methods available:

- Thumb test
- Pocket penetrometer test
- Plasticity test
- Torvane shear test
- Sedimentation test

The following is a description of how a simple *thumb penetration test* can be used to help identify the different types.

Type A soil can be readily indented by the thumb, but can only be penetrated with great effort. Type C, on the other hand, are soils that can be penetrated by the thumb with only minimal effort.

While the discussion of the thumb penetration test in Appendix A (as referenced above) addresses only the characteristics of types A and C soils. Type B soil would fall in between those characteristics, meaning that type B soils can be penetrated with moderate effort. Clay can be type A, B, or C soil depending on how much water is in the clay. This test should be conducted on an undisturbed soil sample such as a large clump of soil, as soon as practicable after excavation. Any soil excavated will dry and not reflect the actual condition of the trench media. If the excavation is later exposed to wetting influences (rain, flooding), the classification of the soil must be changed accordingly.

The *pocket penetrometer test* uses a commercially available tool to perform a test similar to that of the thumb test. The penetrometer, a spring-operated instrument, is pushed into the soil sample and an indicator sleeve displays a reading of the required compression force (see Figure 20-3).

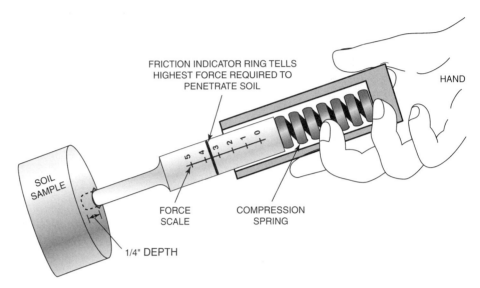

Figure 20-3 Penetrometer test measures how hard it is to penetrate the soil. This gives an idea of how compressive it is.

The instrument is calibrated in either tons per square foot (tsf) or kilograms per square centimeter or kilo Pascals (kPa). It should be pointed out that penetrometers have error rates in the range of +/− 20–40%.

Using the *plasticity test,* you can measure how cohesive the soil is by rolling a moist sample of the soil into a ball. The ball is then rolled into a 1/8" by 2" thread and then held vertically by one end. If the soil remains suspended without tearing, the soil is cohesive (see Figure 20-4).

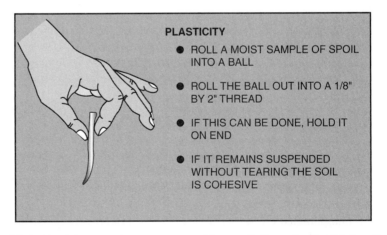

Figure 20-4 Plasticity test gives a measure of how well the soil will stick together.

Once the type of soil has been identified, the maximum allowable slope for the trench can be determined by using Table 20-1 and the following illustrations (see Figures 20-5, 20-6, and 20-7).

Table 20-1 Maximum Allowable Slopes (For Excavations Less Than 20' Deep)

Soil or Rock Type	Maximum Allowable Slopes (H/V)
Stable rock	Vertical (90°)
Type A[1]	3/4:1 (53°)
Type B	1:1 (45°)
Type C	1-1/2:1 (34°)

[1]Exception: Simple slope excavations which are open 24 hours or less and which are 12 feet or less in depth will have a maximum allowable slope of 1/2:1.

Figure 20-5 Maximum allowable slope for type A soil. Exception: Simple slope excavations that are open 24 hours or less and that are 12 feet or less in depth will have a maximum allowable slope of 1/2:1.

Figure 20-6 Maximum allowable slope for type B soil.

In some cases, the excavation is done in such a manner that a bench is formed (see Figure 20-8). It may be used to reduce the amount of excavation, to allow for additional piping that will be placed in the same trench, but it must be at a different elevation or used simply to allow workers a location to place lightweight supplies while the work is being performed. In any case, unshored benches are allowed as described in the following figures. NOTE: They are not allowed in type C soils.

Figure 20-7 Maximum allowable slope for type C soil.

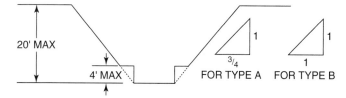

Figure 20-8 Benched excavation for type A and B soils. Bench height is the same for both types.

All excavations 8' or less in depth which have unsupported vertically sided lower portions will have a maximum vertical side of 3-1/2' (see Figure 20-9).

All excavations 20' or less in depth which have vertically sided lower portions that are supported or shielded will have a maximum allowable slope of 3/4:1 for Type A, 1:1 for Type B, and 1-1/2:1 for Type C. The support or shield system must extend at least 18" above the top of the vertical side (see Figure 20-10).

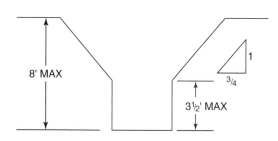

Figure 20-9 When 8' or less in depth, the bench height is reduced to a 3-1/2' max.

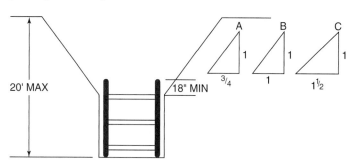

Figure 20-10 Shielded or supported vertically sided lower portion.

In situations where there are different types of soil at different elevations, each layer would follow the above maximum allowable slopes as long as the harder soils are on the lower elevations. For example, if a type C soil were located above a type A soil, the trench would require a slope similar to that of Figure 20-11.

The type of soil on the lower location will determine the slope of the entire trench (see Figure 20-12).

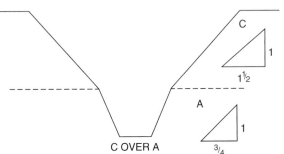

Figure 20-11 Maximum allowable slope for multilayered soils with harder soil on the bottom.

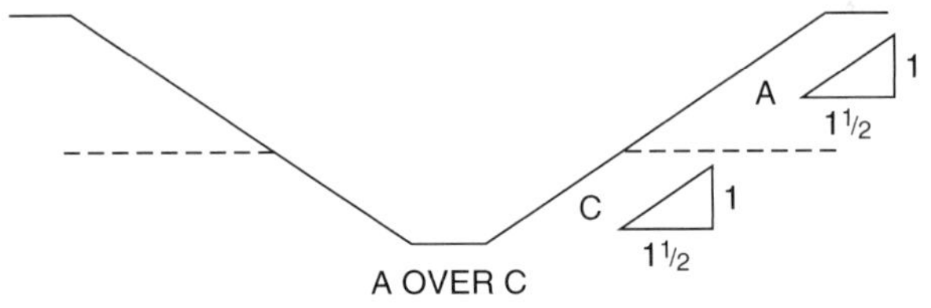

Figure 20-12 Maximum allowable slope for multilayered soils with softer soil on the bottom.

Atmospheric Hazards in Trenches

As stated earlier, a person can suffocate in minutes. While most workers can visualize suffocation because of a cave-in, few think about the possibility of suffocation because of a lack of oxygen. Many think that because there is no roof on the trench it is not a confined space, but that is a deadly misconception. Table 20-2 shows a list of common gases that can be found on jobsites, all of which are heavier than air.

Table 20-2   Common Gases that are Heavier than Air

Argon	Butane
Carbon Dioxide	Chlorine
Refrigerant	Ethane
Hydrogen Chloride	Hydrogen Sulfide
Methyl Ethyl Ketone	Nitrogen Dioxide
Sulfur Dioxide	Propane

Because these gases are heavier than air, they will sink to the bottom of the trench and displace any oxygen-rich air. Workers need approximately 19% oxygen in the air that they breathe in order to stay alive. OSHA requires that the air in a workspace must have 19.5 to 23.5% oxygen. Because many of the above gases are odorless and colorless, a worker would have no warning of the danger; that is why a competent person should check the oxygen levels before anyone enters a trench. For example, carbon dioxide is found in the exhaust gases from a combustion engine, so if workers were in a trench that had a piece of equipment, like a generator, the fumes would eventually fill up the trench. The Table 20-3 an OSHA-approved trench check sheet. It is normally filled out before anyone is allowed to work in the trench.

IN THE FIELD

Provide stairways, ladders, ramps, or other safe means of access within 25' of the workers in all trenches 4' or deeper.

Trench Safety Check Sheet

Table 20-3   Trench Safety Check Sheet

Project:		Date:	Weather:	Soil Type:
Trench Depth:	Length:	Width:	Type of Protective System:	

Yes	No	N/A	**Excavation**
			Excavations and protective systems inspected daily and before start of work by competent person.
			Competent person has authority to remove workers from excavation immediately.
			Surface encumbrances supported or removed.
			Employees protected from loose rock or soil.
			Hard hats worn by all employees.
			Spoils, materials, and equipment set back a minimum of 2' from edge of excavation.
			Barriers provided at all remote excavations, wells, pits, shafts, etc.
			Walkways and bridges over excavations 6' or more in depth equipped with guardrails.
			Warning vests or other highly visible PPE provided and worn by all employees exposed to vehicular traffic.
			Employees prohibited from working or walking under suspended loads.
			Employees prohibited from working on faces of sloped or benched excavations above other employees.
			Warning system established and used when mobile equipment is operating near edge of excavation.
			Utilities
			Utility companies contacted and/or utilities located.
			Exact location of utilities marked when near excavation.
			Underground installations protected, supported, or removed when excavation is open.
			Wet Conditions
			Precautions taken to protect employees from accumulation of water.
			Water removal equipment monitored by competent person.
			Surface water controlled or diverted.
			Inspection made after each rainstorm.
			Hazardous Atmosphere
			Atmosphere tested when there is a possibility of oxygen deficiency or build-up of hazardous gases.
			Oxygen content is between 19.5% and 23.5%.
			Ventilation provided to prevent flammable gas build-up to 20% of lower explosive limit of the gas.
			Testing conducted to ensure that atmosphere remains safe.
			Emergency response equipment readily available where a hazardous atmosphere can or does exist.
			Employees trained in the use of personal protective and emergency response equipment.
			Safety harness and lifeline individually attended when employees enter deep confined excavation.

Signature of Competent Person

Date

Figure 20-13 The builders transit has the ability to pivot vertically.

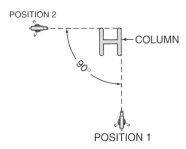

Figure 20-14 Vertical alignment from two 90° planes. Positions 1 and 2 should be a distance equal to or greater than the height of the column.

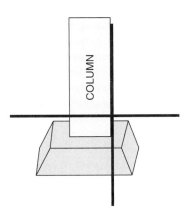

Figure 20-15 Align the vertical crosshair with the side edge near the bottom of the column.

IN THE FIELD

Use a supporting brace to hold the vertical position as seen from the first sighting position; then, a second brace will hold the item properly for all directions when the object has been plumbed from the second sighting position.

USES OF A BUILDERS TRANSIT

As mentioned in Chapter 18, the builders transit is very similar to the builders level with the exception of one added feature. The telescope on the builders transit can pivot about a horizontal axis which allows the telescope to point up or down. There is also a scale so that angles can be measured in the vertical plane (see Figure 20-13).

This added feature allows the builders transit to be used for a number of different tasks. This chapter will focus only on two of the most common: (1) *vertical alignment* and (2) *setting points in a straight line*. Other applications of the transit may be obtained from the instruction book that accompanies your instrument or from the manufacturer's website.

Vertical Alignment

Occasionally items such as building walls, columns, and other tall objects may have to be checked for plumb. *Plumb* is a term that indicates whether an item is exactly vertical or at a 90 degree angle to a level horizontal plane. The transit tripod will need to be setup at two locations that are approximately 90 degrees from each other so that the object can be checked in two separate vertical planes. Doing so will ensure that the object will appear to be plumb from any view. Figure 20-14 shows a steel column being checked for plumb.

Example 1

Determine whether the column in Figure 20-11 is plumb or not.

At position 1, the transit should be setup and leveled as described in Chapter 18. The telescope can then be aimed at one side of the column. The locking lever that holds the telescope level can now be released so that the telescope can be aimed as described in Figure 20-15.

Once the crosshair is aligned with the edge of the column, lock the horizontal clamp so that the transit cannot turn in the horizontal plane. Now the telescope can be tilted up and down to observe the entire length of the column (see Figure 20-16).

If the crosshair follows the edge of the object from top to bottom, the object is plumb in a direction perpendicular to the line of sight from the telescope. The transit can now be moved to position 2 and using the same steps just mentioned, the column can be checked in a second vertical plane.

SETTING POINTS IN A STRAIGHT LINE

When a sewer line is installed, it is generally in a straight line. Knowing the two endpoints of the line, a long string can be used to aid in the marking of any intermediate points. However, this string can become a trip hazard on a busy construction site, and with a little planning, can be avoided all together. In Chapter 19, the tripod was set up so that all points along the projected path could be seen. If the transit can be located, as described in Figure 20-17, then not only can the elevations be measured, but the centerline of the pipeline can also be identified.

Example 2

Use a transit to locate the centerline stakes of the projected sewer line and calculate the elevations (see Figure 20-18).

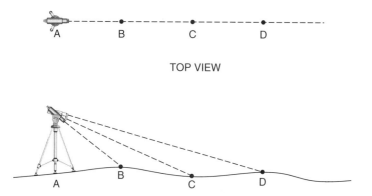

TOP VIEW

Figure 20-17 Tripod centered over point A allows the center line to be marked at other points.

Once points B and D (see Figure 20-18) have been marked with stakes, the transit and tripod can be set up at point A. Point A can be anywhere as long as the vertical crosshair is in line with both stakes B and D. After the transit has been leveled, the instrument height and elevation of point B can be determined. The horizontal clamping screw should be tightened so that the transit will remain in line. After the rod reading has been recorded in Table 20-4, the vertical locking lever can be released and a small nail can be driven into the top of the stake at point B.

Figure 20-16 Align the vertical crosshair with the side edge, near the top of the column.

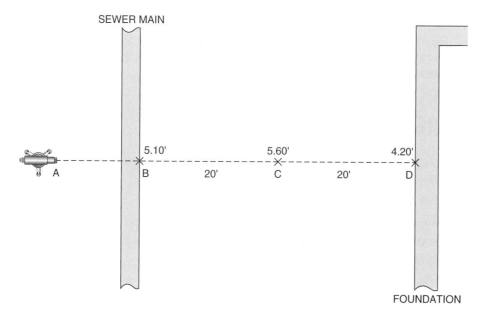

Figure 20-18 The transit, at point A, is used to mark the centerline and measure elevations.

A tape measure can now be used to locate the distance from point C to point B, and with the aid of the vertical crosshair, a nail can be driven into the ground. Again, a small nail can be used to locate the exact location of the centerline. Remember to bring the telescope back to zero in the vertical plane and secure the

Table 20-4   Use of a Builders Transit

Location	Reading	Elevation	Height of Instrument
B	5.10	100	105.10
C	5.60	105.10 − 5.60 = 99.5	
D	4.20	105.10 − 4.20 = 100.9	

vertical locking lever before taking any elevation readings. The readings can again be recorded in Table 20-4. The same procedures used to locate point C can be used to locate and record the elevation of the stake at point D or any other intermediate points along the path.

PROPER STORAGE AND CARE OF BUILDERS TRANSIT AND LEVEL

The builders transit and the builders level are precision instruments and should be treated with care. Anyone not familiar with the instruments should refer to the instruction manual or receive training before attempting to use them. If properly maintained, these instruments will give years of good service and remain reliable. The following are some common guidelines to follow when working with leveling instruments:

- Always make sure the tripod is securely placed on the ground with its feet spread and pressed into the ground.
- Do not overtighten the leveling and clamping screws.
- If the instrument is not in use, cover it with a waterproof cover to protect it from rain and dust.
- Never rub dust or dirt off the lens. Blow the dust or dirt off or use a soft bristled brush.
- When moving from one location to the next, loosen the leveling screws to take the pressure off of the tripod plate.
- Carry the tripod under your arm with the level in front of you.
- Always store the equipment in its carrying case, especially when transporting it in a vehicle.

REVIEW QUESTIONS

1 List four different types of soil classifications in order from hardest to softest.

2 List four gases that are heavier than air.

3 How do the builders level and the builders transit differ?

4 What is the maximum distance a ladder should be from the workers in a deep trench?

5 Who should fill out the trench safety check sheet?

Offsets

The student will:

- Calculate piping lengths for various offsets.
- Calculate piping lengths for parallel offsets.
- Calculate piping lengths for rolling offsets.

TERMINOLOGY AND CONSTANTS

When a pipeline shifts to one side or another as it continues in a common direction, the shift is called an *offset*. Offsets are referred to in terms of the angle of the fittings used to bring about the shift in location. Offsets of 90°, 45°, and 22-1/2° are the most often used in piping systems. Regardless of the angle used, the common terms *offset, travel,* and *run (advance)* will stay the same as will the basic equation used. What will change is the *constant (multiplier)* used, which shows how they are related.

Figure 21-1 shows the meanings of the terms offset (0), run (R), and travel (T). The term *run* (R) is sometimes referred to as advance because it is always measured in the same direction that the pipe is running or advancing. The *travel* is the diagonal distance that has to be traveled in order to get the required offset. The angle made by the fitting will always appear as the one from the centerline of the run to the centerline of the travel. The angle used will determine the ratio of the offset to the run. A low angle fitting, such as a 22-1/2°, will require a longer run than a 45° fitting in order to get the same amount of offset. When using a 45° fitting, the offset will be equal to the run. If both the offset and the run are known, then the Pythagorean theorem can be used to determine the travel. Remember that the Pythagorean theorem is $A^2 + B^2 = C^2$, where A and B are two sides of a right triangle and C is always the longest side. In this case, C in the Pythagorean theorem has to be the travel.

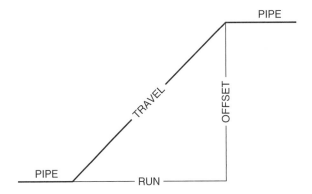

Figure 21-1 Commonly used terminology for piping offsets.

Example 1

If the 45° offset is 10", what is the travel?

$$R = 0 = 30"$$

$$T^2 = R^2 + 0^2$$

$$T^2 = (10)^2 + (10)^2$$

$$T^2 = 100\ in^2 + 10\ in^2$$

$$T = \sqrt{200\ in^2}$$

$$T = 14.142\ in^2$$

Using the Pythagorean theorem requires that both the run and the offset be known, but in many cases they are not. Usually, only the offset and the angle of the fitting used are known. When this is the case, Table 21-1 can be used to determine both run and travel. *The Pipefitter's Handbook* is an excellent reference for additional information concerning piping allowances.

Table 21-1

Angle of Fitting	To Find	Multiply	By
22-1/2°	Travel	Offset	2.613
	Run	Offset	2.414
45°	Travel	Offset	1.414
	Run	Offset	1.000
60°	Travel	Offset	1.155
	Run	Offset	.577
72°	Travel	Offset	1.051
	Run	Offset	.325

Now using the multiplier for a 45° fitting from Table 21-1, determine the travel if the 45° offset is 10".

$$T = 1.414 \times (10")$$

$$T = 14.142 \ in$$

MATH REVIEW

Because the remainder of this chapter deals with converting back and forth from fractions to decimals, now is the time for a quick review of two methods that can be used. The first method uses Table 21-2. With this table, any inch measurement from 0 to 11-7/8" (in 1/8" increments) can be converted into a decimal part of one foot (1/4" = .020'). The chart can also be used in reverse: 0.552' = 6-5/8".

The second method deals with converting a decimal value in inches into a fraction that is normally found on a tape measure. In the previous example, the answer was found to be 14.142 inches and now the .142 needs to be converted into a fraction commonly found on a tape measure (1/16, 1/8, 1/4, 1/2). To do so, simply pick the smallest fraction that is needed for accuracy (for example, 1/16) and multiply .142 by 16. The answer will be the whole number that becomes the numerator in the fraction.

$$.142 \times 16 = 2.3$$

2.3 *needs to be a whole number so round to* 2

Therefore $\dfrac{2}{16} \Rightarrow \dfrac{1}{8}$

So now 14.142 can be written as 14-1/8".

Figure 21-2 shows a common problem. A technician wants to run a pipe vertically upward through a wall, but a beam is located directly under the wall. A plumb bob is used to establish the alignment of the lower pipe so that the exact offset distance can be measured. Once 0 (offset) is known, T (travel) and R (run) can be determined.

Table 21-2   Decimal Values (inches to feet)

Whole Value (inches)	0"	1/8"	1/4"	3/8"	1/2"	5/8"	3/4"	7/8"
			Decimal Value of a Foot (Fractional Parts (inches))					
0"	.000	.010	.020	.031	.041	.052	.063	.073
1"	.083	.094	.104	.115	.125	.135	.146	.156
2"	.167	.177	.187	.198	.208	.219	.229	.240
3"	.250	.260	.271	.281	.292	.302	.313	.323
4"	.333	.344	.354	.365	.375	.385	.396	.406
5"	.417	.427	.438	.448	.458	.469	.479	.490
6"	.500	.510	.521	.531	.542	.552	.563	.573
7"	.583	.594	.604	.615	.625	.635	.646	.565
8"	.667	.677	.688	.698	.708	.719	.729	.740
9"	.750	.760	.771	.781	.792	.802	.813	.823
10"	.833	.844	.854	.865	.875	.885	.896	.906
11"	.917	.927	.938	.948	.958	.969	.979	.990

To use this table, select the whole inch value from the column on the left and the fractional value on the top of the table, and read the equivalent value as a decimal part of one foot.

For example, to find the decimal value in feet for 7-3/8", enter the table at 7" on the left and 3/8" at the top, where the column and the row intersect is the value of .615'.

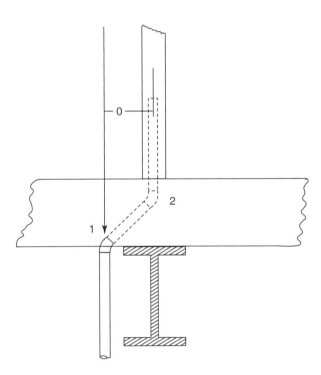

Figure 21-2 Using an offset to miss a beam in a wall.

A 2" PVC pipe running down through the middle of a wall must be offset to miss a beam. Two 45° fittings are used.

Example 2

A 2" PVC pipe running down through the middle of a wall must be offset 16" to miss a beam. If two 45° fittings were used, what would be the required run and travel?

Using the data from Table 21-1:

$$Run(R) = Offset(O) \times 1.000$$

$$R = 16" \times 1.000$$

$$R = 16"$$

$$Travel(T) = Offset(O) \times 1.414$$

$$T = 16" \times 1.414$$

$$T = 22.624"$$

Remember that since the 16" offset was measured from the centerline of the pipe, the calculated run and travel are center-to-center dimensions. In order to calculate the end-to-end measurement for the travel section, the fitting allowance for a 2" 45° PVC fitting must be known. If the fitting allowance (or take-off) were 1-1/2", then the travel end-to-end measurement would be 19.624" [22.624"– (2 × 1-1/2")] or 19-5/8".

Example 3

A 2" PVC pipe running down through the middle of a wall must be offset 16" to miss a beam. If two 60° fittings were used, what would be the required run and travel?

Using the data from Table 21-1:

$$Run(R) = Offset(O) \times .577$$

$$R = 16" \times .577$$

$$R = 9.232"$$

$$Travel(T) = Offset(O) \times 1.155$$

$$T = 16" \times 1.155$$

$$T = 18.48"$$

Offsets may also be used in horizontal piping. Figure 21-3 shows a sewer line that drops to the sewer main at the final connection.

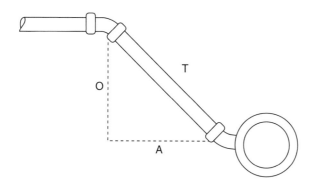

Figure 21-3 4" hubless cast-iron sewer line connecting into the sewer main using a 45° fitting.

Example 4

If the sewer line is 4" hubless cast iron and if the pipe must drop 6.8' in the final run, what would be the necessary end-to-end measurement for the travel?

$$T = 1.414 \times 6.8'$$

$$T = 9.6'$$

$$.6 \, ft \times \frac{12 \, in}{1 \, ft} = 7.2 \, inches$$

$$.2 \, in \times 8 = 1.6 \Rightarrow \frac{2}{8} \Rightarrow \frac{1}{4}$$

$$T = 9 \, ft - 7\frac{1}{4}\,"$$

Table 21-3 shows that the take-off for a 4-1/8" bend is 3-1/8". Therefore, the piece to be cut (assuming 10 foot sections are available) is calculated as follows:

Table 21–3   Hubless Cast Iron 1/8 Bend

Size (Inches)	Dimensions in Inches	
	D (±1/8)	R
1 1/2	2 5/8	2 3/4
2	2 3/4	3
3	3	3 1/2
4	3 1/8	4
5	3 7/8	4 1/2
6	4 1/16	5
8	5	6
10	5 1/16	7

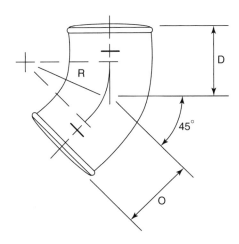

$$T_{end\text{-}to\text{-}end} = 9'7\frac{1}{4}\," - 2 \times \left(3\frac{1}{8}\,"\right)$$

$$T_{end\text{-}to\text{-}end} = 9'7\frac{1}{4}\," - 6\frac{1}{4}\,"$$

$$T_{end\text{-}to\text{-}end} = 9'1"$$

PARALLEL OFFSETS

Parallel offsets are used when two or more pipes are run parallel to each other and an offset must be made. When done correctly, the job will have a very professional and neat appearance. It will also save space and ensure enough clearance for insulation. Figure 21-4 shows two pipes that have a constant center-to-center distance (S) between them. In order to maintain this distance, the offset elbow connecting pipes D and E must be shifted up by an amount H. H is calculated so that all of the fittings will lie on a line known as the *parallel angle*. The parallel angle is always 1/2 of the angle of the fittings used.

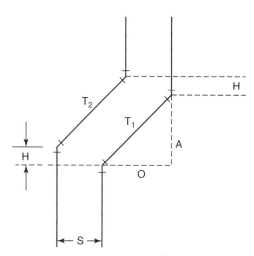

Figure 21-4 Two parallel lines separated by a constant distance (S). Notice how all of the elbows line up on the parallel angle lines.

Using this information along with Figure 21-4, an equation that shows the relationship between S, H, and the fitting angle can be expressed. Table 21-4 shows the multiplier value worked out for some of the more common fitting angles. It is also useful to notice that the travel is the same for both pipes, which also means that the value H appears on both the top and the bottom of the figure.

$$H = S \times \left[\tan\left(\frac{\textit{fitting angle}}{2} \right) \right]$$

$$H = S \times \left[\tan\left(\frac{45°}{2} \right) \right]$$

$$H = S \times \left[.414 \right]$$

Table 21-4

Fitting Angle	To Find	Multiply	By
22.5°	H	S	.199
45°	H	S	.414
60°	H	S	.577
72°	H	S	.727
90°	H	S	1.000

Example 5

If the pipe on the left is a 2" PVC vent stack and the pipe on the right is a 4" PVC waste, calculate the travel center-to-center distance of the two pipes and the dimension of H if the offset is 15" and the distance between pipe centers is 9". Refer to Figure 21-4.

$$T_1 = T_2$$

$$T_2 = 1.414 \times (15")$$

$$T_1 = T_2 = 21.21"$$

$$H = S \times (\tan 22.5°)$$

$$H = 9" \times (.414)$$

$$H = 3.73"$$

Thus, if the center of the two-inch 45° ell on the left is 3-3/4" above the center of the 4" ell on the right, and if the upper four-inch 45° ell is placed 3-3/4" below the upper two-inch 45° ell, the two lines will be equally spaced throughout the offset.

PARALLEL TURNS

Similar methods can be used to solve the problem of two pipes running parallel that have to make a 90° turn. The turn is usually completed with two 45° elbows and the travel is calculated much like the 45° offset. The outer pipe will have its elbows shifted up by the value H, which is calculated in the same manner as before. Once the travel for pipe two (outer) has been calculated, the travel for the inner pipe can be calculated by simply subtracting 2H from the travel of the outer pipe (see Figure 21-5).

Example 6

Using Figure 21-5, if O = 18", and S = 8", calculate T_1 and T_2:

$$T_2 = 1.414 \times (18")$$

$$T_2 = 25.45"$$

$$H = 8" \times (.414)$$

$$H = 3.312"$$

Thus,

$$T_1 = T_2 - 2 \times (3.312")$$

$$T_1 = 25.45" - 6.624"$$

$$T_1 = 18.83"$$

$$T_1 = 18\frac{13}{16}"$$

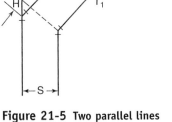

Figure 21-5 Two parallel lines making a 90° turn are separated by a constant distance S. Notice how the travel of the outer pipe is increased by 2H.

ROLLING OFFSETS

Up until this point, offsets have only taken place in one plane at a time, meaning that if a pipe was running up a wall, it stayed in contact with the wall at all times or if the pipe was laying on the floor, it stayed flat on the floor at all times. But what if it were necessary to move in both directions at the same time? This type of movement is called a *rolling offset*. Think of a rolling offset as a pipe that enters through the bottom corner of a box and must exit out the top of the far side (see Figure 21-6).

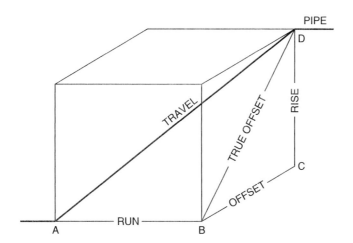

Figure 21-6 Isometric view of rolling offset.

On most jobsites, the technician will know how much to shift the pipe in the horizontal plane and how high it will have to go up or down in the vertical plane, but how will the resulting diagonal (or travel) be calculated and when should the change in direction begin? Looking at Figure 21-7, there are two right triangles of importance: triangle ABD and triangle BCD. Triangle BCD is made up of the dimensions typically known, the offset and the rise. Knowing these two sides and the Pythagorean theorem, the third side (*true offset*) can be calculated. The true offset is the side opposite the fitting angle in the triangle ABC. Since the run, travel, and true offset are in the same plane, they will all be in their true shape and size. This means that both the travel and the run can be calculated using Table 21-1.

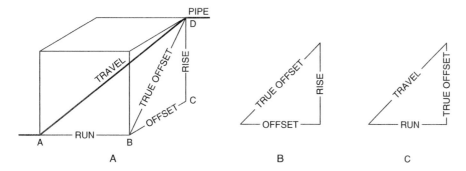

Figure 21-7 Isometric view of rolling offset.

Example 7

Figure 21-7 shows using two 45° fittings to accomplish a rolling offset. If the rise and the offset are both 24", determine the travel and the run.

offset = 24 "

rise = 24 "

therefore

True offset =24 "× (1.414)   (*see Figure* 21-07B)

True offset = 33.94 "

Run = *True offset* × (1.00)

Run = **33.94"**

Travel = 33.94 "× 1.414

Travel = **47.99"**

Let's look at the same type of problem, only this time the rise and offset won't be equal.

Example 8

Figure 21-8 shows using two 45° fittings to accomplish a rolling offset. If the rise is 18" and the offset is 36", determine the travel and the run.

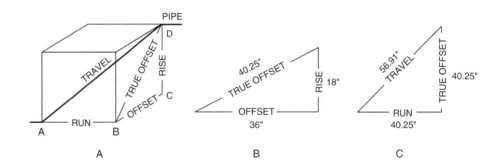

Figure 21-8 Isometric view of rolling offset.

offset = 36 "

rise = 18 "

this time the angle of triangle b is not known, so use the Pythagorean Theorem

True offset = $\sqrt{\left(18^2 + 36^2\right)}$ (*see Figure* 21-07B)

True offset = 40.25 "

Since triangle c is a 45° triangle standard *multipliers can be used*

Run = *True offset* × (1.00)

Run = **40.25"**

Travel = 40.25 "× 1.414

Travel = **56.91"**

Tank Capacities, Volume, and Weight of Water

The student will:

- Calculate the volumes and areas of cylindrical and rectangular tanks.
- Calculate the weight of water from a known volume.

TANK CAPACITIES

Cylindrical and rectangular tanks are the most common shape containers encountered in the plumbing industry. It is frequently necessary to calculate the volume of water a tank can hold, either to determine the amount of water available for some specific purpose or application or to determine the weight of water (and tank) so that structural integrity of the building can be ensured.

Cylindrical Tank

Figure 22-1 shows a cylindrical tank with diameter D and radius R. The length of the tank is L. The *volume* of a cylindrical tank can be found by first finding its cross-sectional area and then multiplying by its length. Since the cross-section of a cylinder is a circle, the formula for *area of a circle* is pi (π) times the radius squared (πR^2). *Pi (π) equals 3.14* and is a constant in the formula. The *radius* is the point from the center of the circle to its outer edge or 1/2 the diameter. The formula is as follows:

$$V = (\text{cross-sectional } \textit{Area of circle}) \times \textit{length}$$

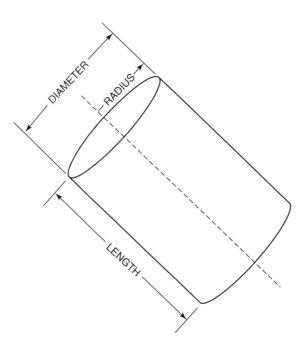

Figure 22-1 Cylindrical tank.

The volume of the tank in gallons is given by the following equation:

$$V = \pi R^2 L \times (7.48)$$

where

V = *volume (gallons)*

R = *radius of the tank (feet)*

L = *length or height of the tank (feet)*

7.48 = *number of gallons per ft^3*

Example 1

Referring to Figure 22-1, determine the capacity (in gallons) for a tank with a diameter of 4' and a length of 8'. Be sure to use consistent units to make these calculations. Be sure that all the dimensions are in feet.

$$V = \pi R^2 L \times (7.48)$$

$$V = 3.14 \times \left(\frac{4\ ft}{2}\right)^2 \times (8\ ft) \times (7.48)$$

$$V = 752\ gallons$$

Rectangular Tank

The volume of a rectangular tank is calculated in the same manner as the cylindrical tank except that the cross-sectional area is a rectangle instead of a circle. The formula for calculating the *area of a rectangle* is the width times the height ($A = W \times H$); when this is combined with the length, the formula is as follows:

$$V = (\text{cross-sectional } Area\ of\ rectangle) \times length$$

The volume of the tank in gallons is given by the following equation:

$$V = W(H)L \times (7.48)$$

where

$V = volume\ (gallons)$

$W = width\ of\ the\ \text{tank}\ (feet)$

$H = height\ of\ the\ \text{tank}\ (feet)$

$L = length\ of\ the\ \text{tank}\ (feet)$

$7.48 = number\ of\ gallons\ per\ ft^3$

Other forms of these equations are possible, so always make sure the units are consistent (feet, inches, yards) before attempting to use a different one.

Example 2

Referring to Figure 22-2, determine the capacity (in gallons) if the tank has the following dimensions: W = 6', H = 5', and L = 8'?

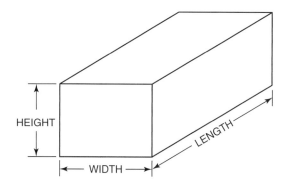

Figure 22-2 Rectangular tank.

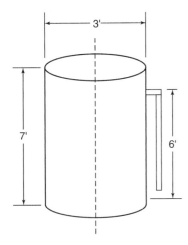

Figure 22-3 Vertical cylindrical tank with an overflow.

$$V = W(H)L \times (7.48)$$

$$V = 6' \times 5' \times 8' \times 7.48$$

$$V = 1795.2 \ gallons$$

Example 3

If an overflow opening is placed in a tank, the maximum (steady state) volume the tank can hold is determined by the height from the bottom of the tank to the overflow. Figure 22-3 shows a 7' vertical tank, 3' in diameter, with an overflow installed. The distance from the top of the tank to the invert of the overflow is 12". What is the capacity of the tank?

$$V = \pi R^2 L \times (7.48)$$

$$V = 3.14 \times (1.5)^2 \times (7 \ ft - 1 \ ft)(7.48)$$

$(7 \ ft - 1 \ ft)$ **Was used for the height because that is the maximum depth**

that can be retained in the tank.

$$V = 317 \ gallons$$

Example 4

What is the capacity (in gallons) of the bathtub in Figure 22-4 if the average width is 27", the depth is 18", and the average length is 51"? The invert of the overflow is located 4 1/4" below the rim of the tub. NOTE: Average values are used because bathtubs have sloped sides.

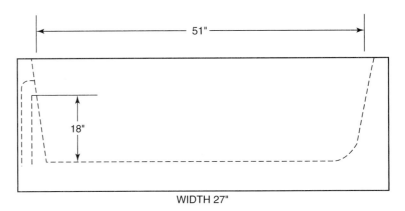

Figure 22-4 Standard residential bathtub with an overflow.

First change all of the dimensions to feet.

$$W = \frac{27"}{12" \ per \ ft} = 2.25 \ ft$$

$$H = 18" - 4.25" = 13.75"$$

$$\frac{13.75"}{12" \ per \ ft} = 1.15 \ ft$$

$$L = \frac{51"}{12" \text{ per } ft} = 4.25 \text{ ft}$$

$$V = (2.25') \times (1.15') \times (4.25') \times (7.48)$$

$$V = 82.26 \text{ gallons}$$

SURFACE AREA

The *surface area* of a tank is frequently required to calculate the amount of insulation or paint needed to cover the tank. The formula for the area of containers is simply the sum of the areas of all of the sides. Cylindrical tanks have two end caps (which are circles) and the barrel portion (which is a rectangle rolled up), so its formula is as follows:

Area of a cylinder = 2(*ends*) + *barrel portion*

$$A = 2(\pi R^2) + 2\pi R(L)$$

(2*πR* is the perimeter of the circle)

The formula for the *area of a rectangular tank* is simply the area of the 6 sides (top, bottom, 2 ends, 2 sides).

Area of a rectangular tank = 2(*W*)(*H*) + 2(*L*)(*H*) + 2(*L*)(*W*)

$$A = 2 \times [(W)(H) + (L)(H) + (L)(W)]$$

Example 5

Calculate the area of the tank in Example 1.

Area of a cylinder = 2(*ends*) + *barrel portion*

$$A = 2(\pi R^2) + 2\pi R(L)$$

$$A = 2 \times [(3.14 \times (2')^2] + 2 \times (3.14 \times 2' \times 8')$$

$$A = 25.12 \text{ ft}^2 + 100.48 \text{ ft}^2$$

$$A = 125.6 \text{ ft}^2$$

Example 6

Calculate the area of the tank in Example 2.

Area of a rectangular tank = 2(*W*)(*H*) + 2(*L*)(*H*) + 2(*L*)(*W*)

$$A = 2 \times [(W)(H) + (L)(H) + (L)(W)]$$

$$A = 2 \times [(6' \times 5') + (8' \times 5') + (8' \times 6')]$$

$$A = 2 \times [(30 \text{ ft}^2) + (40 \text{ ft}^2) + (48 \text{ ft}^2)]$$

$$A = 2 \times [118 \text{ ft}^2]$$

$$A = 236 \text{ ft}^2$$

Volume and Weight of Water

If the volume of a quantity of water is known, then it is possible to calculate how much it weighs. Sometimes it may be more convenient to convert from one unit of measure to another before the weight calculation is performed so some conversion formulas are handy to know. Water weighs 8.33 pounds per gallon or 62.5 pounds per cubic foot (at average temperature). See Table 22-1 for some commonly used conversions. Many of these can be found in the *Pipefitter's Handbook* and other quick-reference guides.

Table 22-1   Volume and Weight Conversion Chart

To Find	Multiply	By
VOLUME (gallons)	VOLUME (cubic inches)	.00433
VOLUME (cubic inches)	VOLUME (gallons)	231
VOLUME (cubic inches)	VOLUME (cubic feet)	.00058
VOLUME (cubic feet)	VOLUME (cubic inches)	1728
VOLUME (cubic feet)	VOLUME (gallons)	7.48
WEIGHT(pounds)	VOLUME (cubic ft of water)	62.5
WEIGHT(pounds)	VOLUME (gallons of water)	8.33
WEIGHT(pounds)	VOLUME (cubic inches of water)	.0361

Example 7

A tank 24" × 24" × 48" has a volume of how many gallons?

$$Volume = 24" \times 24" \times 48"$$

$$Volume = 27,648 \ in^3$$

$$Weight = 27,648 \ in^3 \times .0361 \qquad (from \ Table \ 22\text{-}1)$$

$$Weight = 998 \ lbs$$

Example 8

What is the volume of water needed to fill a pipe that has a 6" inside diameter and is 362' long?

$$V = \pi R^2 L$$

$$V = 3.14 \times \left(\frac{3"}{12}\right)^2 \times (362 \ ft) \qquad (Note: \ radius = 3" \ converted \ to \ ft)$$

$$V = 71.04 \ ft^3$$

or

$$V = 71.04 \ ft^3 \times 7.48 \frac{gallons}{ft^3} = 531 \ gallons$$

or

$$V = 71.04 \ ft^3 \times 1728 \frac{inches^3}{ft^3} = 122,757 \ inches^3$$

Example 9

How much would the pipe in Example 8 weigh when filled with water if the pipe weighs 6.1 pounds per foot?

$$\text{Weight of water} = 531 \text{ gallons} \times 8.33 \frac{\text{lbs}}{\text{gallons}}$$

$$\text{Weight of water} = 4423 \text{ pounds}$$

$$\text{Weight of pipe} = 362 \text{ ft} \times 6.1 \frac{\text{pounds}}{\text{ft}} = 2208 \text{ } lbs$$

$$\text{Total weight} = 4423 + 362$$

$$\text{Total weight} = 6631 \text{ } lbs.$$

Example 10

A water heater has an empty weight of 103 lbs and has a tank size that is 14" in diameter, and 5' high. How much would the tank weigh when completely full of water?

$$Capacity \text{ of } \text{tank} = \pi \times R^2 \times L \times (7.48)$$

$$= 3.14 \times \left(\frac{7"}{12}\right)^2 \times 5' \times 7.48$$

$$Capacity \text{ of } \text{tank} = 40 \text{ gallons}$$

$$\text{Weight of water} = 40 \times 8.33 \frac{\text{pounds}}{\text{gallon}}$$

$$\text{Weight of water} = 333 \text{ lbs}$$

$$\text{Total weight} = 333 + 103$$

$$\text{Total weight} = 436 \text{ pounds}$$

23

Ratios and Proportions

The student will:

- ⊗ Apply ratios and proportions for solving problems.

RATIOS

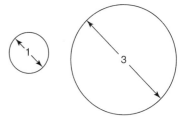

Figure 23-1 **The ratio of the two diameters is 1:3.**

A *ratio* is a linear relationship between two quantities. It is usually written as two numbers separated by a colon. The ratio of 1:3 would be read as "one to three" and means that the whole is made up of 1 part of one thing and three parts of another. A ratio can also be written as a fraction. For example, the ratio of 1/3 is the same as saying 1:3 and should also be read as "one to three." Ratios are not expressed in units. Because two objects are being compared using the same measure ratios are unitless; the units cancel out of the ratio.

Example 1

Figure 23-1 shows two circles with diameters of one and three. The ratio between the two diameters is written 1:3. The values 1 and 3 are called the *terms of the ratio*. Since a ratio can be expressed as a fraction, it may also be solved like one. The terms can be reduced without changing the ratio by simply dividing both numbers by the same divisor. For example:

8:16   (*if both are divided by 2, becomes*)

4:8   (*if both are divided by 4, becomes*)

1:2

But be careful. The relationship in Figure 23-1 deals with the diameters; in this case, the ratio of the areas would be squared (Area = (πR^2)).

So, now instead of

d:D it would be $d^2:D^2$

1:3 *becomes* 1:9

This shows that a 3" pipe has 9 times the area of a 1" pipe.

A common example of a ratio is the scale used on a drawing. If the scale is 1/4" = 1' 0", the ratio can be expressed in several ways:

1. 1/4":1'
2. 1":4'
3. 1":48"

PROPORTIONS

A *proportion* is an equation with a ratio on each side. It is a statement that two ratios are equal. Two examples of how proportions can be written are:

1. 1:2 = 4:8
2. 1/2 = 4/8

In this proportion, 1 and 8 are called *extremes,* 2 and 4 are called the *means,* and it is read "one is to two as four is to eight." When one of the four numbers is unknown, cross multiplication may be used to find the unknown number. This is called *solving the proportion*.

For example: 1:6 = 13:X

The products are equal, thus,

1(X) = 6(13)

X = 78

So 1:6 = 13:78.

This method is an excellent way of determining values not listed in charts. Table 23-1 allows the user to quickly change from Fahrenheit degrees to Celsius degrees.

Table 23-1   Farenheit and Celsius Comparison

Fahrenheit (F)	Celsius (C)
68°	20°
104°	40°
140°	60°
176°	80°

Example 1

Calculate the Celsius equivalent for 115° F.

If 104° F is 40° C and 140° F is 60° C, then the Celsius equivalent for 115° F must be between 40° C and 60° C. Set up the proportion and then solve. Remember ratios are unitless.

$$104:40 = 115:X$$
$$104X = 115 \times 40$$
$$104X = 4600$$
$$X = 44.23$$
$$115° \text{ F is } 44.23° \text{ C}$$

This assumes that the chart values are linear, which they are. Other job-related problems can take many forms as the following examples show.

Example 2

A 4" pipeline of hubless soil pipe is 180' long. If a hanger is required every five feet and at each end, how many hangers are required?

The proportion is set up as:

$$1:5 = X:180$$
$$5(X) = 1(180)$$
$$X = \frac{180}{5}$$
$$X = 36 \text{ hangers } (\textit{plus} \text{ one for the end of the line})$$

Therefore, a total of 37 hangers will be needed.

Example 3

If a person uses 25.3 gallons of hot water in one day, how much is used in a month (30 days)?

$$1:5 = X:180$$
$$5(X) = 1(180)$$
$$X = \frac{180}{5}$$

Example 4

A low-consumption water closet uses 1.0 gallon per flush. If the consumption for older model fixtures is four gallons, how much water is saved per day and per year if the water closet is used four times per day?

Daily savings:

1 flush:water saved = 4 flushes:total saving

$$1:(4 - 1.0) = 4:X$$

$$1:3 = 4:X$$

$$X = 12 \text{ gallons saved per day}$$

Annual savings:

$$1:12 = 365:X$$

$$X = 12(365)$$

$$X = 4380 \text{ gallons saved per year}$$

Example 5

The 6" building drain was to be installed with a minimum slope of 1/8" per foot. If the length of the drain was 250' and the total drop in elevation was 32', did it meet the minimum?

$$X:1\,ft = 32:250$$

$$250X = 32$$

$$X = .128$$

$$.128 \times 8 = 1.02 \Rightarrow \frac{1}{8}$$

Since the slope of the drain was 1/8" per foot, then YES, the installation meets the code.

Example 6

The scale of a drawing is 1/8" = 1'0". A building on the drawing measures 8" × 11-1/2".

What is the actual size of the building?

First side:

$$\frac{1}{8}:1 = 8:X$$

$$\frac{1}{8}X = 8$$

$$X = 64\,ft$$

Second side:

$$\frac{1}{8}:1 = 11\frac{1}{2}:X$$

$$\frac{1}{8}X = 11\frac{1}{2}$$

$$X = 92\,ft$$

Thus, the building is 64' × 92'.

IN THE FIELD

Remember to always think about the results of your calculations and ask yourself, "Was that what I was expecting to get?" Math errors can happen, but if you have an estimate of what the answer should be, the error can be caught. For example, since the Celsius equivalent in Example 1 should have been somewhere between 40° C and 60° C, if the calculated answer was 35° C, then clearly there was an error and the problem should be reworked.

Check the initial setup for the proportion to be sure the units are consistently used. Do not be misled by the seeming simplicity of the methods of proportions as discussed. Many times, the conversion factors can simply be multiplied without the proportional setup, but it is good to start with the proportion form. This method can clear the air when the initial information is more complicated and you are less likely to divide when you should multiply and vice versa.

LEARNING OBJECTIVES

The student will:

- ⊗ Describe drainage terms.
- ⊗ Explain the details of installation of storm drains.
- ⊗ Calculate the sizing of the components of the storm drainage system.

DEFINITIONS

Area Drain

An *area drain* is a receptor designed to collect surface or storm water from an open area (see Figure 24-1). If the area drain receives water from an area of 100 square feet or less, it may be connected to the foundation drain; otherwise it must be discharged separately.

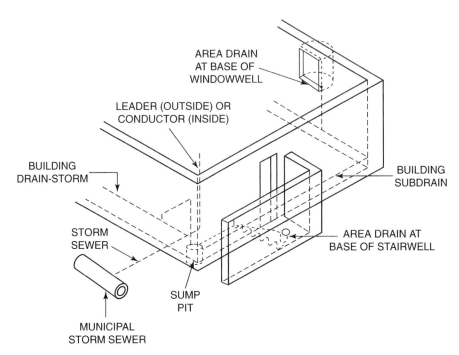

Figure 24-1 Storm drainage system. Notice how the subsoil drain feeds into a sump pit which is pumped out while a conductor brings water down from the roof.

Building Sewer (Combined)

A *building sewer* is a building drain that conveys sewage and storm water and/or other drainage. The building sewer begins at a point 3' or 5' from the outside of the building. (In many jurisdictions, combination sanitary and storm systems are not permitted as the storm water provides too great a load on the treatment plant.)

Building Storm Drain

A *building storm drain* is a building drain that conveys storm water or other drainage, but no sewage. The building storm drain is the lowest piping in the drainage system and is at least 3' or 5' (depending on code) from the outside of the building. It is used to convey water away from the building (see Figure 24-2).

Building Storm Sewer

A *building storm sewer* is the part of the drainage system located outside of the building that conveys storm water or other drainage, but no sewage. See Figure 24-2.

Conductor

A *conductor* is the water drainage pipe located inside the building. It extends from the roof to the building storm drain, combined building sewer, or other means of disposal. See Figure 24-3.

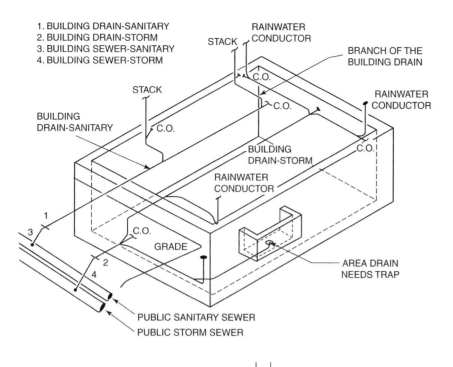

1. BUILDING DRAIN-SANITARY
2. BUILDING DRAIN-STORM
3. BUILDING SEWER-SANITARY
4. BUILDING SEWER-STORM

Figure 24-2 Building storm drain. Notice how this building has separate sanitary and storm drainage systems.

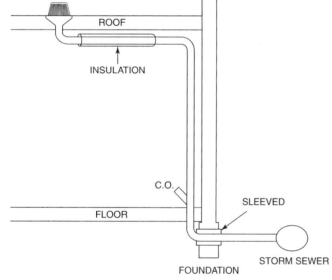

Figure 24-3 Conductors are storm water drain pipes located on the inside of the building.

Downspout or Leader

A *leader* or *downspout* is an exterior vertical drainage pipe for conveying storm water from roof or gutter drains. See Figure 24-4.

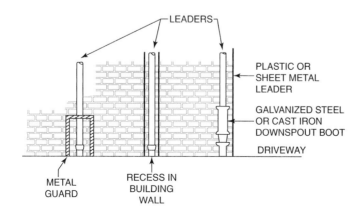

Figure 24-4 Leaders or downspouts are located on the outside of the building and must be protected from damage. Courtesy of Plumbing-Heating-Cooling Contractors–National Association.

Foundation Drain

A *foundation drain* is a drain installed to receive any water that may collect near the building foundation. It then conveys that water away from the building to a suitable disposal point. See Figure 24-10.

Roof Drain

A *roof drain* is a drain installed to receive water collecting on the surface of a roof, and then to discharge it into a leader or a conductor (see Figure 24-5). A strainer is required, even on high-rise buildings. Because of birds or updrafts, leaves, twigs, and dirt can accumulate in roof drains and quickly clog a trap. In cases where the roof is easily accessible, a vandal resistant hood cap may be required.

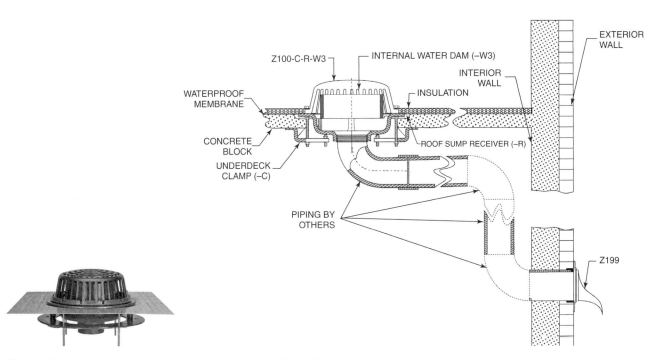

Figure 24-5 Sectional view of a roof drain. Courtesy of Zurn Industries, LLC.

Scupper Drain

A *scupper drain* is a drain or opening for a secondary drainage system and is usually located along any part of the structure that would cause water to dam and collect on the rooftop (see Figure 24-6). Many flat-top roofs have a relatively short vertical wall that goes along the perimeter of the building; this wall is commonly called a *parapet wall.*

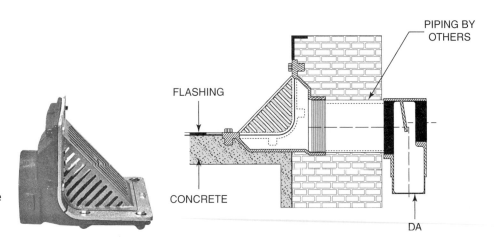

Figure 24-6 Scupper drain discharging along the outside of the building. Courtesy of Zurn Industries, LLC.

Sewage

Sewage is any liquid waste containing animal or vegetable matter in suspension or solution, and may include liquids containing chemicals in solution.

Sewage Pump

A *sewage pump* is a permanently installed mechanical device, other than an ejector, for removing sewage or liquid waste from a sump.

Storm Sewer

A *storm sewer* is a sewer used for conveying rainwater, surface water, condensate, cooling water, or similar liquid wastes.

Subsoil Drain

A *subsoil drain* is a drain that collects subsurface or seepage water and conveys it to a place of disposal. See Figure 24-7.

Figure 24-7 Some subsoil drains can be on both the inside and outside of the foundation. Usually the piping on the inside of the foundation is used for venting radon gas.

Sump

A *sump* is a tank or pit that receives sewage or liquid waste; it is located below the normal grade of the gravity system and must be emptied by mechanical means. See Figure 24-8.

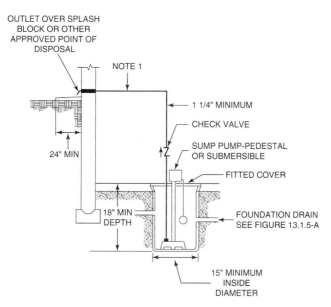

Figure 24-8 A sump for a foundation drain. Note that minimum diameter and depth varies by code. Courtesy of Plumbing-Heating-Cooling Contractors–National Association.

Sump Pump

A *sump pump* is a pump that is a permanently installed mechanical device, other than an ejector, for removing liquid waste from a sump. See Figure 24-9.

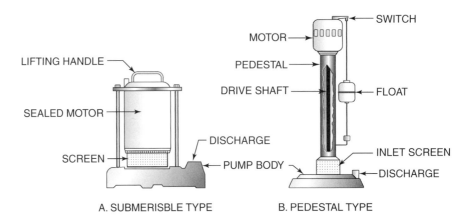

Figure 24-9 Two types of sump pumps. Courtesy of Plumbing-Heating-Cooling Contractors–National Association.

RATE OF RAINFALL

Rainfall is described quantitatively in inches per hour. That is, if a certain rate of water accumulation continued for one hour, a depth of that many inches would collect.

The U.S. Weather Bureau has compiled data for many years at many locations in the nation and published this data in a report known as Technical Paper No. 40.

A study of this report confirms what you might expect: rain can occur at higher rates for short periods of time and extremely heavy rains occur less frequently than lighter rains. These two facts are recognized by reporting the data in these ways: rainfall is observed for various periods, from 30 minutes to 24 hours; and storm intensity is reported in terms of years between the recurrences of a storm of certain intensity. Thus, a return period of 2 years indicates a lesser storm than a 10-year return period, and 25, 50, and 100-year return periods indicate storms of increasing severity. Storms that exceed the assumed rate will not usually produce structural problems, but may result in short-duration flooding or other nuisances.

Table 24-1   Rainfall Rates for Cities

State and City	Primary Storm Drainage 60-minute Duration 100-year Return		Secondary Storm Drainage 15-minute Duration 100-year Return	
	Inches/Hour	GPM/sq ft.	Inches/Hour	GPM/sq ft.
Alabama, Mobile	4.5	0.047	10.1	0.105
California, Los Angeles	2.0	0.021	4.3	0.045
Washington DC	4.0	0.042	8.6	0.089
Florida, Tampa	4.2	0.044	10.1	0.105
Indiana, South Bend	2.7	0.028	6.0	0.062
North Carolina, Charlotte	3.4	0.035	8.1	0.084

For a complete listing, see Table A.1 from the 2006 *National Standard Plumbing Code.*
Courtesy of Plumbing-Heating-Cooling Contractors–National Association.

SYSTEM DESIGN

Residential structures and similarly sized buildings usually have sloped roofs that divert rainwater into gutters and downspouts, while larger buildings are usually equipped with flat roofs and roof drains. In either case, leaders (or downspouts) are used to connect the roof drains or gutters to a suitable point of disposal. Sometimes the leaders are connected to a storm (or combined) sewer system, but most often on small buildings, the water is discharged on the ground, which should be sloped away from the building structure.

If any portion of the building space is below grade, such as a basement, cellar, or crawl space a foundation drain must be provided around the perimeter. The foundation drain will allow any accumulated water to be conveyed away from the structure. The drain usually consists of a perforated or open-joint drain tile or pipe positioned in a bed of gravel or some other approved porous material (see Figure 24-10).

IN THE FIELD

2006 *National Standard Plumbing Code*: When discharged at grade, the point of discharge shall be at least 10' from any property line and shall not create a nuisance.

IN THE FIELD

Form-A-Drain is a relatively new product that combines a form for the foundation and a continuous subsoil drain. Instead of conventional framing that must be removed after the foundation has set, Form-A-Drain remains to serve as a conduit to allow removal of underground water or to vent radon gas (see Figure 24-11).

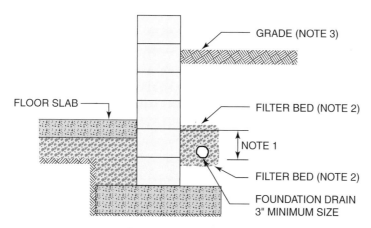

Figure 24-10 A foundation drain outside of a footing.
Courtesy of Plumbing-Heating-Cooling Contractors–National Association.

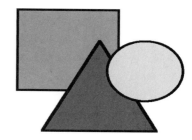

Figure 24-11 Form-A-Drain® is used to convey water or radon gas. The outer course of piping is drained either to grade or into a sump pit. This material provided by CertainTeed Corporation and used with permission.

Traps in Storm Drainage Systems

Storm drain openings are required to be trapped unless the openings are located in areas that are normally approved for vent terminals. Since most roof drains are located far enough away from any building openings, there is usually no need for a trap. However, any floor drain or any other receptor within the building will require a trap regardless of whether it connects to a storm or a combined drainage system. (Consult local codes.)

To guarantee collection of rainwater as it falls on the roof, roof drains must be placed sufficiently close.

TYPES OF STORM DRAIN DESIGN

Two basic storm drainage design options have been used for many years: (1) conventional and (2) controlled-flow.

Conventional

The concept for the *conventional system* is that the water drains away as it falls on the roof with minimal build-up or delay. This type is designed to maintain an open flow of air throughout the entire system; usually the piping will only be 1/2 to 2/3 full. The design is commonly left to the plumbing contractor.

Figure 24-12 Controlled flow roof drain. Notice how the opening gets smaller toward the top of the drain fitting. Courtesy of Jay R. Smith Mfg Co.®

Controlled-Flow

In the early 60s, a number of big buildings with large footprints and huge parking lots were built and rainwater that would normally be absorbed into the ground was being diverted into combined storm drains. Thus, a huge inrush would cause an overload to the municipal sewage system and possible environmental problems. As a result, *controlled-flow system* drains were designed to drain the storm water slowly through orifices or restrictions in the special roof drains (see Figure 24-12).

One of the determining factors for the rate of flow through an opening is the height of water above the drain opening: the higher the water, the faster the flow. Because of this, the controlled-flow roof drain has a parabolic opening that becomes more restrictive as the water level rises. The system is designed to pond water on the roof for controlled release. For short period storms, this reduces the amount of water delivered to the storm system over a given period of time. This also results in smaller conductors and storm drain piping. However, during heavy storms, the roof structure may be subjected to heavier loads and significant water depth accumulation on the roof. Because this system requires various engineering decisions, the method will not be discussed in depth here.

Siphonic

Siphonic roof drainage systems are engineered systems that were developed in Europe in the late 1960s and have been used successfully for many years. In 2005, the ANSI/ASME A112.6.9-2005 standard was approved which establishes minimum requirements and provides guidelines for the proper design, installation, examination, and testing of siphonic roof drains in the United States.

During periods of high rainfall and peak loads, the siphonic roof drain has a special air baffle that limits the flow of air into the system (see Figure 24-13).

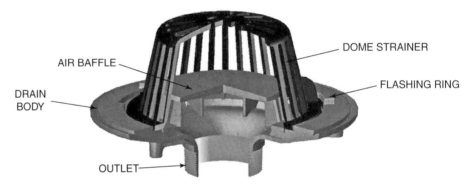

COMPONENTS OF A SIPHONIC ROOF DRAIN

Figure 24-13 Cutaway of siphonic roof drain. Courtesy of Jay R. Smith Mfg Co®.

Under light loads, the drain acts like a conventional storm drain, letting in a small amount of air with the water. The piping will remain only partially full. If the water level on the roof continues to rise with greater rainfall intensity, the air baffle will start to limit the amount of air intake. As the pipe starts to become full the flow will start to pulse, causing suction on the pipe inlet (see Figure 24-14).

This decrease in interior pressure, combined with atmospheric pressure on the roof, causes an increase in the rate of flow of the water.

Once the pipe is full, the system operates as a siphon. Since a siphon works on the basis of differential pressure and not gravity, no slope is required to create or maintain flow; thus, the horizontal runs can be completely level and flow will continue.

Because siphonic roof drain systems allow full-bore flow, smaller piping can be used. This will reduce construction costs and allow for pipes to be installed in confined locations. Since the system allows horizontal runs to be level, the depth of excavation for buried pipes is reduced, as is the need for a high ceiling space for overhead installations.

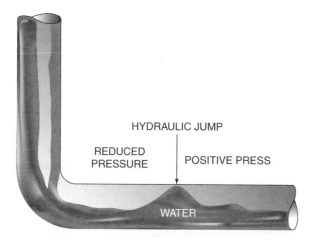

HYDRAULIC JUMP

REDUCED
PRESSURE   POSITIVE PRESS

WATER

Figure 24-14 As the fast traveling water transitions from vertical to horizontal, it meets with the slower moving water in the horizontal section and creates a slug of water that completely fills the pipe. This causes a reduced pressure at the pipe inlet and helps suck more water into the pipe.

It should be pointed out that the system is an engineered system, so it must be installed exactly as described by the engineer. However, the design can be very flexible, so even though the engineer may have to revamp the original design, it is possible to make modifications when necessary.

Design Considerations

No matter which type of system is designed, storm drain stack design limits allow greater flow than sanitary stacks. This is because hydraulic and pneumatic problems are less likely to cause problems with storm stacks. Because of this, when connecting to a combined system, storm drains should connect at least 10 feet downstream from a sanitary stack connection to minimize pressure fluctuation problems (see Figure 24-15).

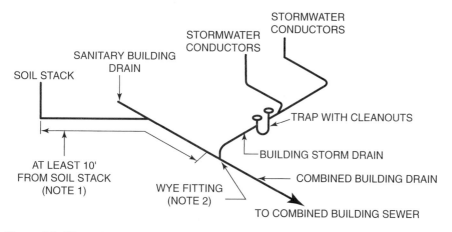

SOIL STACK

SANITARY BUILDING
DRAIN

STORMWATER
CONDUCTORS

STORMWATER
CONDUCTORS

TRAP WITH CLEANOUTS

BUILDING STORM DRAIN

COMBINED BUILDING DRAIN

AT LEAST 10'
FROM SOIL STACK
(NOTE 1)

WYE FITTING
(NOTE 2)

TO COMBINED BUILDING SEWER

Figure 24-15 Notice how storm drainage system has a running trap (which is not permitted in all areas) and is tied into a sanitary system.
Courtesy of Plumbing-Heating-Cooling Contractors–National Association.

Figure 24-16 Expansion joint to allow for thermal growth.
Courtesy of Zurn Industries, LLC.

To connect a roof drain in the center of a roof bay (usually the low point) to the conductor installed next to a column, horizontal offsets are usually required. Such offsets are desirable because they provide flexibility for the piping, which is necessary to compensate for dimension changes that come with temperature changes and the settling of buildings.

If an offset is not possible expansion joints are available to provide for the necessary dimension changes (see Figure 24-16). Also see Chapter 25 for further information.

It is not uncommon for the rainwater entering a storm drain to be cooler than ambient temperature, which may make it necessary to insulate any horizontal segments of roof drain piping that pass through a conditioned space; otherwise condensation can occur under some conditions. The weather (outdoor temperature and humidity) extremes in your area will dictate whether such insulation is required.

SIZING STORM DRAINS

Once the building type and roof design have been determined by the owner or architect, it is necessary to select the appropriate storm drainage arrangement as discussed earlier. One additional item of information is required before the size of the system can be determined: the *local design rainfall rate*. With the rates obtained from Table 24-1 or some other source (local custom, code, or designer's decision), sizes can be established.

Sloped Roof Drainage

Buildings with a sloped roof are usually designed with gutters along the lower edge. The roof area that will cause rainwater to flow into that section of the drainage system usually determines the size of the gutter and downspout. Be sure to use the actual roof area, not the projected horizontal area, because under certain wind conditions, the rain could be falling perpendicular to the roof and this maximum water accumulation must be drained by the gutter system.

Table 24-2 shows the semi-circular gutter sizes required as a function of roof area.

Table 24-2   Size of Roof Gutters

Diameter of Gutter in Inches	Maximum Projected Roof Area for Gutters 1/16 inch slope (greater slopes do not increase the capacity of the gutter)	
	Square ft	Gallons per minute (GPM)
3	170	7
4	360	15
5	625	26
6	960	40
7	1380	57
8	1990	83
10	3600	150

Table reproduced from Table 13.6.3 from the 2006 *National Standard Plumbing Code*.
Based on a maximum rate of 4 inches per hour.

Example 1

What gutter size is required for a sloped roof that measures 40' × 38' and the rainfall rate is 4"/hr? Figure 24-17 illustrates the building.

Area of one sloped side = 40' × 38'
Area = 1520 sq. ft.
Select the gutter size from Table 24-2.

The table shows that a 7" gutter will serve up to 1380 square feet, so it would not be adequate. An 8" gutter can serve up to 1990 square feet, so the 8" gutter is selected. A gutter of equivalent area may be selected from the various commercial shapes that are available.

Another option would be to slope the gutter from the midpoint of the eave to the outermost ends of the building and have a leader at each end. This would allow the water on the left half of the building roof to drain out through the left side gutter and

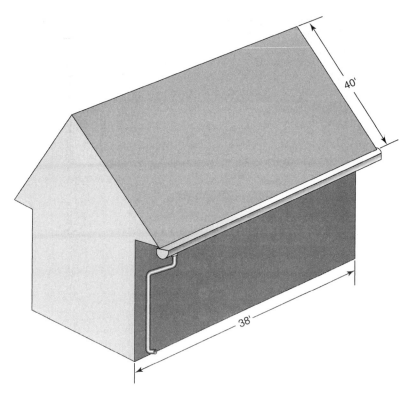

Figure 24-17 Typical sloped roof construction. If two downspouts are used, the gutter should be sloped from the midpoint to the ends so that only half of the storm water will run to that downspout.

downspout, while the right side of the building roof would drain out through the right side gutter and downspout. With this arrangement, a 6" gutter would be satisfactory since only one half of the roof is feeding into that portion of the gutter. The one drawback is that it would require an additional downspout per side of the building.

Vertical Conductors and Leaders

The size of the downspout is determined from Table 24-3.

This table shows the area of roof that can be drained by conductors or downspouts of various diameters. Conductors or leaders of rectangular shape may be sized for the same cross-sectional area as the round pipe provided that the ratio of the sides does not exceed three to one. For this example (area = 1520 sq. ft.), a 3" leader is satisfactory. If the gutter were drained on each end as discussed above, a 3" diameter leader would still have to be used on each end because a 2" diameter leader is only good for 545 sq. ft. at the 4" per hour rate of rain.

Flat Roof Drainage

Flat roof drainage design is accomplished by controlled-flow, continuous-flow, or siphonic systems. Since both the controlled-flow and the siphonic-flow systems require various engineering decisions, only the continuous-flow system will be discussed. Roof drains should be securely fastened to the roof deck and adequate slope needs to be provided on horizontal lines unless the system dictates otherwise. For a long-lasting installation, be sure to use adequate hangers and supports that will maintain a secure roof drain system.

Continuous-Flow Method

The *continuous-flow method* is designed to remove the water as quickly as possible using large conductors. In order to function properly, an air space must be maintained throughout the system.

Table 24-3 Size of Vertical Conductors and Leaders

Nominal Diameter (inches)	Flow Capacity (GPM)	Allowable Projected Roof Area at Various Rates or Rainfall per Hour (Sq. Ft.)					
		1"	2"	3"	4"	5"	6"
2"	23	2,180	1,090	727	545	436	363
3"	67	6,426	3,213	2,142	1,607	1,285	1,071
4"	144	13,840	6,920	4,613	3,460	2,768	2,307
5"	261	25,094	12,547	8,365	6,273	5,019	4,182
6"	424	40,805	20,402	13,602	10,201	8,161	6,801
8"	913	87,878	43,939	29,293	21,970	17,576	14,646
10"	1655	159,334	79,667	53,111	39,834	31,867	26,556
12"	2692	259,095	129,548	86,365	64,774	51,819	43,183
15"	4880	469,771	234,886	156,590	117,443	93,954	78,295
		7"	8"	9"	10"	11"	12"
2"	23	311	272	242	218	198	182
3"	67	918	803	714	643	584	536
4"	144	1,977	1,730	1,538	1,384	1,258	1,153
5"	261	3,585	3,137	2,788	2,509	2,281	2,091
6"	424	5,829	5,101	4,534	4,080	3,710	3,400
8"	913	12,554	10,985	9,764	8,788	7,989	7,323
10"	1655	22,762	19,917	17,704	15,933	14,485	13,277
12"	2692	37,014	32,387	28,788	25,910	23,554	21,591
15"	4880	67,110	58,721	52,197	46,977	42,706	39,146

Courtesy of Plumbing-Heating-Cooling Contractors–National Association.

The first requirement is to determine the *equivalent roof area* to be drained. The equivalent area includes the actual roof area plus half the area of any walls that extend above the roof and can contribute water to the roof surface. For a freestanding building with equal parapet walls all around, only half the parapet can contribute water to the roof under any kind of wind and rain. Thus, the added area for this type building is equal to the parapet wall area times 50% (because only half of the water can add to the roof water) and times 50% again because opposite walls will reduce the amount of water striking the roof.

Example 2

A 60' × 90' building with 4' parapet walls requires a storm drain system. How many 3' roof drains are needed if the rainfall rate is 3"/hr?

Actual roof = 60' × 90'
Actual roof = 5400 sq. ft.
Wall area = 4' (60' + 90' + 60' + 90')
Wall area = 1200 sq ft
Wall area contributing to roof load = .5 × 1200 sq. ft.
Wall area contributing to roof load = 600 sq. ft.
Effective load of wall = .5 × 600 sq. ft.
Effective load of wall = 300 sq. ft.
Total equivalent roof area = 5400 sq. ft. + 300 sq. ft.
Total equivalent roof area = 5700 sq. ft.
Area of roof to be serviced by 3" vertical drain (at 3"/hr) = 2142 sq. ft. (Table 24-3). Thus, the number of 3" roof drains required is:

$$\frac{5700 \ sq. \ ft.}{2142 \ sq. \ ft.} = 2.66$$

So, three 3" roof drains will be needed.

IN THE FIELD

The 2006 *International Plumbing Code* rates a 3" drain (3"/hr) at 2030 sq ft. Only 2 roof drains would be required if that code were being applied.

If a drainage system receives the discharge from a continuous or intermittent flow source such as an air conditioner condensate drain, a sump pump, or any other approved source, then that quantity can either be converted into an equivalent square footage or the entire drainage system should be sized basis on the gallons per minute as shown in Table 24-1. The *International Plumbing Code* states that this type of flow could be converted at a rate of 1 gallon per minute (gpm), which is equal to 96 square feet (at a rate of 1"/hr), the same as 1 gpm being equal to 24 square feet (at a rate of 4"/hr).

Example 3

If a sump pump and an air-conditioning system deliver 18 gpm to the storm drainage system, what size vertical line should be used to connect this flow to the system?

$$Equivalent\ roof\ area = 18\ gpm \times \left(\frac{96\ sq\ ft}{1\ gpm} \right)$$

$$Equivalent\ roof\ area = 1728\ sq\ ft$$

Therefore, according to Table 24-3, a 2" diameter pipe would be sufficient. Of course, following the second column until it meets the row for a 2" pipe will verify that a 2" vertical conductor is rated for up to 23 gpm.

Horizontal Piping

Horizontal piping in the storm system must be sized from Table 24-4.

Notice that the chart is divided vertically into four different pipe slopes: 1/16 inch/foot, 1/8 inch/foot, 1/4 inch/foot, and 1/2 inch/foot.

Example 4

Using Table 24-5 and Figure 24-18, determine the size piping required for the building storm drain. Figure 24-18 shows a building that is broken into six equal sections, each of which is 1600 sq. ft. If the building where located in Washington DC, determine the size for each section of pipe based on the local rainfall. Assume all horizontal runs at 1/4" per foot slope.

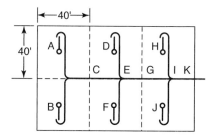 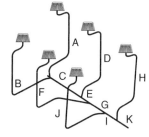

Figure 24-18 Notice how this large flat roof has been divided into six smaller sections. Each section drains into its own roof drain.

According to Table 24-1, Washington DC has a rainfall rating of 4"/hr and 0.042 gpm/sq. ft. The gpm for each area can be determined by multiplying the area being served times the rate/area (0.042).

Example 5

A slightly different example is shown in Figure 24-19 where one roof is at a different elevation than another. Using Table 24-6 and Figure 24-19, determine the size piping required for the building storm drain.

Table 24-4   Size of Horizontal Storm Drains (for 1"/hr to 6"/hr Rain Fall Rates)

Size of Drain (inches)	Design Flow of Drain (GPM)	Allowable Projected Roof Area at Various Rates or Rainfall per Hour (Sq. Ft.)					
		1"/hr	2"/hr	3"/hr	4"/hr	5"/hr	6"/hr
Slope 1/16 inch/foot							
2							
3							
4	53	5,101	2,551	1,700	1,275	1,020	850
5	97	9,336	4,668	3,112	2,334	1,867	1,556
6	157	15,111	7,556	5,037	3,778	3,022	2,519
8	339	32,629	16,314	10,876	8,157	6,526	5,438
10	615	59,194	29,597	19,731	14,798	11,839	9,866
12	999	96,154	48,077	32,051	24,039	19,231	16,026
15	1812	174,405	87,203	58,135	43,601	34,881	29,068
Slope 1/8 inch/foot							
2							
3	35	3,369	1,684	1,123	842	674	561
4	75	7,219	3,609	2,406	1,805	1,444	1,203
5	137	13,186	6,593	4,395	3,297	2,637	2,198
6	223	21,464	10,732	7,155	5,366	4,293	3,577
8	479	46,104	23,052	15,368	11,526	9,221	7,684
10	869	83,641	41,821	27,880	20,910	16,728	13,940
12	1413	136,002	68,001	45,334	34,000	27,200	22,667
15	2563	246,689	123,345	82,230	61,672	49,338	41,115
Slope 1/4 inch/foot							
2	17	1,636	818	545	409	327	273
3	50	4,813	2,406	1,604	1,203	963	802
4	107	10,299	5,149	3,433	2,575	2,060	1,716
5	194	18,673	9,336	6,224	4,668	3,735	3,112
6	315	30,319	15,159	10,106	7,580	6,064	5,053
8	678	65,258	32,629	21,753	16,314	13,052	10,876
10	1229	118,292	59,146	39,431	29,573	23,658	19,715
12	1999	192,404	96,202	64,135	48,101	38,481	32,067
15	3625	348,907	174,454	116,302	87,227	69,781	58,151
Slope 1/2 inch/foot							
2	24	2,310	1,155	770	578	462	385
3	70	6,738	3,369	2,246	1,684	1,348	1,123
4	151	14,534	7,267	4,845	3,633	2,907	2,422
5	274	26,373	13,186	8,791	6,593	5,275	4,395
6	445	42,831	21,416	14,277	10,708	8,566	7,139
8	959	92,304	46,152	30,768	23,076	18,461	15,384
10	1738	167,283	83,641	55,761	41,821	33,457	27,880
12	2827	272,099	136,050	90,700	68,025	54,420	45,350
15	5126	493,379	246,689	164,460	123,345	98,676	82,230

Courtesy of Plumbing-Heating-Cooling Contractors–National Association.

A 60' × 100' building has two roof drains on the upper elevation and two roof drains on a lower 20' × 100' section. The piping is shown in the isometric view and sizing is displayed in the following table. The short horizontal sections associated with the roof drains themselves are sized the same as vertical pipe. Assume 4"/hr rainfall.

SECONDARY ROOF DRAINS

Secondary roof drains are required on any building that has parapet walls or any other construction that extends above the roof. If the primary roof drain were to become clogged or were undersized for the amount of rainfall at that moment, a

Table 24-5   Pipe Sizes to Accompany Figure 24-18

Conductor	Horizontal Drain	Area Served (Sq. Ft.)	Load @ 4"/hr (gpm)	Pipe Size V=verti. H=hori.
A		1600	67.2	3" (V)
B		1600	67.2	3" (V)
	C	3200	134.4	5" (H)
D		1600	67.2	3" (V)
	E	4800	201.6	6" (H)
F		1600	67.2	3" (V)
	G	6400	268.8	6" (H)
H		1600	67.2	3" (V)
	I	8000	336	8" (H)
J		1600	67.2	3" (V)
	K	9600	403.2	8" (H)

NOTE: The horizontal sections could be sized differently at different slopes.

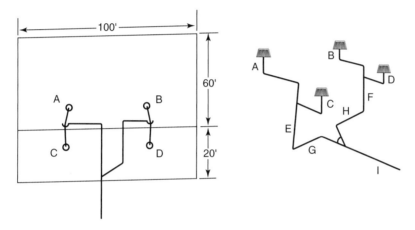

Figure 24-19 Bi-level roof. Notice how the piping is sized based on the area being discharged into it.

Table 24-6   Pipe Sizes to Accompany Figure 24-19

Conductor	Horizontal Drain	Area Served (Sq. Ft.)	Load gpm	Pipe Size	Slope	Table Used
A		3000	125	4"		24.3
B		3000	125	4"		24.3
C		1000	41.7	3"		24.3
D		1000	41.7	3"		24.3
E		4000	166.7	5"		24.3
F		4000	166.7	5"		24.3
	G, H	4000	166.7	5"	1/4 in/ft	24.4
	I	8000	333.3	8"	1/4 in/ft	24.4

dangerous amount of water could accumulate on the rooftop. This extra weight could weaken or collapse the roof structure. In order to avoid this possibility, the secondary drainage system should be designed and constructed separately from the primary roof drain. The secondary roof drain is usually either a scrubber or a roof drain with an internal dam. The height of the scrubber or the height of the dam should be set to the maximum height at which the storm water could be allowed. At this height, the primary storm drainage system has either become plugged, otherwise nonfunctional, or simply undersized for the current rainfall. The 2006

National Standard Plumbing Code dictates that the secondary system should be designed based on a 100-year, 15-minute storm—considerably more severe than the 100-year, 60-minute storm. On the other hand, the 2006 *International Plumbing Code* states that the secondary drainage system should be sized identically to the primary system. Both codes do agree on the fact that the primary and secondary systems should not share any common piping.

REVIEW QUESTIONS

1. What is the difference between a leader and a conductor?

2. What is the maximum area an area drain can serve and still be allowed to discharge into a foundation drain?

3. Give an example of where a roof drain may require a trap.

4. According to the *International Plumbing Code,* what is the minimum diameter allowed for a sump pit?

5. Using Table 24-3, what size conductor would be required for a 15,000 square feet area assuming a 2"/hr rainfall?

6. Using Table 24-4, what size horizontal pipe (sloped at 1/8 in/ft) would be required for a 42,000 square foot area assuming a 5"/hr rainfall?

7. Using Table 24-4, what size horizontal pipe (sloped at 1/2 in/ft) would be required for a 42,000 square foot area assuming a 5"/hr rainfall?

CHAPTER

25

Energy and Temperature, Piping Expansion, Heat Transfer, Insulation, Humidity, and Condensation

The student will:

- Discuss several types of insulation and their application.
- Explain the difference between temperature and heat.
- Describe the different methods of heat transfer.
- Discuss the relationship between temperature, humidity, and comfort.
- Describe the effects of humidity and condensation within a building.

TEMPERATURE

All substances are made up of molecules. Molecules are too small to be seen by even the most powerful microscopes, but these tiny ingredients determine the material and its properties. Within the substance, the molecules are constantly in motion; the more heat energy added to the substance, the faster they move. Theoretically, if enough heat energy were removed from the substance, all of the molecules would stop moving. This condition would happen at a temperature known as *absolute zero*. Absolute zero is measured as either 0°Rankine (R) (imperial) or 0°Kelvin (K) (metric). 0°Rankine is equal to –460°Fahrenheit (F), while 0°Kelvin is equal to –273°Celsius (C). To put this in perspective, water freezes at 32°F and 0°C. Absolute zero is essential when relationships between substances at various temperatures and pressures are involved, but it is too low on a scale to be used in normal day-to-day calculations. In the United States, temperatures are commonly measured in degrees Fahrenheit (F), while Canada and the rest of the world use the degrees Celsius or Centigrade scale. Because today's industry is a global market, a technician needs to know how to convert back and forth between these two scales.

The formula to convert °Fahrenheit to °Celsius is:

$$F = \left(\frac{9}{5}\right)C + 32$$

The formula to convert °C to °F is:

$$C = \left(\frac{5}{9}\right)(F - 32)$$

When a temperature is needed, it is typically measured with a *glass tube thermometer*, a *thermocouple*, or even an *infrared thermometer* (see Figure 25-1).

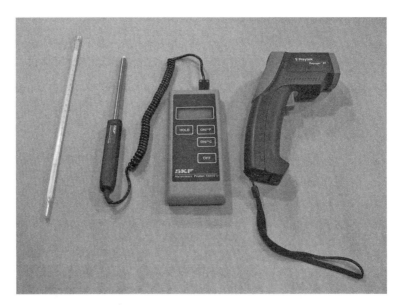

Figure 25-1 Different instruments for measuring temperature: glass tube thermometer, thermocouple, and infrared thermometer.

The glass tube thermometer is probably the most common form of thermometer; it uses a glass tube containing mercury or alcohol. Mercury is a metal that is in liquid form between the temperatures of about –40°F to 675°F (–40°C to 357°C). A thermocouple is a device that measures temperatures very rapidly. It consists of two small wires of different metals joined at one end. When the joint at that end is placed in contact with a material, a very small voltage is developed. This voltage is then read by a voltmeter and converted into a suitable number. Thermocouples should be checked for calibration from time to time by submersing the jointed end into a bath of ice water. A mixture of ice and water will always be at 32°F under normal atmospheric pressure. The infrared thermometer is a noncontact device that has an optic sensor that picks up emitted, reflected, and transmitted energy. This energy is then translated into a temperature reading through electronics. It should be noted that if the infrared thermometer uses a laser, its purpose is for aiming the device, not for reading the temperature.

ENERGY

Energy is the ability to do work. Energy appears in many forms, but for the purpose of this discussion we are concerned with heat energy. Heat energy actually is a form of *kinetic energy* (energy contained in items in motion). Under some circumstances, heat energy can also be considered as potential (e.g., steam under pressure can be thought of as contained potential energy). Examples of heat energy being added or removed every day are:

- Water will boil when sufficient heat is added.
- Water will freeze when sufficient heat is removed.
- Pipe will expand when heated and contract when cooled.
- Road pavements will buckle in very hot weather due to expansion of the material.
- A wire fence will sag in hot weather and pull taut in cold weather.

One of the best ways to distinguish the difference between heat energy and temperature is to look at how water is affected by it. Assume one pound of chopped ice is placed in a pan and the pan is gently heated. The ice absorbs heat and the molecular activity increases. The more active molecules become liquid water. The temperature of the ice-water mixture will stay at 32°F until all the ice melts. Note that during the change from ice to water, heat is added to the mixture without changing the temperature. Heat energy that causes a change in state—solid to liquid or liquid to gas—is called *latent heat*. The heat that changes the state of ice to water is called the *latent heat of fusion*. The heat energy that is required to raise one pound of water one degree Fahrenheit is known as a *BTU (British Thermal Unit)*. It takes 144 BTUs of heat to change one pound of ice into one pound of water.

After the ice is melted, the temperature of the water will increase as more heat is added. All of the heat added that raises the temperature of the water from 32°F to 212°F is known as *sensible heat*. It takes 180 BTUs (212°F–32°F) of heat to change one pound of water from 32°F to 212°F. When the temperature of the water reaches 212°F, the water starts to boil. *Boiling* is a violent process of vaporization where the liquid and the vapor coexist at the same temperature. As heat is added to the water at 212°F, the temperature will not rise as the water changes from liquid to vapor. This heat is called the *latent heat of vaporization*. It takes 970 BTUs of heat to change one pound of water to one pound of steam. After all the water has changed to vapor (steam), the addition of more heat will increase the temperature of the steam provided that it was contained without an increase in pressure.

The graph (see Figure 25-2) shows the relationship between temperature and amount of heat energy and changes of state for ice, water, and steam.

Looking back at the example of heating water, it should be clear now that even though the water and the steam are both at the same temperature, the steam has 970 BTUs more heat energy. Thus, the temperature of a substance and its state (solid, liquid, or gas) are good indicators of how much heat energy is contained in a substance.

IN THE FIELD

WARNING: Do not point the laser directly at your eyes. Because thermocouples and infrared thermometers react quickly to a temperature change, they can save time. While the thermocouple must come into contact with the substance being measured, the infrared thermometer does not. However, reflective surfaces can cause inaccurate readings with infrared thermometers.

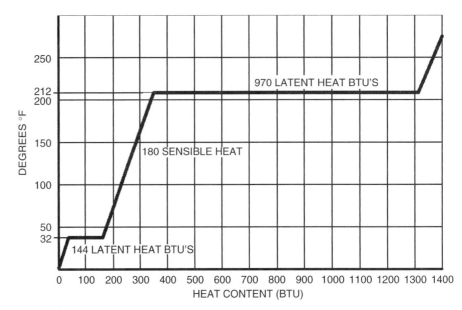

Figure 25-2 Relationship between temperature and amount of heat energy in changes of state for ice, water, and steam.

PIPING EXPANSION

All piping materials expand with an increase in temperature. The reason for this change is that the molecular activity of the material increases as the material temperature increases and this greater molecular activity results in more space between molecules. Most materials contract when cooled because reduced molecular activity results in less space between molecules. Metal and non-metal piping materials expand when heated, but non-metallic materials usually expand more than the metals.

If a pipe section is installed so that it is completely restrained at the ends (see Figure 25-3) and a heated fluid is passed through the pipe, the section will try to increase in length according to the new temperature. Since it is completely restrained, large forces may develop within the material that can cause the pipe to bend, the restraints to break, or pipe joints to break.

To guard against these problems any pipe that can be heated (or cooled) significantly should be installed so that dimensional changes can take place without any damage to the building, piping, or any other connected device (such as fixtures, pumps, valves, hangers, etc.).

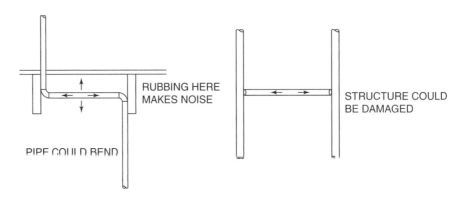

Figure 25-3 Picture of pipe expanding due to temperature, but restrained.

COMMON SITUATIONS

Hot water distribution or hot water or steam heating lines should have adequate clearance when passing through floor or wall openings.

Plastic pipe should not be hung with supports which could abrade, cut, or distort the pipe.

Pipe should not be restrained on the ends as the pipe could be bent, noises could be generated, or the structure damaged (Figure 25-3).

Freezing of Water

When exposed to temperatures above 36°F, water usually reacts to temperatures like other materials—it expands when heated and contracts when cooled. Below 36°F, however, it expands slightly and when it changes from liquid at 32° to solid at 32°, it expands volume about 10%. This expansion causes a very large pressure increase of the contained fluid. The freezing water can develop extremely high forces up to 450,000 psi.

Pressure Protection

To be sure that destructive forces cannot develop as a result of thermal effects, excessive fluid pressures must be prevented by using relief valves, expansion tanks, or blowout diaphragms or discs.

Figure 24-4 shows a typical water heater installation with an expansion tank.

IN THE FIELD

Most codes require that an expansion tank be installed near a water heater to protect the piping system from increased pressures generated due to the heating of the water (see Figure 25-4).

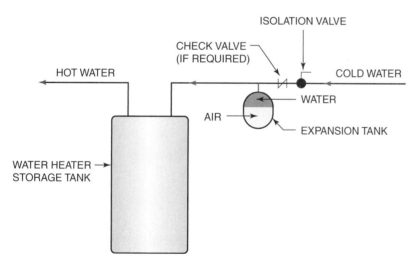

Figure 25-4 Typical water heater installation with an expansion tank.

COEFFICIENT OF THERMAL EXPANSION

The numerical value that describes how much a certain material will expand for a certain temperature change is called the *coefficient of thermal expansion*.

The scientific form of this coefficient is given in decimal form as change of length per unit length per degree Fahrenheit. Thus, the same decimal number would express the change in inches per inch, feet per foot, yards per yard, etc. This decimal number is very small, however, so for practical piping work, we use the unit *inches per 100 feet per 100 degrees Fahrenheit*.

Table 25-1 lists the coefficients in both of the above-described forms for several common piping materials. Note, for example, that the *scientific* form for cast iron is 5.9×10^{-6} (.0000059) inches for every inch of length, for every degree Fahrenheit change. In some instances, the other form, 0.71 inches per 100 feet per 100°F, is much more convenient to use. Also, note that nonmetallic materials expand from 4 to 9 times more than their metallic counterparts.

Table 25-1 Coefficients of Thermal Expansion

Material	Coefficient in/in/°F	Expansion in/100 ft/100°F
Metallic		
Brass (red)	9.2×10^{-6}	1.1
Brass (yellow)	10.0×10^{-6}	1.2
Cast Iron	5.9×10^{-6}	0.71
Copper	9.3×10^{-6}	1.1
Steel	6.4×10^{-6}	0.77
Nonmetallic		
ABS	$55. \times 10^{-6}$	6.6
CPVC	$35. \times 10^{-6}$	4.2
PB	$75. \times 10^{-6}$	9.0
PE	$80. \times 10^{-6}$	9.6
PVC	$30. \times 10^{-6}$	3.6
PEX	$90. \times 10^{-6}$	

Example 1

210' of 2" copper tube is contained in a hydronic heating system. When the tube was installed, the temperature was 60°F. The pipe was tested with 140°F water. How much expansion occurred?

$$Expansion = \frac{length}{100}\left(\frac{temperature\ change}{100}\right)(coefficient\ of\ expansion)$$

$$Temperature\ rise = 140°F - 60°F$$

$$Temperature\ rise = 80°F$$

$$Expansion = \left(\frac{210\ ft}{100\ ft}\right)\left(\frac{80°F}{100}\right)(1.1)$$

$$Expansion = 2.1 \times .8 \times 1.1$$

$$Expansion = 1.85"$$

Thus, the 2" copper tube will expand nearly 2" in 210' with an 80°F temperature rise. Notice that the size of copper tube does not enter into the calculation. Thus, any copper tubing of the same original length undergoing the same temperature change will expand by the same amount.

Example 2

How much will a 380' PVC storm drain contract when 35°F water flows through it, if the job was installed in 60°F weather?

$$Expansion\ (+)\ or\ Contraction\ (-) = \left(\frac{length}{100}\right)\left(\frac{temperature\ change}{100}\right)(coefficient\ of\ expansion)$$

$$Temperature\ drop = 60°F - 35°F$$

$$Temperature\ drop = 25°F$$

$$Contraction = \left(\frac{380}{100}\right)\left(\frac{25}{100}\right)(3.6)$$

$$Contraction = 3.42"$$

Note that this small temperature change results in the pipe shortening by nearly 3-1/2". If this temperature change is not distributed throughout the run, it could easily result in a cement joint coming apart. A marginal joint could fail in a short time, but even well-made joints could separate after repeated stress cycles.

Example 3

An 80' concrete floor contains steel radiant heat piping. What will be the change of length of the steel pipe when the temperature changes from 50°F to 150°F?

$$\text{Expansion} = \frac{length}{100}\left(\frac{temperature\ change}{100}\right)(coefficient\ of\ \text{expansion})$$

$$\text{Temperature rise} = \ 150°F - 50°F$$

$$\text{Temperature rise} = 100°F$$

$$\text{Expansion} = \left(\frac{80\ ft}{100\ ft}\right)\left(\frac{100°F}{100}\right)(.77)$$

$$\text{Expansion} = .8 \times 1 \times .77$$

$$\text{Expansion} = .62"\ or\ about\ \frac{5}{8}"$$

Since the concrete will expand about the same amount, no significant forces are encountered.

METHODS OF CONTROL

Expansion and contraction may be controlled by using expansion loops or expansion devices. These devices permit the dimensional change to take place without damaging the piping, the connected equipment, or the building.

Field Constructed

Figure 25-5 shows the *expansion loop*. The turns in the loop can be threaded fittings, weld fittings, or bent pipe sections. As the pipe expands and contracts along its length, the expansion loop will flex inward or outward as the dotted lines indicate.

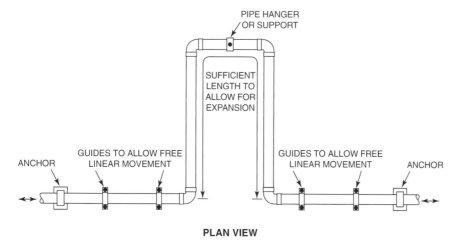

PIPE HANGER OR SUPPORT

SUFFICIENT LENGTH TO ALLOW FOR EXPANSION

ANCHOR   GUIDES TO ALLOW FREE LINEAR MOVEMENT   GUIDES TO ALLOW FREE LINEAR MOVEMENT   ANCHOR

PLAN VIEW

Figure 25-5 Expansion loop.
Courtesy of Plumbing-Heating-Cooling Contractors—National Association.

Figure 25-6 shows a *swing joint*. Swing joints, where a thrust from one-line results in turning a nipple in an ell or tee, can also accommodate expansion. This is not desirable for lines with significant pressure, however, as the turning of the thread will eventually cause leaks.

Figure 25-6 PVC joints used in the landscaping industry. Top fitting allows for minor axial expansion. Bottom fitting allows for minor angular expansion.

Mechanical Expansion Joints

Expansion joints, like the one shown in Figure 25-7, are mechanical devices arranged to permit the movement of connected piping.

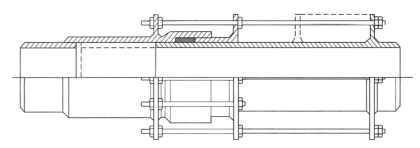

Figure 25-7 Mechanical expansion joint.
Courtesy of Plumbing-Heating-Cooling Contractors—National Association.

One type of expansion joint consists of a smooth-sided tube that is arranged to slide in packing, but these devices require frequent maintenance.

Another version uses an accordion bellows to take up the movement. This type does not require packing maintenance, but sometimes fails with fatigue cracks in the bellows. Any expansion provision usually requires some sort of anchor to be associated with it to control where the pipe moves and where it stays put! You should realize that most small systems, besides operating at minimal temperature changes, contain enough changes of direction that expansion forces are not a problem (see Figure 25-8). It is important to install expansion joints per the manufacturer's instructions.

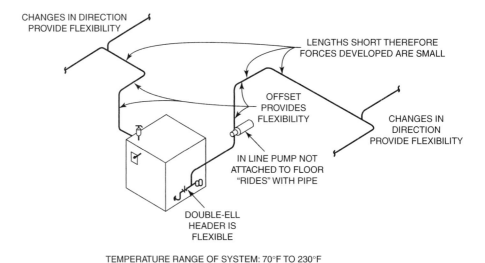

Figure 25-8 A small system with relatively short runs. Expansion per run is small.

Figure 25-9 shows a larger system in which the main trunk line has an anchor point and an expansion loop.

The anchor point will minimize the shifting due to thermal expansion, which will protect the shorter runs and stationary equipment. The expansion loop will accommodate the thermal expansion, thereby protecting the long truck line. Temperature changes can produce damaging conditions in a pipeline. If in doubt, check with an experienced piping designer before installing a line that will undergo significant temperature changes.

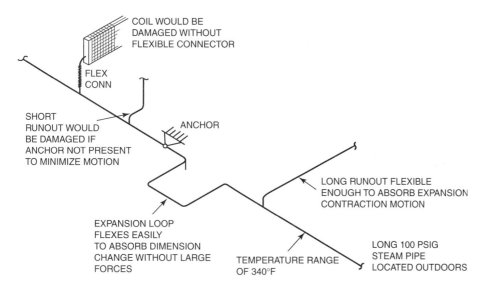

Figure 25-9 Portion of a large system showing an anchor, an expansion loop, and loads.

HEAT TRANSFER

You may have noticed that if the thermometer touches the bottom of a pan, the temperature is much hotter than the water. You may even measure temperatures higher than 212°F. How is this possible? The answer is simple: heat flows from hot to cold and the amount of heat depends on the temperature difference and how well a substance conducts heat. If you think about it, the heat had to be transferred from the burner to the pan and then from the pan to the water. Therefore, the burner had to be hotter than the bottom of the pan or the pan would not get hot. This type of heat transfer is called *conduction*. As the heat was added to the bottom of the pan, the molecules in the metal pan started to move faster; as the molecules moved, they collided with other molecules, thereby passing the energy along to the next group of molecules. Materials that transfer heat quickly are known as *good conductors of heat* while materials that do not transfer heat well are called *insulators*. If you were to pick up a hammer that had been left outside on a cold winter night, the head would feel much colder than the handle. Both the metal head and the handle would measure the same temperature, but because the metal conducts heat faster it would appear to be cold to your touch.

If you go back to the example of boiling water on the stove, you may have noticed that as the water started to heat, it also started to swirl. If you looked closely, you could see the currents in the water. The water closest to the bottom of the pan heated up first and expanded, and as the heated water started to rise to the top of the pan, the colder, more dense water came in to take its place. This referring of heat through flowing currents is called *convection*. Convection is only possible in a gravity field where lifting of a heated fluid and sinking of a cooled fluid takes place. Convection is the mechanism used in gravity hot water heating systems—water heated in the boiler rises throughout the system and water cooled in the radiators sinks back to the boiler. All the details of the piping are arranged to aid this natural flow in the following ways:

1. Supply piping connects to the top of the boiler
2. Return piping connects at the bottom of the boiler
3. Air must be removed from all high points of the system
4. Piping is sloped to prevent air locking

Convection heat transfer also occurs in air. The most common example is the use of baseboard heat. The air in contact with the baseboards will heat and rise toward the ceiling and the cooler air at the floor flows toward the space that the hotter air just left. This cooler air is then warmed and the process continues until the room is filled with warmed air. Chimneys also operate on the principle of convection. Hot gases from the fire at the base of the chimney start upward because they are lighter. The colder air from the room comes in to take their place and this causes a draft.

The last form of heat transfer to consider is radiation. *Radiation* is the transfer of energy by electromagnetic waves and rays. While conduction and convection require a physical substance to act as a medium, radiation does not. The largest source of radiation energy is the sun. We are either trying to block that energy, such as when we want to cool a house in the summer time, or we are trying to absorb the energy, such as when we want to heat water with solar panels. One of the most basic principles to keep in mind is that light colored or mirrored objects tend to reflect the solar energy while darker, rougher objects tend to absorb the solar energy.

A good example of technology that is reflecting the solar energy is the low *e coatings* on windows being produced today. The low e coating refers to *emissivity*. Emissivity is a comparison of how well a material absorbs the radiant energy compared to a perfect radiator. If a material has a low emissivity rating, it reflects more energy than it absorbs. Standard clear glass has an emittance of 0.84, which means that 84% of the energy is absorbed and only 16% is reflected. A window with a low e coating of .04 would only absorb 4% of the energy while 96% is being reflected back.

INSULATION MATERIALS AND PROCEDURES

An *insulator* is a material that is a poor conductor of heat. One way of describing how an insulator works is to point out that it slows down the flow of heat from hot to cold, but does not stop it. The temperature of the insulation itself will vary between the two extreme temperatures. Insulation can come in many different sizes, shapes, and types, but the following eight configurations of insulating materials are frequently seen in plumbing and heating work. The temperature range, within which the term *thermal insulation* will apply, is from −100°F to 1500°F. All applications below −100°F are termed *cryogenic* and those above 1500°F are termed *refractory*.

Insulation Properties

When selecting the insulation for an application, it is important to consider the properties of the insulation. Usually manufacturers will only publish properties that are common to most applications. If there are special conditions such as a chemical corrosive environment, you may have to contact the manufacturer to get further information. Below is a table with a brief description of some of the more common thermal, mechanical, and chemical properties (see Table 25-2).

INSULATION MATERIALS

Blanket

Blanket is made of fibrous materials in various widths, usually sold in rolls of standard length. This form is best to insulate irregular or bulky objects as well as stud and joist spaces in buildings.

Formed Fibrous

Formed fibrous insulation is made of fiberglass or similar material and formed into tubular shapes for different pipe sizes and different wall thicknesses (usually 1/2", 1", 1-1/2", 2").

Table 25-2 Thermal Properties

Temperature limits	Upper & lower temperatures within which the material retains all its properties.
Thermal conductance "C"	The rate of heat flow for the actual thickness of a material.
Emissivity "E"	Important when the surface temperature of the insulation must be regulated, as with moisture condensation or personal protection.
Thermal resistance "R"	The overall resistance of a "system" to the flow of heat.
Mechanical/chemical properties alkalinity (pH or acidity)	Important when corrosive atmospheres are present. Also, insulation must not contribute to corrosion of the system.
Appearance	Important in exposed areas and for coding purposes.
Chemical resistance	Important when the atmosphere is salt or chemical laden.
Coefficient of expansion and contraction	Enters into the design and spacing of expansion/contraction joints and/or the use of multiple-layer insulation applications.
Combustibility	One of the measures of a material's contribution to a fire hazard.
Dimensional stability	Important when the material is exposed to atmospheric and mechanical abuse such as twisting or vibration from thermally expanding pipe.
Fire retardancy	Flame-spread and smoke-developed ratings should be considered.
Resistance to fungal or bacterial growth	Important in food or cosmetic process areas.
Toxicity	Must be considered in food processing plants and potential fire hazard areas.

Corrugated

Corrugated insulation is usually made of asbestos paper in alternate smooth and corrugated layers. No longer made, but frequently encountered in existing work, this material was formed into tubes for various pipe sizes and was also made in flat form to insulate large objects (such as boilers).

Foam

Factory-formed insulation is made of various compounds and in two shapes: tubular for pipe covering or flat for large surfaces. *Foam* is produced by combining certain chemicals that react together in appropriate forms and releases a gaseous, bubbly product that takes the shape of the form.

Foamed-in-Place

Foamed-in-place involves mixing two liquid chemicals in a nozzle and delivering the product into a cavity that is to be filled with insulation. The two chemicals react to produce a gas-filled, cellular material that has excellent insulating

IN THE FIELD

Do not disturb or attempt to remove asbestos material. Only persons with the training and necessary equipment should perform this type of work.

properties, and if the procedure is done properly, the material will fill the cavity completely.

Slab or Block

Slab or block is cork, shredded wood, binder, and similar fibrous or granular materials that are formed into flat sheets to insulate large objects.

Spray-on

Spray-on is a fibrous or cellular material mixed with an adhesive and then sprayed onto its final position. This material is used to cover irregular shapes such as structural steel in order to aid in fireproofing.

Insulation Board

Made from fibrous materials and pressed into boards, *insulation board* can be used for sheathing, sound insulation, or roof decking. Some of these materials are flammable and require appropriate precautions, especially if you are working around them with open flames. Other materials will not produce a fire hazard, but the insulation can be damaged by flame, so care must be observed. Determine the characteristics of any insulating materials on the job in advance so you are aware of the necessary precautions.

APPLICATION

Insulation is applied for one or more of the following reasons:

1. To control heat loss (or heat gain) and reduce the operating cost of a system.
2. To retard freezing, i.e., reduce the possibility of freezing during short-term exposure to freezing temperatures.
3. To prevent condensation by keeping moist air away from cold pipes.
4. To control noise by muffling the sounds associated with rushing water.
5. To protect against heat and flame deformation in case of a fire by insulating the surface of steel beam, for example.
6. To prevent injury by shielding temperature extremes from body contact and also forming a padded or resilient surface to protect against impact injury.

The last consideration is important for protection of persons who do not have sensation or feeling in parts of their body.

SAFETY

Some insulation materials encountered in existing work contain asbestos. Asbestos fibers have been deemed harmful to your health under certain circumstances. Consult OSHA, NIOSH, or local safety codes for guidance in working with asbestos. Notify your supervisor at once if you suspect the presence of asbestos. It is strictly prohibited for anyone who is not trained and properly equipped to handle asbestos in any way. In general, use great care and practice good personal hygiene when working with any material that is fibrous or is irritating to your skin. If fibers irritate your hands or forearms, they probably should be kept out of your lungs!

Use long-sleeved shirts, gloves, and face filters when working with any fibrous insulating material. After working with such products, wash yourself carefully and thoroughly. Be sure to launder insulation-fouled clothing separately from other clothing.

HUMIDITY AND CONDENSATION

Moisture in a space is often described as being taken up by the air or supported in the air, but at a given temperature, a certain amount of water vapor can exist in a space regardless of whether any other gases are present or not. Thus, it is proper to say that water vapor is a component of air.

DEFINITIONS

These concepts will help describe the action of water vapor as a component of air. They will also be used with a psychrometric chart to see how they are all related.

Condensation

Condensation is the change of state from gas to a liquid due to a reduction in temperature.

Dew Point

Dew point is the temperature at which the moisture in the air will begin to condense.

Dry-Bulb Temperature

Dry-bulb temperature is the temperature of the air as read on a dry conventional thermometer.

Enthalpy

Enthalpy is the amount of heat content, measured in BTUs per lb, in a substance usually compared to air at 0°F or water at 32°F.

Humidity

Humidity is the amount of water vapor in the air, usually measured by weight in grains. There are 7000 grains to one pound.

Relative Humidity

Relative humidity is a measure of the amount of water in the air compared to the maximum amount of water the air could hold if the temperature and pressure were to remain the same. It is expressed as a percentage.

Saturated Air

Saturated air is air that contains its maximum amount of moisture at the given temperature. Saturated air has a dry-bulb temperature equal to the dew point.

Wet-Bulb Temperature

Wet-bulb temperature is the temperature measured by a thermometer whose bulb is covered by a wetted wick and exposed to a current of rapidly moving air. If the humidity is greater than zero and less than 100%, this temperature is higher than the dew point and cooler than the dry-bulb temperature. If the air being measured were at saturation, then the dry-bulb temperature would equal the dew-point temperature, which would also equal the wet-bulb temperatures.

MEASURING DRY-BULB AND WET-BULB TEMPERATURES

If the dry-bulb and wet-bulb temperatures are known, much information can be deduced about an air mass. These temperatures can be measured with either a *sling psychrometer* (Figure 25-10) or a *digital relative humidity sensor*.

IN THE FIELD

The Department of Energy's website (http://www.eere.energy.gov/buildings/info/components/envelope/insulation.html) is an excellent source of information concerning green construction.

Figure 25-10 Digital relative humidity and temperature meter and a sling psychrometer.

Figure 25-11 Sling psychrometer. The wet bulb has a piece of cloth that must be soaked with water before use.

The *sling psychrometer* (see Figure 25-11) is an assembly of two thermometers on a small frame. One thermometer has a water-soaked wick around its mercury bulb while the other thermometer is left dry. Both are mounted on a frame that has a swivel handle. The assembly is whirled around several times by holding the handle and then the thermometers are read. The plain thermometer gives the dry-bulb temperature while the wick-encased thermometer gives the wet-bulb temperature.

Table 25-3 shows relative humidity as a function of dry-bulb temperature and wet-bulb depression. *Wet bulb depression* is the difference between dry-bulb and wet-bulb readings. The sling psychrometer has two sliding scales that can be used to calculate the relative humidity based on wet-bulb and dry-bulb temperatures.

Example 3

What is the relative humidity of an air mass if the dry-bulb temperature is 76°F and the wet-bulb is 66°F?

The wet-bulb depression is
76°F − 66°F = 10°F

Using Table 25-2, where the vertical column for wet-bulb depression of 10°F crosses the horizontal row for a dry-bulb temperature of 76°F, a relative humidity reading of 59% is obtained. This means that the air in this example only contains 59% of the maximum amount of moisture that it could possibly hold.

Psychrometric Chart

The *psychrometric chart* is a very useful tool to help see how the above-mentioned characteristics are related. If any two characteristics are known, then the others can be determined. At first the chart may appear to be confusing, but Figure 25-12 shows how each set of lines are oriented.

The skeleton psychrometric chart shows what each line represents.

When all of the scales and corresponding reference lines are combined, a completed psychrometric chart is produced (see Figure 25-13).

Example 4

Using the psychrometric chart, determine the relative humidity, the number of grains of moisture per pound of dry air, and the dew point temperature if the dry-bulb temperature is 75°F and the wet-bulb temperature is 61°F.

Find 75°F on the bottom scale and follow the line up vertically. Also, find 61°F on the curved line on the left-hand side of the chart. The 61°F line will slope downward to the right until it crosses the 75°F vertical line. Mark this intersection. The intersection falls somewhere between the 40% and the 50% relative humidity lines. It can be interpolated to be approximately 45%.

By following the horizontal line closest to the intersection to the right, it can be determined that there are approximately 58 grains of moisture per pound of air.

If the horizontal line were followed to the left, it would meet the dew point scale at 52°F.

So what does this mean? Simply, any surface cooler than 52°F will have condensation on it. The chart also points out that if the dry-bulb temperature were to increase, the relative humidity would decrease, but the dew point would stay the same. This tells a technician that if condensation is a problem, either the moisture must be removed or the surfaces below 52°F must be insulated from the moist air. When Figure 25-13 is studied more closely, several other conclusions can be observed:

1. Relative humidity cannot exceed 100%.
2. Relative humidity is higher when dry-bulb temperature and dew point come closer together.
3. The higher the dry-bulb temperature, the more moisture it takes to saturate the air.

Table 25-3 Psychrometric Table Relative Humidity

WB[1] Depression

DB[2] temp °F	1	2	3	4	5	6	7	8	9	10	11	12	13	14	15	16	17	18	19	20	21	22	23	24	25	26	27	28	29	30
32	90	79	69	60	50	41	31	22	13	4																				
36	91	82	73	65	56	48	39	31	23	14	6																			
40	92	84	76	68	61	53	46	38	31	23	16	9	2																	
44	93	85	78	71	64	57	51	44	37	31	24	18	12	5																
48	93	87	80	73	67	60	54	48	42	36	31	25	19	14	8	3														
52	94	88	81	75	69	63	58	52	46	41	36	30	25	20	15	10	6	0												
56	94	88	82	77	71	66	61	55	50	45	40	35	31	26	21	17	12	8	4											
60	94	89	84	78	73	68	63	58	53	49	44	40	35	31	27	22	18	14	6	2										
64	95	90	85	79	75	70	66	61	56	52	48	43	39	35	31	27	23	20	16	12	9									
68	95	90	85	81	76	72	67	63	59	55	51	47	43	39	35	31	28	24	21	17	14									
72	95	91	86	82	78	73	69	65	61	57	53	49	46	42	39	35	32	28	25	22	19									
76	96	91	87	83	78	74	70	67	63	59	55	52	48	45	42	38	35	32	29	26	23									
80	96	91	87	83	79	76	72	68	64	61	57	54	51	47	44	41	38	35	32	29	27	24	21	18	16	13	11	8	6	4
84	96	92	88	84	80	77	73	70	66	63	59	56	53	50	47	44	41	38	35	32	30	27	25	22	20	17	15	12	10	8
88	96	92	88	85	81	78	74	71	67	64	61	58	55	52	49	46	43	41	38	35	33	30	28	25	23	21	18	16	14	12
92	96	92	89	85	82	78	75	72	69	65	62	59	57	54	51	48	45	43	40	38	35	33	30	28	26	24	22	19	17	15
96	96	93	89	86	82	79	76	73	70	67	64	61	58	55	53	50	47	45	42	40	37	35	33	31	29	26	24	22	20	18
100	96	93	90	86	83	80	77	74	71	68	65	62	59	57	54	52	49	47	44	42	40	37	35	33	31	29	27	25	23	21
104	97	93	90	87	84	80	77	74	72	69	66	63	61	58	56	53	51	48	46	44	41	39	37	35	33	31	29	27	25	24
108	97	93	90	87	84	81	78	75	72	70	67	64	62	59	57	54	52	50	47	45	43	41	39	37	35	33	31	29	28	26

[1]Web Bulb [2]Dry Bulb

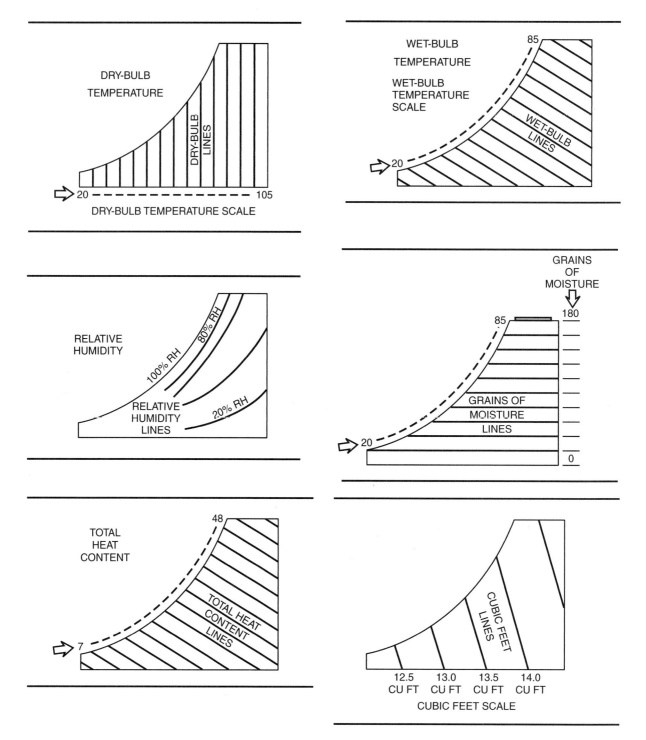

Figure 25-12 Skeleton psychrometric chart showing what each line represents. Courtesy of RSES.

Condensation

Condensation is the change of state of a gas to a liquid accompanied by the release of the heat of vaporization. Water condensation takes place at any location that is cooler than the dew point of the air mass.

Condensation becomes a significant problem in buildings if there are cold areas in walls, windows, ceilings, or other areas where moisture can condense to a liquid. Such water can do damage to the building structure, finish materials, or even form ice that can push apart walls or ceilings.

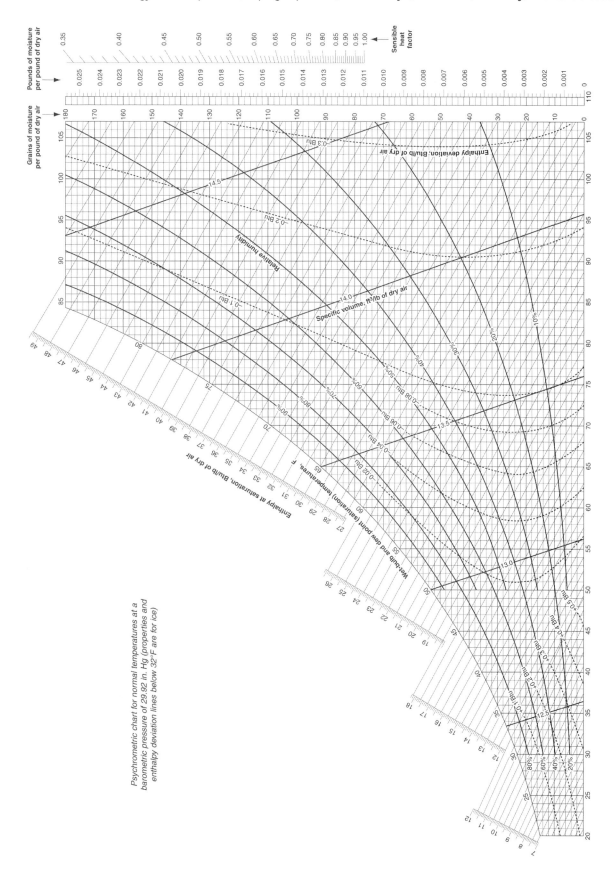

Figure 25-13 Psychrometric chart. Courtesy of RSES.

Condensation may also form on cold water pipes, chilled water pipes, water closet tanks, and storm drain conductors.

Condensation can be controlled by insulating and sealing the cold surfaces. The easiest way to do this for a wall or ceiling is to place insulation within the construction and place a vapor barrier between the insulation and the warm surface. The insulation raises the temperature of the surface of the insulation, which prevents condensation from occurring on that surface; the vapor barrier prevents the moisture from passing into the wall structure where it could condense in cold regions within the wall structure. Similarly, insulation on any cold surface will raise the temperature of the surface exposed to the room and, provided the insulation is made vapor-tight, condensation will be prevented.

Compressed air systems are especially subject to condensate formation inside the system. When air is compressed, the volume of the receiver is packed over and over again with air at ambient pressure until the desired pressure is reached. All the components of the air, including the water vapor, are packed into the receiver. If air at 20% relative humidity is compressed to only 60 psig, the air in the tank is about five times as dense as the air at the compressor inlet. At this density, the air in the tank is saturated ($5 \times 20\% = 100\%$).

Thus, if the initial air is more than 20% relative humidity or if the final pressure is higher than 60 psig (air at 105 psig is about eight times as dense as inlet air), condensation in the tank is sure to result. Likewise, condensation will form at any cool points in the distribution piping. Thus, low-point drains must be provided and in critical cases, refrigerated dryers must be used so the air in the pneumatic lines will not become saturated.

Comfort

Some general remarks about heating and cooling comfort are in order:

1. An *air-conditioning system* operates to control temperature in warm seasons. In most cases, relative humidity conditions are improved as a consequence of the temperature control, but in some situations, humidity must be controlled also.
2. High moisture content in warm air is very uncomfortable as it inhibits perspiration, the evaporative cooling process of the body.
3. In cold weather, the heated air inside a building tends to have a low relative humidity. Moisture may be added to raise the relative humidity level by installing humidifiers.

Low relative humidity produces the following conditions:

1. People suffer from dry skin and respiratory tract dryness.
2. Hygroscopic materials (wood, food, clothing, etc.) dry out rapidly.
3. Static electricity becomes a problem as a source of nuisance shocks and can also cause data loss problems with computers.
4. When space heating is required, evaporation is enhanced, producing discomfort in the human body.

REVIEW QUESTIONS

1 What is the difference between temperature and heat?

2 What are the three types of heat transfer?

3 What is dew point?

4 What is a sling psychrometric used for?

5 What is a BTU?

CHAPTER 26

Water Treatment

LEARNING OBJECTIVES

The student will:

- Describe some common contaminants and the contaminant groups as defined by the Safe Drinking Water Act.
- Describe treatment methods for various water problems.

PUBLIC WATER SUPPLIES AND CONTAMINANTS

Typically water systems are divided into two main categories, public and private. A public water system is one that supplies water to at least 15 service connections or regularly serves an average of at least 25 people each day for at least 60 days per year. A private water system is any system that supplies water for human consumption that does not meet the description of a public water system. Both systems frequently need some form of water conditioning, which is a general term applied to any treatment or process for improving the quality of water delivered to a building. Such processes may accomplish one or more of the following effects:

- Reduce minerals or acids
- Remove suspended solids
- Eliminate bacteria in the water to levels considered desirable, unobjectionable, and safe for human consumption

The Environmental Protection Agency (EPA) has established standards for drinking water for all public water systems. These standards are divided into two types, primary and secondary. *Primary standards* are health-based and enforceable while *secondary standards* deal with the aesthetic quality of water such as color, odor, and taste. Water with contaminants above the secondary standards may not be pleasant to drink, but will not cause any health problems. The EPA allows the local regulatory agencies to administer secondary standards.

Primary standards are usually set up as either Maximum Contaminant Level Goals (MCLGs), Maximum Contaminant Levels (MCLs), or Treatment Technique (TT) Requirements.

Maximum Contaminant Level Goals are set by the EPA and maintain that the level of contaminants, if consumed for a lifetime, result in no known or anticipated adverse effect on the health of a person. This is a goal and not an enforceable standard. Once a maximum goal is set, the EPA works toward setting a Maximum Contaminant Level. The *MCL* is the highest level of a contaminant allowable in the water. The level is set as close to the MCLG as economically feasible using the best available technology, treatment techniques, and other means EPA has available.

When there is no reliable method that is economically and technically feasible to measure a contaminant at particularly low concentrations, a *Treatment Technique* (TT) is set rather than an MCL. A *treatment technique* is an enforceable procedure or level of technological performance which must be followed by public water systems to ensure control of a contaminant.

Table 26-1 shows the six contaminant groups and a brief description, along with a few common examples of that type of contaminant. A complete list can be found at http://www.epa.gov/safewater/dwh/index.html.

Table 26-1   EPA Regulated Contaminants

Group	Description of Group Components	Examples of Contaminants
Microorganisms	Bacteria, viruses, and protozoa, some of which cause disease	Legionella, total coliforms, and turbidity
Inorganic chemicals	Natural occurring and man-made metals and minerals	Arsenic, Antimony, Cadmium, Beryllium, Copper, Lead, Nitrate
Organic chemicals	Man-made chemicals	Benzene, Chlordane, Carbon Tetrachloride
Radionuclides	Naturally occurring radioactive chemicals	Radium 226, Uranium, Alpha particles
Disinfectants	Chemicals added to water to disinfect	Chlorine, Chlorine dioxide
Disinfection by-products	Chemicals formed when a disinfectant such as chlorine is added to water containing organic matter	Bromate, Chlorite, Haloacedic acids

When water falls as rain, it absorbs carbon dioxide, which forms a mild acid. As the water migrates through the earth—depending on the nature of the strata—it dissolves various chemicals and may pick up bacterial growth, dissolved gases, or suspended solids. These conditions can range from nuisances to serious problems resulting in harmful bacteria levels, offensive odors, fixture staining, highly acidic, or highly basic (alkali) chemical characteristics.

Treatment for these problems begins with identifying the condition to be corrected. A laboratory analysis is required for this determination. Table 26-2 lists

Table 26-2 **Water Conditioning Identification and Treatment**

Symptom	Possible Cause	Treatment Procedure
	TURBIDITY	
Muddy or dirty water.	The water contains sediment, organic substances, or bacterial organisms which cloud the water.	Sand filter or some other fine particle filter if the turbidity exceeds 10 parts per million (ppm).
	ACIDITY	
Green stains appear on the fixture under a dripping faucet because the acid in the water is decomposing the copper and brass in the piping.	The water contains an acid that is formed by absorbing carbon dioxide. A pH reading of the water that shows a pH value less than 6 often indicates the need for special treatment.	A soda ash solution may be added to the water through the use of a chemical feed. Chlorine may also be added to the soda ash solution.
		A neutralizing tank may be used to reduce water acidity. The tank is filled with limestone, which neutralizes the acid. This method may produce hardness.
	HARDNESS	
As the hardness increases, the need for more soap for suds is needed.	The water contains minerals in solution. These minerals are usually calcium and magnesium. Treatment is desirable if the hardness exceeds 50 ppm.	A resin softener or a reverse osmosis softener is normally used to correct hardness.
Human skin feels rough after washing with hard water.		
Scale forms on the inside of supply piping. Dishes may appear streaked or dirty after washing. Bathtub ring is very common with hard water.		
	IRON	
Red staining of fixtures or an iron taste is noted.	Dissolved iron.	A softener will remove a small amount of iron, but a manganese filter is usually required for higher concentrations.
	SULFUR ODOR AND/OR FLAVOR	
Water smells and tastes like rotten eggs and is unsuitable for cooking.	The water contains high concentrations of sulfur, bacteria, hydrogen sulfide gas, and possibly iron particles. If testing shows more than 1 ppm of sulfur content, treatment is required.	A chlorinator and filter are used for treatment. The chlorine mixes with the sulfur and forms particles that are then removed by the filter.
Steel, copper, and iron decomposition is found in the pump and the piping.		
If a combination of iron and sulfur is found in the water, the water may develop fine black particles, turning silverware black.		
	OTHER FLAVORS	
Water tastes of salt, metal, or chlorine.	The water contains high mineral content or may contain an excessive amount of chlorine.	Salty tastes are treated with a reverse osmosis softener.
		Metallic tastes may be corrected using the hardness or iron treatment methods.
		Chlorine may be removed using a charcoal or activated carbon filter.

typical problems and probable causes so that an effective solution to the problem conditions can be identified. Once the problem is defined, the appropriate equipment can be selected by consulting with manufacturers of such equipment.

BASIC TREATMENT METHODS

The basic treatment methods for certain conditions include the following:

- Filters are used to remove suspended particles that cause turbidity (cloudiness in the water). These filters include sediment and filter systems, pressure filters, and diatomaceous earth filters.
- Neutralizing tanks or chemical feeders are used to control acidity.
- Softeners (using Zeolite) or reverse osmosis devices are used to minimize hardness.
- Manganese oxidizing filters and charcoal filters are used to remove dissolved iron and/or hydrogen sulfide gases dissolved in the water.
- Chlorination, ultraviolet radiation, and ozone injection are disinfection methods used to kill bacteria and other harmful organisms.

The details of operation of these devices are described below.

FILTERS

Figure 26-1 shows a sediment and filter system. The water intake (from a lake or reservoir) is introduced through an inlet opening equipped with a fine-mesh screen. A float is used so that the water is taken from near the surface of the lake, thus minimizing the turbidity.

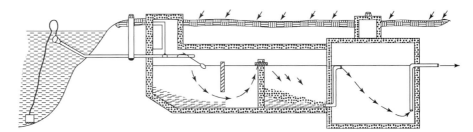

Figure 26-1 Sediment and slow sand filtration of surface water.

The water is drawn into a settling chamber and may receive a chemical treatment to aid in the clarifying process. Certain chemicals will cause finer sediment particles to cluster together to form a combination of particles that is heavier and tends to settle to the bottom more quickly. The water then passes to the next chamber and must go through layers of sand and gravel before exiting out into the final holding tank. The flow through this portion is relatively slow due to the fact that gravity alone is at work, so a large surface area of filter is required. Then the water is taken to a storage tank and pumped from the tank to the distribution system.

Pressure Filter

Pressure filters are used in the portion of the piping from the pump to the system. Figure 26-2 shows a typical arrangement. The filter tank contains layers of sand and gravel as the filtering media. Somewhat higher flow rates are possible with this filter type. These filters remove only relatively large particles. Cleaning either of the

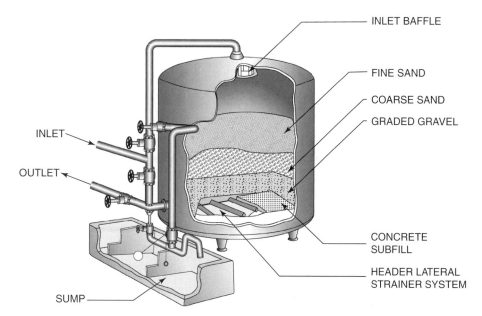

Figure 26-2 Pressure filter.
Courtesy of American Society of Plumbing Engineers.

above filter types involves backwashing the media at a rate sufficient to loosen the foreign material, but not so rapidly as to wash out the filtering media (the sand).

Figure 26-3 illustrates the regular operation and backwash cycles for a pressure filter. Backwashing can be performed manually by reversing certain valves and continuing the backwash for a certain time period or the filter can be equipped with automatic valves and time clocks which initiate and control the backwash cycle.

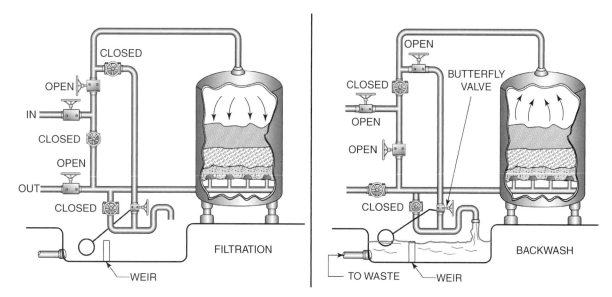

Figure 26-3 Operating and backwash cycles.
Courtesy of American Society of Plumbing Engineers.

Diatomaceous Earth Filter

Diatomaceous earth filters (see Figure 26-4) are composed primarily of fossilized remains of diatoms, a type of hard-shelled algae, which coat a fabric to form the filtering material. Once coated, this assembly makes an effective filter that is lighter

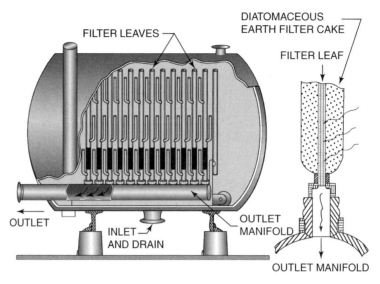

Figure 26-4 Diatomaceous filter.
Courtesy of American Society of Plumbing Engineers.

and less bulky than the other two types previously discussed, but it is more expensive and has a higher pressure drop than the others (see Figure 26-4).

Neutralizing Tank or Chemical Feeder

Acidity is controlled with a neutralizing tank or a chemical feeder. The chemical placed in the *neutralizing tank*—limestone or marble—is dissolved by the acid, which thereby removes the acid component in the water. Eventually, the limestone or marble is depleted and must be replaced.

Figure 26-5 shows a neutralizing tank. The flow through the chemical must be slow enough to permit the chemical reaction to take place, so the tank must be sized carefully for the flow rate required. Backwashing is also required to clean out any foreign material that collects in the chemical structure in the tank.

Chemical feeders are pumps or other devices that supply the desired chemical to the water at a rate that is proportional to system flow. Whatever chemical is used, it should be handled with great care to avoid injury or overtreating the water. Figure 26-6 shows a manually controlled chemical feeder while Figure 26-7 shows an automatic feeder.

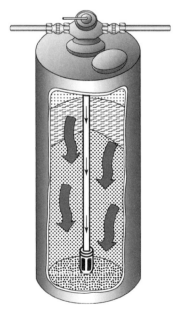

Figure 26-5 Neutralizing tank.

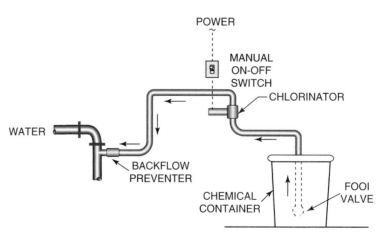

Figure 26-6 Manually controlled chemical feeder.
Courtesy of American Society of Plumbing Engineers.

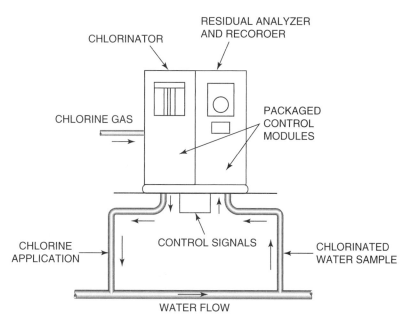

Figure 26-7 **Automatically controlled chemical feeder.**
Courtesy of American Society of Plumbing Engineers.

SOFTENERS

Hardness is a condition wherein water contains scum-forming ions, principally calcium and magnesium. When such water passes through a Zeolite mineral bed, the calcium and magnesium are held by the Zeolite bed and *sodium ions* are released into the water. The sodium ions are much more soluble than the others and by remaining in solution, the water is much more effective with soaps. Thus, personal bathing, clothes washing, and dishwashing are greatly improved. In addition, scale formations do not develop in water heaters, so the life of such equipment is extended.

When the sodium ions in the softener are depleted, regenerating the softener restores them. Regeneration involves backwashing to remove any dirt material caught by filtering action and recharging the Zeolite by introducing a brine solution (sodium chloride). The rich sodium chloride displaces the calcium and magnesium ions, which are washed out along with the excess chloride ions, until all excess brine is removed. Softeners used to be regenerated manually on a weekly to monthly basis, but automatic softeners are now available that can be regenerated daily if needed. Because the operation can vary from manufacturer to manufacturer, it is important to follow the recommended operating procedure very closely.

REVERSE OSMOSIS

For persons who cannot tolerate sodium ions (anyone on a low salt diet), *reverse osmosis* is another method of removing unwanted matter from water. A special filter membrane is used that has the property of permitting small molecules (water) to pass through, but larger molecules (hardness components) cannot. Thus, water will pass to the clean side, but the larger molecules will stay on the inlet or dirty side. These units have very small capacity, so they are useful only in special circumstances. Typically, in systems where the water quality is required to be at a high level, plastic piping is preferred so additional minerals are not picked up from metal piping.

MANGANESE OXIDIZING

To clear up a hydrogen sulfide problem (smell of rotten eggs), an *oxidizing filter* (manganese) or *chlorinator* and filter is required (see Figure 26-8). The offensive material is oxidized and filtered out. Backwashing is required and occasional recharging of the manganese may be needed.

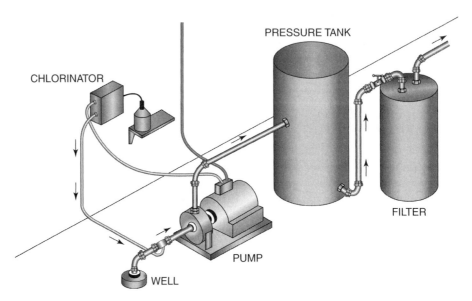

Figure 26-8 Chlorinator and filter.

Manganese filters also remove dissolved iron. Water containing iron is unpleasant to drink and will produce red stains on fixtures.

ACTIVATED CHARCOAL FILTER

If a water supply has an offensive taste, the problem is probably hydrogen sulfide or iron. *Activated charcoal filters* will hold many ingredients and should be included in any process to improve taste.

CHLORINATION, ULTRAVIOLET RADIATION, AND OZONE INJECTION

Adding chlorine to the water supply in order to kill disease-causing organisms is called *chlorination*. The three most common organisms are virus, bacteria, and protozoa, each of which has varying resistance to chlorine. When chlorine is first added to a water supply, the chlorine will react with ferrous iron and hydrogen sulfide. All of the chlorine is used up, so no disinfection has occurred at this stage. As more chlorine is added, chloroorganic compounds are formed from the ammonia and other organic compounds in the system. Any additional chlorine added would be what is referred to as *Free Chlorine Residual*. A level of 0.2 parts per million (PPM) of free chlorine is required to disinfect the entire distribution system. The temperature and the pH level of the water affect the amount of chlorine required. As the temperature decreases and/or the pH level increases, more chlorine will be required. Another important factor is how long the chlorine will be in contact with the water and the distribution system. There must be sufficient time for the chlorine to kill the organisms. The length of time that chlorine is required to be in contact with water before the water reaches the first tap in the distribution system is called *contact time*. Contact time may be anywhere from 15 minutes to 1 hour depending on the concentration of chlorine and the temperature and pH level of the water.

Article 10.9 of the 2006 *National Standard Plumbing Code* states: "The piping shall be disinfected with a water-chlorine solution. During the injection of the disinfecting agent into the piping, each outlet shall be fully opened several times, until the concentration of not less than 50 parts per million chlorine is present at every outlet. The solution shall be allowed to stand in the piping for at least 24 hours."

ULTRAVIOLET (UV) RADIATION

Ultraviolet light has shorter wavelengths than that of visible light. There are three types of ultraviolet rays: *UVA, UVB,* and *UVC.* The UV spectrum is usually considered to be in the 400 nm to 160 nm range. Of that range, only the shortest section, UVC, is effective in killing organisms. The amount of organisms killed is affected by the UV intensity and the amount of exposure time. This dosage is usually expressed in microwatt-seconds per square centimeter (mw-s/cm^2). Typically, either a high intensity or a long exposure time is required. The *kill rate* is normally expressed as a percentage of microorganisms killed; a 99% kill rate is usually the highest rate listed, but an 80% or 90% kill rate is more commonly used. Only microorganisms that have direct exposure to the UV rays will be killed.

Ultraviolet purifiers have some disadvantages. First, there is no residual disinfection, meaning any contamination downstream will not be affected. Secondly, the light source will have to be wiped clean occasionally so that the UV light can reach the water. Some manufacturers have a wiper built into the unit specifically for this reason (see Figure 26-9). Lastly, the unit requires power and the bulbs have a finite life, generally lasting approximately one year.

OZONE

Ozone (O$_3$) is a gas that is very effective in killing organisms—approximately 10 times more powerful than chlorine. Unlike chlorine, ozone does not leave any odor or taste in the water. However, ozone has a short life and must be generated at the point of injection.

Ozone is injected into the water as bubbles. The amount of ozone transferred to the water is a function of the bubble size and the amount of time they are in contact. Any excess gas must be destroyed before it reaches the atmosphere. Methods used to dispose of ozone include granular activated carbon, catalytic ozone destruction, and ultraviolet light.

A problem water supply should be tested to determine the problem before prescribing a solution. Once the problem is definitely known, review Table 26-2 for solutions. For additional information on water quality treatment, you may wish to contact the Water Quality Association (address available at your local library).

SWIMMING POOLS AND SPAS

Swimming pools and spas are a particular example of systems that require water treatment. These pool systems usually include a circulating pump, filter, and a heater to maintain water temperature.

Check instructions for any swimming pool or spa that you install or service.

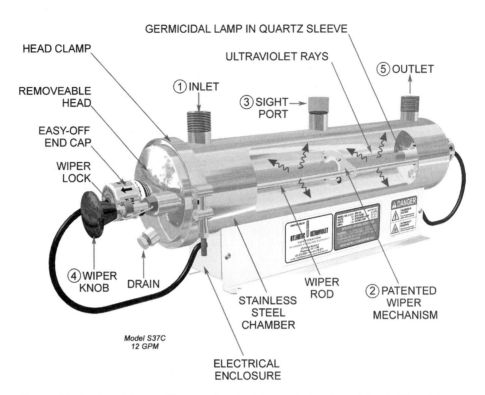

Figure 26–9 Ultraviolet purifier. Reprinted with permission from Atlantic Ultraviolet Corporation, http://www.ultraviolet.com.

REVIEW QUESTIONS

1 What are some common metals and mineral contaminants found in water?

2 What is a common treatment for water that smells like rotten eggs?

3 Softeners (using Zeolite) or reverse osmosis devices are used to minimize _____.

4 When discussing water quality, what is "turbidity"?

5 When water falls as rain, it absorbs _____, which forms a mild acid.

CHAPTER

27

Heat Transfer in Water Heaters—Solar, Stratification, Multiple Heaters, and Recirculation

The student will:

- Describe the different types of water heaters.
- Explain how heat is transferred in water heaters.
- Calculate heat energy.
- Explain different installation methods.

GAS-FIRED WATER HEATERS

The selection of water heaters is usually based on two factors: the type of fuel available and the related energy costs associated with that fuel. Both gas- and oil-fired water heaters convert the potential energy of the fuel into thermal energy of a fire through the process of combustion. Each heater has a combustion chamber near the bottom that allows the gases of combustion to transfer their heat energy to the bottom of the tank (see Figure 27-1 and 27-2).

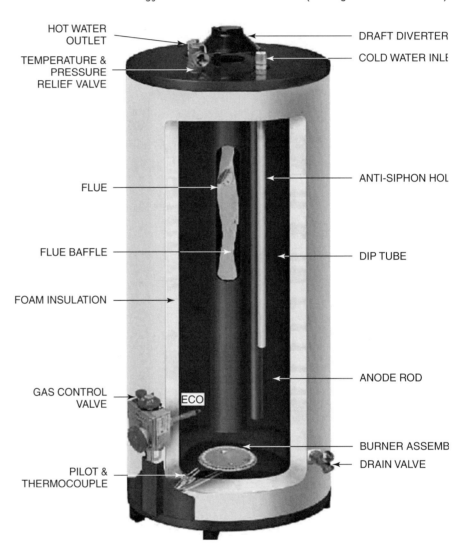

Figure labels:
- HOT WATER OUTLET
- TEMPERATURE & PRESSURE RELIEF VALVE
- FLUE
- FLUE BAFFLE
- FOAM INSULATION
- GAS CONTROL VALVE
- ECO
- PILOT & THERMOCOUPLE
- DRAFT DIVERTER
- COLD WATER INLE
- ANTI-SIPHON HOL
- DIP TUBE
- ANODE ROD
- BURNER ASSEMB
- DRAIN VALVE

Figure 27-1 Gas-fired water heater.
Courtesy of Rheem Manufacturing Company, Water Heater Division.

As these gases of combustion pass through a flue pipe and they continue to pass heat into the tank. The flue pipe contains baffles to slow down the gases, allowing more time for the heat to be transferred. The heat from the flue gases is passed through the tank and into the water by the process of conduction. As the water near the surface of the tank warms, it rises and cooler water is brought in to replace it. Through the process of convection, the contents of the tank are heated and when the set point of the thermostat is reached, the burner will shut off. During the period of time when the burner is not firing, convection currents tend to move the hot water toward the top of the tank and the cooler water toward the bottom. This layering of water can cause the temperature at the top

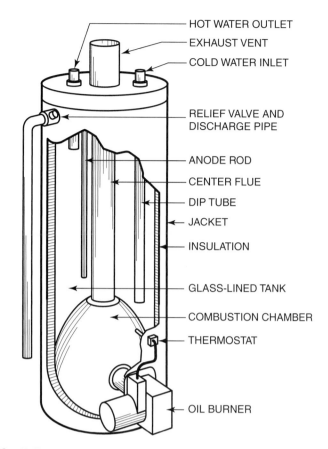

HOT WATER OUTLET

EXHAUST VENT

COLD WATER INLET

RELIEF VALVE AND
DISCHARGE PIPE

ANODE ROD

CENTER FLUE

DIP TUBE

JACKET

INSULATION

GLASS-LINED TANK

COMBUSTION CHAMBER

THERMOSTAT

OIL BURNER

Figure 27-2 Oil-fired water heater.

of the tank to be several degrees hotter than what the thermostat reads near the bottom.

ELECTRIC WATER HEATERS

The biggest difference between electric water heaters and fuel-fired water heaters is that electric water heaters have no need for combustion air, a combustion chamber, or a flue pipe. The heaters are usually inserted directly into the water which means there is no energy loss up the chimney (see Figure 27-3).

Notice how in both the gas-fired and electric water heaters, cold water is delivered through the dip tube to the bottom of the heater and hot water to the building is taken from the top of the tank. With this arrangement, nearly constant hot water temperature will be delivered to the hot water distribution piping until the heater contents have been almost completely replaced by cold water.

There are some low-input models that have the electrical elements wrapped tightly around the outer surface of the tank, but they are not common. Because the heating element is in direct contact, the heat energy is transferred by conduction to the water.

The electric heater industry has developed a clever answer to the problem of long recovery time. A heating element and thermostat are placed high in the tank. This element is switched on when the top of the tank is cool. This upper element only heats the upper 20% of the tank, so about 20% of the tank is heated quickly.

To limit the load on the electric line to one heating element only, the lower element is automatically switched off when the upper element is on. As soon as the upper section is warmed, the upper element is switched off and the lower element

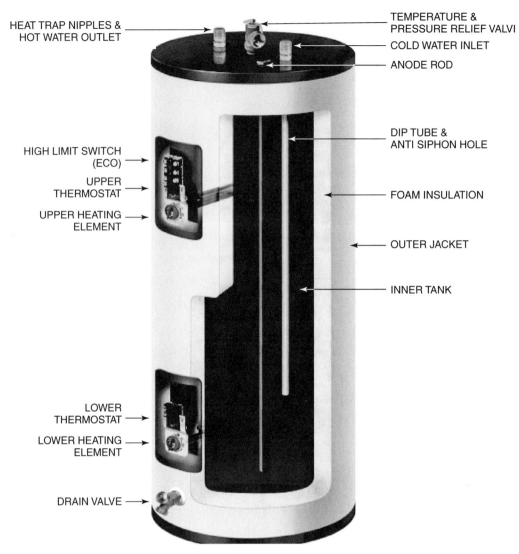

Figure 27-3 Electric water heater.
Courtesy of Rheem Manufacturing Company, Water Heater Division.

is permitted to operate under control of the lower thermostat. Thus, some hot water is available in a short time even if the tank contents become cold with a temporary heavy draw off.

IMPORTANT: An electrically energized heating element that is not submerged in water will overheat and fail in a matter of seconds. As the water surrounding the heating element heats up, convection currents move that water toward the top of the tank and cooler water comes in to take its place. This process continues until the entire contents are heated to the thermostat's set point, at which time the heater is de-energized.

SOLAR WATER HEATERS

In some regions of the world, including parts of the United States, solar radiation is becoming increasingly popular for heating domestic hot water. It is expected that solar installations will continue to be made, so this manner of heating water will become more and more important in the energy balance of the nation and the world. Solar energy could offer an excellent solution wherever fossil fuels or electric energy costs are high or where an installation is remote from conventional energy sources. Solar water heaters are typically divided into two different types: *direct and indirect*. Indirect types use a separate loop, usually filled with a mixture of water and antifreeze, to transfer solar heat energy from the collector to a heat exchanger. The heat exchanger then transfers the heat energy from the antifreeze mixture into the water storage tank. Because one loop has antifreeze and the other has potable water, a double wall heat exchanger must be used. Direct-type water heater systems circulate the potable water through the solar collector so the solar energy is applied directly to the potable water. There is no need for a heat exchanger. Direct-type water heater systems must have some method of draining the solar collector or freezing temperatures may cause damage to the system. The solar collector is required to be rated for full system pressure.

Figure 27-4 shows a diagram of a typical indirect solar water heating system. Cold water enters the bottom of the insulated storage tank through a shutoff valve and

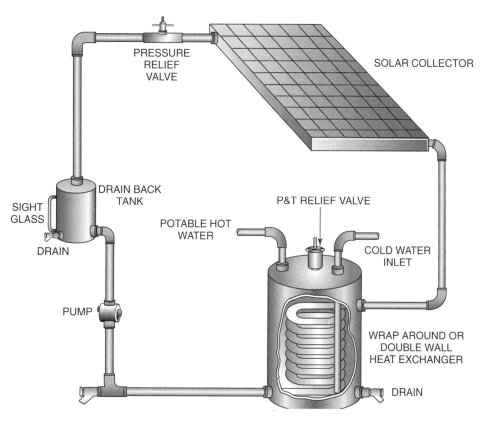

Figure 27-4 Drainback active indirect solar water heating system.

dip tube. Hot water leaves from the top of the tank through a *tempering valve* that maintains the supply to the house in a general temperature range. Remember that a tempering valve is not designed for scald protection.

A mixing valve will be required at the hot water discharge line to limit the temperature to a maximum of 140°F.

The tank is equipped with the solar coil heat exchanger, sensing controls, a combination temperature and pressure relief valve, and an electric element for backup. The solar system consists of collectors mounted to receive the rays of the sun, interconnecting piping, circulator pump that causes the fluid to move through the system, the heat transfer fluid, and the accessories (valves, compression tank, and relief valve) to ensure satisfactory operation.

In the *drainback system,* the pump only runs when the differential controller senses that the solar collector is hotter than the tank temperature. When the pump shuts off, water is drained from the solar collector back to the drainback tank by gravity. The tank should be large enough so it can hold the entire contents of the collector when the pump is off, but still have enough in the tank so the pump has a constant supply. Since the tank acts as an expansion tank and air purge control, there is no need for a vacuum breaker or air vent.

Figure 27-5 shows a diagram of a typical differential-controlled active direct system. It uses a differential temperature controller to sense the temperature difference between the collector and the tank. When there is anywhere from 8–20°F difference, the pump will turn on and water will circulate from the bottom of the tank to the collector and then back into the tank. The check valve in this line prevents hot water from rising out of the tank and into a cold collector if the pump is not running. Whenever a freeze condition occurs, appropriate measures should be taken to drain the solar collector.

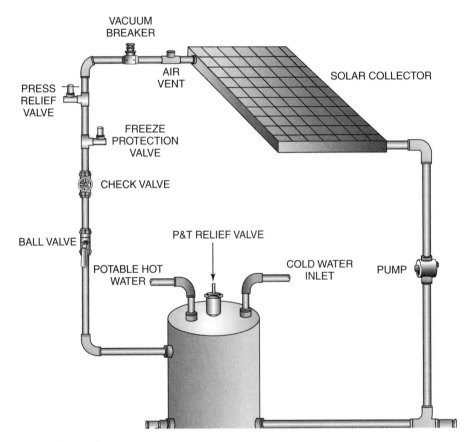

Figure 27-5 Active direct solar water heating system.

Energy Losses

Water heater efficiency is now rated by what is known as an *energy factor*. The energy factor is a percentage of how much of the purchased fuel energy is converted into useable hot water. If you compared energy factors for electrical, gas, and oil heaters with the same size tank, you may find numbers like .92, .60, and .55 respectively. As the size of the tank increases, however, the energy factor decreases for all three. At first glance, the electrical heater would appear to be best, but electric water heaters are typically more expensive to operate. To get a better picture, let's look at some of the energy losses and how they are being minimized.

For natural gas, the losses include 9% for the heat of vaporization of the water vapor that goes up the chimney, 5% to heat the nitrogen in the air to about 450°F,

3% to 5% for the heat for the excess air delivered to the combustion space, and 5% to 10% in jacket losses from the heater enclosure. Similar losses take place for oil firing. Other losses occur when the heater is not firing: room air drawn up the warm chimney, gas burned in the standing pilot flame, and jacket losses into the room from the stored hot water.

When these losses are examined, it is tempting to consider how to recover most of them. The higher-efficiency appliances increase efficiency by working on those items. The problems and costs, however, become formidable:

1. To recover the heat of vaporization, you must condense the water vapor. Such condensate is usually corrosive and requires a stainless steel final heat exchanger. Also, the condensed liquid water must be drained to some place of disposal.
2. The same extended heat recovery system for water vapor will also recover some of the heat for the going-along-for-the-ride nitrogen and excess air.
3. To reduce the heat lost to excess air, we must reduce the excess air. This effort requires a power burner at much increased initial cost and increased complexity of the appliance.
4. Jacket losses are reduced by increasing the amount or quality of insulation, which will increase the cost of the unit.
5. Standing-pilot flame losses are eliminated by using systems that ignite the pilot burner only at the beginning of each heating cycle. This is an option that adds cost and complexity.
6. Off-cycle chimney loss is reduced by using automatic flue dampers. This option also adds cost and complexity.

Another problem that is becoming more common is the lack of combustion air. Today's houses and buildings are much tighter than in the past. This means it is harder for the heater to draw enough combustion air, causing problems like soot or flame roll out. The topic of combustion air for gas-fired appliances is covered in Chapter 15 in this book.

One way manufacturers are dealing with this problem is with *sealed combustion* or *power vented systems* (see Figure 27-6). These systems are increasingly being used because they are highly efficient. Sealed combustion means that outside air is brought directly into the water heater and exhaust gases are vented directly outside. The combustion air is totally separated from the house air. Power vented equipment can use house air for combustion, but flue gases are vented to the outside with the aid of a fan.

For *electric water heaters* there are no flue losses or other fuel-burning problems, but because the energy input to a water heater is considerably less than a fuel-fired water heater, the storage tank must be larger. Let's look at a standard 50-gallon gas-fired water heater and a standard 50-gallon electric water heater.

Example 1

How much energy is required to heat the contents of a 50-gallon water heater, from 60°F to 120°F? (One BTU is required to raise the temperature of one pound of water 1°F.)

Weight of water: $\left(8.33\dfrac{pounds}{gallon}\right) \times (50\ gallons) = 416.5\ pounds$

Change in Temperature: $120°F - 60°F = 60°F$

Energy Required: $416.5 \times 60 = 24,990\ BTU$

A normal firing rate for this size gas water heater would be approximately 40,000 BTU/hr. Even if the heater was only 75% efficient, the output would be around 30,000 BTU/hr. Now, let's compare that to a normal 4500 watt electric water heater.

First, 4,500 watts needs to be converted into BTUs (1 watt is equal to 3.412 BTUs). So, 4,500 watts is approximately $4,500 \times 3.412 = 15,354$ BTUs, which is approximately

Figure 27-6 Power vented gas water heater.
Courtesy of Rheem Manufacturing Company, Water Heater Division.

Figure 27-7 Tankless water heaters take up less space and have less standby losses. Courtesy of Rheem Manufacturing Company, Water Heater Division.

half of the firing rate of a gas heater. This explains why a gas-fired water heater will have a much higher recovery rate when compared to that of an electric water heater. This also explains why the electric water heater will have a larger tank compared to the tank selected for the gas-fired water heater in most applications.

TANKLESS WATER HEATERS

Realizing that the larger the tank, the greater the standby losses is what brought about the idea of a *tankless water heater*. Tankless water heaters, which are sometimes called *on demand water heaters*, have a staged firing rate that is capable of a much higher firing rate than a standard gas heater. For example, one manufacturer has a model that will modulate its firing rate from 19,000 BTUs to 199,000 BTUs depending on the demand. The advantage is that they require minimal space and have no tank to maintain at an elevated temperature.

Installation Methods

Heaters must be installed at least 18" above the finished floor when located where combustible materials are present, such as garages. Also, water heaters must be protected from being struck by vehicles and other motorized traffic.

Earlier, the effects of convective heat transfer within the tank of a water heater were discussed. We will now consider some implications of the convective heating process.

First, the hottest water will be at the top of the tank with the temperature decreasing at lower and lower levels in the tank.

Second, in a well-designed heater, there should be very little mixing of these levels.

Third, if the heat input to the water heats a small quantity of water to a high temperature, the hottest water (referred to above) could be well above the average tank temperature or significantly above the thermostat setting.

If the temperature spread is great enough so that the first water out of the heater would be too hot to be safe, we refer to the heater as being *stratified* or we say that the heater is *stacking*.

Even though these are vague terms, the problem is real enough. There are a couple of conditions that give rise to stacking. A partially scaled heater wherein a smaller amount of water is heated to an abnormally high temperature will give rise to stacking. Or, stacking may occur when a heater is operating at repeated light loads on a minimum amount of new cold water input or where the thermostat is activated by a combination of natural tank cooling and cycles of heating. The heavy insulation of modern heaters can actually exacerbate the stacking problem because there is so little water cooling through the tank wall. The infrequent burner operation of a lightly loaded heater keeps the top zone temperature working higher and higher.

One concept to counter stacking involves the use of special *dip tubes* for the cold inlet. Referring back to Figure 27-1, the newer dip tubes not only break up sediment, but also help eliminate stacking. Another design can be seen in Figure 27-8.

The dip tube is a plastic over metal assembly with many holes drilled in the two overlaying materials. The inlet cold water enters the tank at the level where the holes line up in the double dip tube assembly. The coefficient of expansion of the two materials is such that the incoming water is discharged into the hottest strata in the tank and thereby reduces the hottest temperature.

At typical flow rates in a normal heater, there is enough agitation of the strata to counteract the stacking tendency. In a heater with a return circulator, the contents of the tank are averaged out and stacking cannot occur (see Figure 27-9).

Figure 27-8 Engineered dip tube.

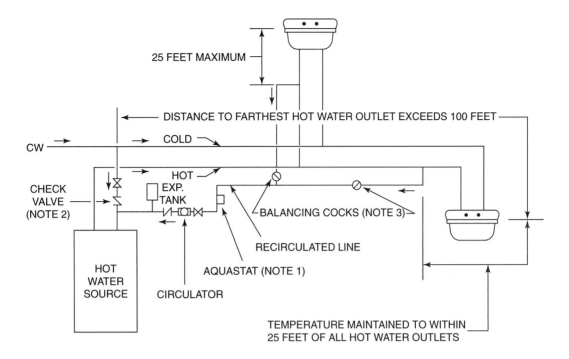

NOTES:
1. THE AQUASTAT OR TIME/TEMPERATURE CONTROLLER STARTS THE CIRCULATOR WHEN THE RECIRCULATED LINE COOL THE CIRCULATOR STOPS WHEN THE LINE TEMPERATURE RISES.
2. THE CHECK VALVE IN THE COLD WATER SUPPLY TO THE HOT WATER SOURCE KEEPS RECIRCULATED HOT WATER FROM FLOWING INTO THE COLD WATER MAIN DUE TO PRESSURE FLUCTUATIONS IN THE DISTRIBUTION SYSTEM.
3. IF THE RECIRCULATION PIPING HAS MORE THAN ONE BRANCH, BALANCING COCKS IN EACH BRANCH PERMIT THE FLOWS TO BE ADJUSTED.
4. A MEANS OF MEASURING THE RETURN TEMPERATURE SHOULD BE PROVIDED.

Figure 27-9 Water heater using a recirculation pump to maintain hot water within 25 feet of the farthest outlet. Courtesy of Plumbing-Heating-Cooling Contractors—National Association.

RECIRCULATION

When the water heater is located a long distance away from the point-of-use, a considerable amount of water can be wasted waiting for the water to get hot. To avoid this, the *National Standard Plumbing Code* requires *recirculating* loops be used on hot water supply systems where developed length of hot water piping from the source of the hot water supply to the farthest fixture supplied exceeds 100 feet. Typically, the circulation loop will allow water to be circulated from the water heater to within 25 feet (or closer) of the point-of-use. A thermostat mounted on the return line of the circulation loop will operate a circulation pump to maintain the desired temperature. Circulating hot water in the lines eliminates the need to run water until the water at the point of usage gets hot. This can save a considerable amount of water. These lines must be well insulated and sized based on the size of the circulating pump and the necessary length of piping.

RELIEF VALVE LOCATION

The *temperature and pressure relief valve* (*T & P*) is required to be located within the top 6" of the heater where it can respond to the hottest water. It should be noted that the sensing tube on the T & P valve must be in contact with the water in order for it to sense the water temperature. The relief valve must connect directly to the

IN THE FIELD

Local energy codes and standards should be consulted prior to installation of recirculation lines. Some standards require that a pump and timer be installed on the line to minimize energy loss.

water heater with no valves between the water heater and the relief valve. Most residential relief valves screw into the water heater and have a threaded outlet. The discharge for the relief valve must be to a location that can do no physical harm to anyone or the building, but it should be located so that it is noticed. Figure 27-10, from the 2006 *National Standard Plumbing Code,* shows the required piping arrangements.

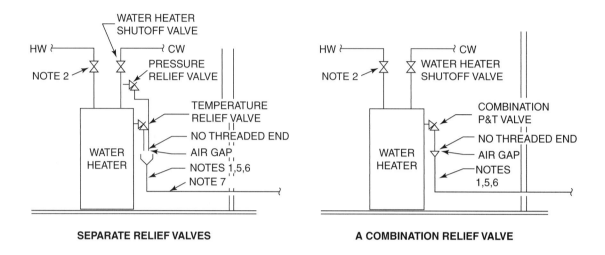

SEPARATE RELIEF VALVES   **A COMBINATION RELIEF VALVE**

NOTES:
1. RELIEF VALVES CAN DISCHARGE TO THE FLOOR WHERE PERMITTED BY THE AUTHORITY HAVING JURISDICTION.
2. A SHUTOFF VALVE IN THE HOT WATER SUPPLY IS PERMITTED.
3. SHUTOFF VALVES MUST NOT ISOLATE RELIEF VALVES FROM THE TANK THAT THEY ARE PROTECTING.
4. THREADS ARE PROHIBITED ON THE ENDS OF RELIEF VALVE DISCHARGE PIPES TO PREVENT THE INSTALLATION OF A PIPE CAP IF THE VALVE BEGINS TO LEAK.
5. DRAINS OR INDIRECT WASTE PIPES THAT RECEIVE THE DISCHARGE FROM RELIEF VALVES MUST BE SIZED ACCORDING TO TABLE 10.16.6, BASED ON THE SIZE OF THE RELIEF VALVE DISCHARGE PIPE.
6. THE INLET TO INDIRECT WASTE PIPING MUST BE ELEVATED TO ESTABLISH SUFFICIENT STATIC HEAD ABOVE THE HORIZONTAL PORTION OF THE DRAIN TO PREVENT SPILLAGE.
7. WHERE TWO INDIRECT WASTE PIPES FROM RELIEF VALVES ARE JOINED, THE CROSS-SECTIONAL AREA OF THE COMMON DRAIN MUST BE EQUAL TO OR LARGER THAN THE SUM OF THE AREAS OF THE INDIVIDUAL WASTE PIPES. REFER TO SECTION 9.3.6.B.

Figure 27-10 Water heater relief valves and discharge piping.
Courtesy of Plumbing-Heating-Cooling Contractors—National Association.

MULTIPLE HEATER INSTALLATIONS

Multiple heater installations are a common solution to the need for high recovery rates. Such multiple installations accomplish the following goals:

1. Permit the use of less costly, high-production quantity heater appliances.
2. Allow for the use of compact equipment that will pass through ordinary doorways with ordinary installation devices such as hand trucks, etc.
3. Permit staged firing during operation.
4. Provide redundancy in case of failure.

In hospitals, nursing homes, and restaurants there is a constant need for hot water. If one of the water heaters should fail, the building would still have a source of hot water through the remaining units even though it will be at a reduced capacity.

Multiple unit installations should use identical model heaters with the same volume, piping connections, and firing rates. The piping details must be carefully designed so that equal (or nearly equal) flow will pass through each individual heater.

Figure 27-11 shows a typical multiple water heater installation. Note that all paths are similar—pipe segments lengths are exactly equal and fitting patterns are the same.

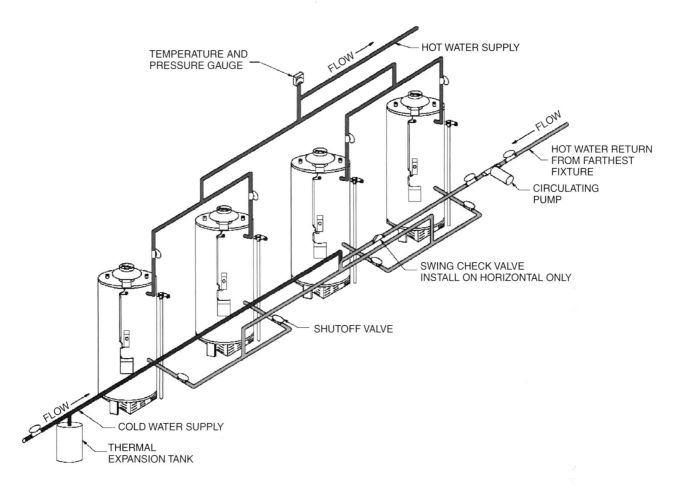

Figure 27-11 **Four water heaters connected in parallel.**
Courtesy of Rheem Manufacturing Company, Water Heater Division.

Each heater is provided with valves and unions located so any heater can be shut off and replaced without shutting down the entire installation. Notice how in Figure 27-9, the recirculating line ties into the incoming cold water supply and the check valve prevents a backward flow into the cold water line.

OPERATING TEMPERATURE OF WATER HEATER

One final note on water heaters relates to the selection of temperature output. Many utilities are encouraging the consumer to dial down water heater settings to 120°F. This practice achieves the desirable effect of reducing standby losses, the heat lost from the tank when hot water is not being used. It may also reduce the potential for scalding. Water heaters are now set at lower temperatures at the factory prior to shipment. The lowering of the water temperature to less than 122°F can allow for the growth of Legionella bacteria. Legionella is the cause of the disease commonly known *Legionnaire's Disease*.

When water temperature is reduced, however, consumers often find they do not have enough hot water. When a lower heater operation temperature is desired, a water heater with a larger volume should be selected.

IN THE FIELD

If scald protection is desired with higher water storage temperatures, pressure-balance or thermostatic valves must be installed for user protection.

REVIEW QUESTIONS

1 Why is it dangerous to set water heaters to temperatures below 122°F?

2 What is the purpose of a dip tube?

3 What is the purpose of a circulation loop in a hot water system?

4 Why is it that both top and bottom heating elements don't come on in some electric water heaters?

5 What does "stacking" refer to when discussing water heaters?

Basic Electricity, Electric Current, and Electric Motors

The student will:

- ⊗ Define basic electrical terms.
- ⊗ Describe various power supply voltages and arrangements for single phase and three phase.
- ⊗ Describe different types of motors.

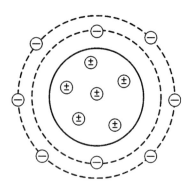

Figure 28-1 Oxygen atom with orbitting electrons.

BASIC ELECTRICITY

n order to understand electricity, you must first understand that materials are made up of atoms. *Atoms* are the smallest particles that the material can be reduced to and still remain the same material. Figure 28-1 shows an atom of oxygen.

The centermost structure is the *nucleus*. There are eight electrons in orbit around the nucleus. An *electron* is a negatively charged particle that rotates around the nucleus. If an atom has the same number of *protons* (positively charged particles) as it does electrons, then it will have a neutral charge. In materials that are good conductors of electricity, the forces holding the outermost electrons are weak, so an external force could easily knock the electron free. In a material like copper, which is a good conductor, the electron that gets knocked loose from one copper molecule will be passed on to the next copper molecule. As the electron is passed from one molecule to another, an electric current is produced. *Electric current* is the rate of flow of electrons, usually measured in amperes. If there is a big external force that dislodges several electrons and the material is a good conductor, the electron flow will be high and show up as high current. If the material is not a good conductor such as rubber or glass, the molecules will hold onto their electrons and there will be no current or electron flow. A device known as an *ammeter* is used to measure the amount of current passing through a wire. Figure 28-2 shows an ammeter placed in a circuit to measure the current flow.

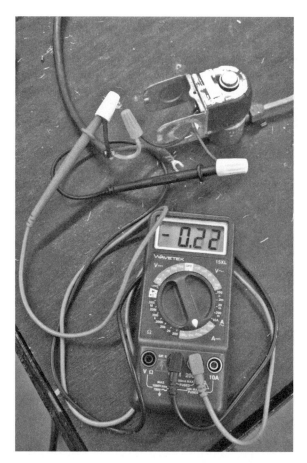

Figure 28-2 Ammeter placed in series with the circuit. Always check to make sure the fuse rating of your meter will not be exceeded.

If electrons are removed from atoms (or molecules), the body is said to be *positively charged*. If extra electrons are added, the body is *negatively charged*. *Static electricity* is when there is a buildup of charges, but the electrons are not flowing. This would be similar to having water being pumped out of a well and into an expansion tank. The water moving through the pipes would be like the electrons flowing through the wires in a circuit. Once the pressure setting is reached, the pump will turn off and the expansion tank will have a positive pressure while the water in the well will be at atmospheric pressure.

Voltage

When one material has more electrons than another, there is said to be a potential difference between the two materials. This potential difference is commonly called *voltage*. Voltage is always the reference of one point compared to another. This would be similar to taking pressure measurements on a pump that is pumping water through a closed system. The discharge line would have a higher pressure than the suction line. Another example would be when you say a car battery has 12 volts, you really mean there are 12 volts from one terminal to the other terminal, not from one terminal to the middle of the case.

Voltage can be generated in a number of ways: a chemical reaction between the acid and lead plates in a car battery; from heating two dissimilar metals in a thermocouple; from pressure in certain crystals; or from magnetism in generators. If the voltage becomes high enough, it can actually jump across an insulator. We have all seen a spark jump across an air gap, such as with a spark plug, but it can also go through plastic or rubber as well. Because this voltage or potential difference can actually cause the electrons to start flowing, it is also commonly called an *electromotive force* or *emf*. The symbol for voltage can be V, E, or emf.

Voltage can be either direct (DC) or alternating (AC). Voltage from a battery is an example of DC where one post of the battery is positive and the other is negative. This concept was shown earlier when the section on cathode protection was discussed in Chapter 14. When the pipe was positively charged, electrons would only flow from the *anode* (negatively charged) to the pipe and thereby protect the pipe. DC voltage is also used as a power supply to emergency lighting when the power is interrupted. The control circuits draw power from the batteries and turn on the emergency lights.

AC voltage is generated from a generator and will change polarity (+ or −) 60 times in one second or at 60 *hertz (Hz)*. Because the AC voltage is constantly changing back and forth, it creates a magnetic field and can be transformed into higher or lower voltages. The concept of transformers will be discussed in Chapter 31. AC voltage is how power is supplied to homes and public buildings, running most of the appliances that we need.

Voltage is measured through a device known as a *voltmeter*. Voltmeters can be purchased in a variety of different styles and voltage ratings. Always make sure that the voltmeter being used is rated for the voltage level being tested or higher. It is also important to make sure that the voltmeter is set to the proper voltage type (AC or DC). This is very important because voltmeters set to read a DC voltage would not read an AC voltage and vise versa. Always know which type of voltage you are working on and only work on energized equipment when you have to check for voltage readings.

Resistance

Earlier it was said that if an electron could not be freed from the molecule easily, then it was a good *insulator*. This insulator, measured in units called *ohms*, would have a high resistance to the flow of electrons. The symbol for ohms is the Greek letter omega (Ω). A good insulator has a high ohm value while a good conductor has a low ohm value. An example of a high resistance would be using a small wire on a very long run. This would be similar to trying to use a garden hose to water

Figure 28-3 Ohmmeter being used to test for continuity on a water heater element.

plants a mile away. If any water came out, it would be moving slowly and have almost no pressure at all. Most equipment manufacturers will indicate the smallest wire size to be used. This will be discussed more in Chapter 31. Figure 28-3 shows an ohmmeter being used to test for continuity on a water heater element. *Continuity* means a complete path from one terminal to the next.

Power

Power is the rate of doing work or using energy. When dealing with AC, power will normally be referred to in terms of *VA* (voltage times amperage). When dealing with DC, power will be referred to in terms of watts (also voltage times amperage). A good way to think about power is to look at a 60 watt light bulb and a 150 watt light bulb. Both produce light using the same 120 volts, but the 150 will produce more light and use a higher current draw. Another way of looking at it is how the horsepower of a pump is calculated. The horsepower of a pump is calculated by multiplying the pressure (which is similar to voltage) times the flow rate (which is similar to current). Both a big pump and a little pump can move water from one place to the next, but the big pump can do it faster, at a higher pressure, or both. Because both are rates of doing work they can be converted from one to another. For example, there are 746 watts in one horsepower. The concept of power will be discussed more in Chapter 31 when sizing transformers and later again in this chapter when discussing motors.

AC VOLTAGES FOR DIFFERENT APPLICATIONS

Depending on the application, there are varying levels of power available. In residential settings, the incoming power will be two hot or energized conductors and a neutral (see Figure 28-4).

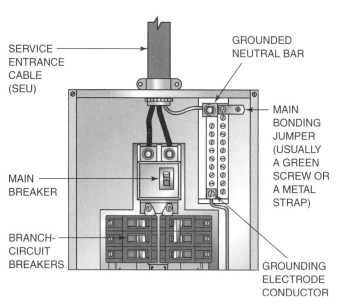

Figure 28-4 Circuit breaker panel for a residential building. Two energized conductors and a neutral are brought into the panel.

This is known as *single-phase power*. Figure 28-5 shows a voltmeter being used to check the incoming power to a residential circuit breaker panel. The two energized conductors, L1 and L2, are distributed so that the L1 goes to half of the circuit breakers and L2 goes to the other half. When an appliance needs the higher voltage (240 V), two consecutive breakers need to be used and both should be physically

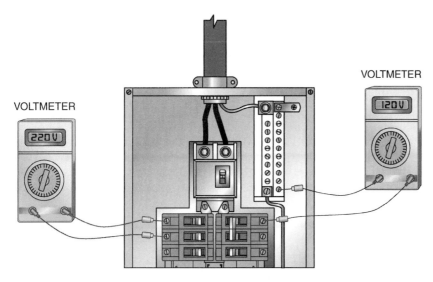

Figure 28-5 Using a voltmeter to check the voltage being supplied through the circuit breakers.

connected so that if one is tripped, both will turn off. If only low voltage is needed, then any one breaker can be used along with the neutral.

In industrial and commercial settings, larger amounts of power and higher voltages are more common. Typically, there will be three energized conductors and a neutral brought into the building. This is commonly referred to as *three-phase power*. Figure 28-6 shows a three-phase circuit breaker panel with some examples of how loads may be connected depending on their power requirements.

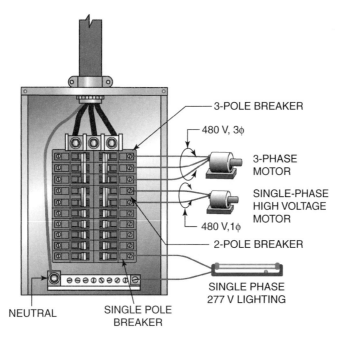

Figure 28-6 Three-phase panel with various loads. Notice 120 V and 240 V is not available in this panel.

Where even larger power requirements are expected, such as in industrial settings, three-phase power may be supplied in what is known as a *four-wire system*. The four-wire system will typically have 480 volts located between energized conductors while 277 volts will exist between any one of the three energized conductors and the neutral. It is important to know that 120 volts or 240 volts cannot be obtained from this arrangement.

ELECTRIC MOTORS

An *electric motor* is a rotating machine that transforms electrical energy into mechanical energy. There are many different types of these motors. Figure 28-7 identifies many parts of a single-phase induction motor. The model illustrated is known as *totally enclosed, fan cooled (TEFC)*. This means that all of the electrical components are enclosed within the housing and there is no free air exchange between the inside of the case and the environment. The external fan cools the housing which in turn helps dissipate the internal heat being generated. The dirtier the motor becomes, the less capable it is to dissipate this heat and overheating may occur. This type of motor is used in an environment that could cause damage to the windings or other internal parts of the motor.

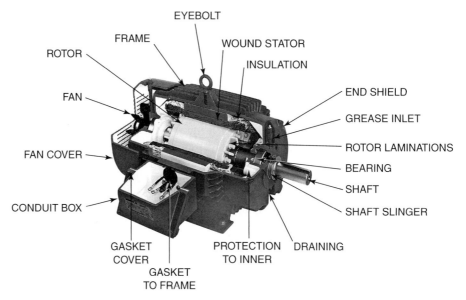

Figure 28-7 Diagram of single-phase electric motor. Copyright © 2007, Baldor Electric. All rights reserved. Used with permission.

- The *armature* is the rotating part that contains the power windings and commutator.
- The *stator* is the stationary part of the motor.
- The *rotor* is the part that turns, but does not contain windings.
- The *capacitor* is a device that aids the starting effort.
- The *starting switch* disconnects the capacitor and/or starts windings when the motor reaches 75% of its rated speed.

The *capacity* of a motor is its ability to do work, which is, its power rating. Some motors will have their power ratings listed on the nameplates in kilowatts (1 kilowatt = 1000 watts) or in horsepower. As discussed earlier, one horsepower is equivalent to 746 watts. If two motors had the same voltage rating, but different horsepower ratings, the larger of the two motors would have a higher current rating than the smaller horsepower motor.

Horsepower can also be expressed in mechanical terms such as torque and rpm, or ft-lbs/min. One horsepower equals 33,000 ft-lbs/min. Think of a horsepower as a large amount of work done every minute. Example 1 shows how the work of one man compares to that of a motor in terms of horsepower.

Example 1

What horsepower is developed if a 220-pound man runs up a flight of stairs for a vertical distance of 10' in 6 seconds?

$$hp = (220 \text{ } lbs \times 10 \text{ } ft) \times \left(\frac{60 \text{ sec}}{\text{min}} \right) \div 6 \text{ sec}$$

$$hp = \left[22{,}000 \frac{ft - lbs}{\text{min}} \right] \div \left[\frac{33{,}000 \frac{ft - lbs}{\text{min}}}{hp} \right]$$

$$hp = 2/3 \text{ } hp$$

Horsepower can also be expressed in hydraulic terms such as pressure and gallons per minute. Example 2 uses a common formula to determine the amount of horsepower generated by a pump.

Example 2

What horsepower is developed if a pump is operating at 50 psi and a flow rate of 10 gallons per minute?

$$HP = \frac{psi \times gpm}{1714}$$

$$HP = \frac{50 \text{ psi} \times 10 \text{ gpm}}{1714}$$

$$HP = \frac{500}{1714} = .29 \text{ } hp$$

Since this is the amount of power being generated, the motor running the pump would have to be slightly larger to account for inefficiencies such as friction, noise, heat, and vibration.

Another important characteristic of motors is the amount of torque they can generate both at start-up and while running. *Torque* is an indication of turning effort and is measured in lbs-feet. For a given horsepower and speed, *full load* torque is determined. However, various motor types have considerable variation in *starting* torque capability.

Horsepower is equal to the product of torque (T) times rpm, divided by a constant.

$$hp = \frac{(T \times n)}{5252}$$

hp = horsepower
T = torque in lbs-ft
N = revolutions per minute
5252 = constant for conversion of units

This means that if you had two motors of the same horsepower rating and one had 2 times the speed of the other the faster turning motor would have half the torque of the slower one. An example of this concept could be where a pump is expected to lift water to the top of a building. If a high rpm motor were stopped in the middle of the process and then expected to restart, it could have problems overcoming all of the weight of the water in the discharge pipe. A lower rpm motor with the same horsepower rating would have the necessary torque needed to restart.

Table 28-1 shows the usual range of horsepowers for standard voltages. Residential equipment is usually a single-phase fractional horsepower motor, but most commercial and industrial motors (larger than fractional horsepower) are three-phase. Frequency is usually 60 Hz.

Table 28-1   Usual Horsepower Values for Specific Voltages

Phase	Motor Voltage (AC)	Recommended Minimum/Maximum Horsepower
Single-phase	110–115	Up to 1
	220–230	Up to 5
Three-phase	110	Up to 5
	220	Up to 200
	440	1 to 1,000

*ASHRAE data book.

MOTOR TYPES

Three-Phase Induction Type

Three-phase motors typically are more efficient and simple in terms of construction than comparable single-phase or DC motors. Unless the driven load has unusual problems, a squirrel cage induction motor should be selected. The motor gets its name because the rotor consists of two metal rings that hold metal bars in a cylindrical shape resembling a cage. Between the metal bars are sheets of laminated metal; this laminated metal gets a current induced into it by the stator windings. These motors are available in a variety of types, but most have a high-starting torque, high-efficiency, rugged construction, and a low cost when compared to either the single-phase or DC equivalents.

The stator windings are connected to the incoming power and a current is induced in the rotor. These two currents oppose each other and produce the torque of the motor. The coils in the motor are arranged in such a manner that a rotating magnetic field is produced in the stator. This magnetic field will rotate around the stator at a constant speed known as *synchronous speed*. The synchronous speed of a motor is determined by the number of poles in a motor and the frequency of the applied power. The number of poles in the motor is determined by how the manufacturer arranges the windings in the stator. The frequency of the applied power is determined by the national grid and is a constant 60 hertz. The formula to calculate synchronous speed is:

$$S = \frac{(120 \times F)}{P}$$

S = synchronous speed in rpm

F = the frequency of the applied AC power in hertz

120 = a constant value

P = the number of stator poles

Because our power suppliers hold frequency constant and the number of poles is constant after a motor was made, changing the rotating speed of a three-phase motor was difficult and expensive. The two most popular motors that had the ability to have their speed changed were the synchronous motor and the wound rotor induction motor. Since the advent of the variable frequency, or variable speed drives, both of these motors are rarely seen. The variable speed drive is a solid state controller that can change the frequency and voltage of the output power, and according to the above equation, the rotating speed of the motor changes with the frequency. Make sure to use a three-phase motor that is rated for use with inverters; otherwise, its life could be dramatically reduced. Figure 28-8 shows both an inverter rated motor and a variable speed drive.

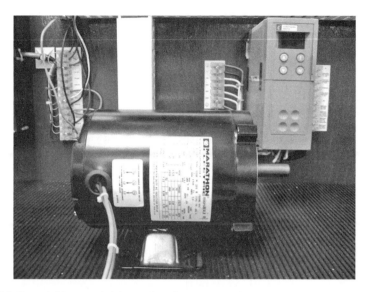

Figure 28-8 Variable speed drive and an inverter rated motor.

Variable speed drives are commonly found on large circulating pumps on hydronic systems for building heating and cooling. These drives allow the pump gpm to be raised and lowered to match the heating and cooling demands of the building, allowing for lower energy consumption.

Single-Phase

Single-phase motors are used where small horsepower is required and where the convenience of single-phase applications is favorable. The major of single-phase motors are fratical horsepower motors. L1 and a neutral are used for low voltage connections while L1 and L2 are used for high voltage connections.

All single-phase motors have a start and run winding which receive power when first energized. The *run winding* is made of heavy gauge insulated copper wire while the *start winding* is made of a finer gauge insulated wire. This difference in wire size causes enough shift in the magnetic fields to produce a starting torque. Because the run winding is made of heavier gauge wire than the start winding, a technician can use an ohmmeter to determine which set of windings is which. If a motor can have its direction changed, it is usually accomplished by swapping the leads of the start winding.

A very common single-phase motor is the split-phase motor. It is simply constructed, reliable, and relatively inexpensive, but has a low starting torque. Because there is low starting torque, applications that call for the motor to start under a light load or no load are preferred. Examples of these applications would be for a spa, jet pump, or a food waste disposal.

The split-phase motor has a start and run winding, and a centrifugal switch mounted on the rotor of the motor. The *centrifugal switch* is a normally closed switch that remains closed until spun quickly enough to cause a set of spring-loaded weights to move outward and allow the switch to open. The centrifugal switch is wired in series with the start windings. When the motor is first energized, both the start and run windings have power applied. When the motor reaches about 70% of its full speed, the centrifugal switch opens and removes power from the start windings. When the motor slows to approximately 40% of its full speed, the centrifugal switch will close again. Figure 28-9 shows how the windings and centrifugal switch are wired together.

When higher starting or running torque is required, usually either a capacitor-start or a capacitor-start capacitor-run motor is used. The motors can easily be

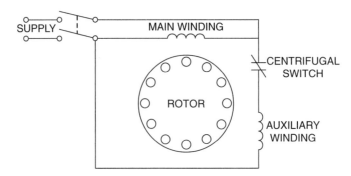

Figure 28-9 Centrifugal switch opens the circuit to the start winding once the motor is up to speed.

identified because they will have one or two cylindrical capacitors mounted on their outer shell. The capacitor creates a bigger phase shift, allowing the motor to generate more torque. They both work in the same manner as the split-phase motor, with the exception that there is a capacitor in series with either the start or run windings (see Figure 28-10).

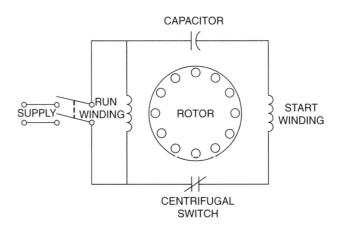

Figure 28-10 Capacitor-start motor.

Another single-phase induction motor that is commonly used is called the AC series or universal motor. It is called the *AC series motor* because the stator field windings are wired in series with the armature windings. The connection is made through a set of brushes and a commutator. Figure 28-11 shows a schematic of how it works.

It is also called a universal motor because it can run off of either AC or DC voltage. Because it has a high starting torque and can adjust to variations in load, it is commonly used for vacuum cleaners, handheld drills, saws, and many other portable appliances.

Shaded Pole Motor

The shaded pole induction motor is generally a small, fractional horsepower motor, typically from 1/250 to 1/2 horsepower. It has a low starting and running torque, so it is typically used in small blowers, ceiling exhaust fans, and small appliances (see Figure 28-12).

It is made up of a squirrel gage rotor, laminated pole pieces, and a heavy loop of copper wire that replaces the start windings and centrifugal switch. The loop of copper wire causes the magnetic field in the pole piece to shift across the face of

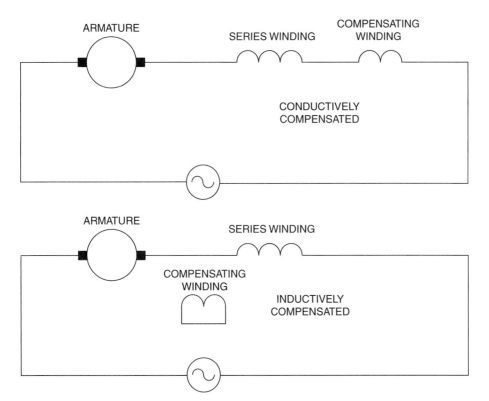

Figure 28-12 The shaded pole motor is commonly found in small appliances as a fan motor.

Figure 28-11 AC series motor.

the pole piece instead of creating one over its entire face. This shift causes the rotor to turn and follow the magnetic field. Table 28-2 gives an overview for the selection criteria of motors.

Table 28-2   Motor Selection Table

Type of Power	Motor Type	Advantages	Identifying Characteristic
AC Polyphase (three-phase)	Squirrel cage induction	High efficiency, lower cost, most rugged	No physical connections to rotor
AC Single-phase	Split-phase	Constant speed and general versatility	Squirrel cage rotor and centrifugal switch
	Capacitor start-induction run	Higher starting torques at lower currents	Squirrel cage rotor, centrifugal switch and capacitor mounted on the motor casing
	Series or universal	Variable speed, very high torque, operates on AC or DC	Brushes and commutator, slows down with increased load

MOTOR CHARACTERISTICS

When selecting a motor replacement, the technician should always try to use the exact same motor or get a recommendation from the equipment manufacturer. If this is not possible, then the technician must select a new motor based on the following motor characteristics:

- Power supply (AC or DC, single- or three-phase, frequency, voltage)
- Horsepower and speed
- Type of application
- Frame size
- Mounting style

Thus far, power supply, horsepower, and speed have been discussed. Now, it's time to address operating characteristics and physical configuration. The NEMA (National Electrical Manufacturers Association) rating listed on the previous motor will give a good idea as to the starting torques and how the starting current is affected. It will typically be listed on the nameplate as one of the letters described in Table 28-3.

Table 28-3 NEMA Ratings and Applications

NEMA Design Rating	Typical Applications
A (High locked rotor torque* and high locked rotor current)	Fans, blowers, centrifugal pumps, and others that require low starting torques
B (Normal locked rotor torque and normal locked rotor current)	Fans, blowers, centrifugal pumps, and others that require low starting torques
C (High locked rotor torque and normal locked rotor current)	Conveyors, stirring motors, and reciprocating pumps, and others that require starting under a load
D (High locked rotor torque and high slip)	Punch presses, elevators, shears, or others that require high peak loads with flywheels

*Locked rotor torque (LRT) is the torque the motor produces when the rotor is stationary and full power is applied to the motor.

Motor frame size is listed by a number on the nameplate. For small motors, frame size numbers are 42, 48, and 56. Manufacturers have standardized these numbers to correspond to key components of the motor such as the following:

- Shaft diameter
- Keyway or flat space on the shaft for pulley mounting
- Distance from the center of the shaft to the mounting surface
- Approximate frame diameter

Matching motor frame numbers greatly increases the chances of finding a suitable replacement for a failed motor. In addition to the physical size and shape of the motor, there are also variations on how the motor is constructed. The following are general types of common constructions.

Open

In the open type, the end bells are open to allow free circulation of ventilation air. These motors cannot be exposed to wet or dirty environments.

Splash Proof

This motor is similar to open end, except the ventilation openings are so arranged that water falling on them will not be drawn into the cooling air passages.

Motors for Hazardous Locations "Explosion-Proof"

These motors are built with very heavy, flanged, cast-iron frames. If a spark occurs within the motor, the flame is quenched before it can get out of the motor frame.

Totally Enclosed

No cooling air is taken inside the motor casing, so all the motor heat must travel to the outside of the motor frame before it can be released to the surroundings.

Hermetic

These motors are contained within a gas-tight vessel. Refrigeration compressors, for example, use the refrigerant to cool the motor windings.

Mountings

Most motors are mounted on four feet that are solidly attached to the stator housing. These feet can be either stamped metal or cast-iron. Cast-iron housings are better for damping vibration, but can easily be damaged if dropped or struck. If vibration is an issue, special spring or rubber mounting feet can be used (see Figure 28-13).

Figure 28-13 **Rubber feet on compressor motor to prevent vibration from being transmitted to equipment housing.**

Other motors are *face mounted*—the end bell is machined flat with tapped or through holes so the motor is bolted directly to the frame of the load. Remember to pay special attention to the frame size when ordering the motor. Most motors are intended to be installed with the shaft horizontal. If the motor must be installed vertically, check with the manufacturer concerning the bearing type and arrangement.

Bearings

Motor bearings are the components in motors that allow the rotor to spin and still be supported by the stationary frame. Sleeve bearings and ball bearing types are the two most common choices of bearings. *Sleeve bearings* are quiet running and can support a larger radial load, but periodical lubrication is required. *Ball bearings* require less service and have less friction than sleeve bearings. They are available in many types and styles, but they are more expensive.

When a motor has failed and needs to be replaced, you should always try to determine why the motor failed. If it was due to normal wear and the motor has lasted several years, you should try to match the existing one as closely as possible (bigger is not always better). If the motor has failed prematurely, then you should consult an expert. Several motor manufacturers have a technical hotline that can assist in finding a suitable replacement. Table 28-4 lists some considerations for selecting an induction motor.

Table 28-4 **Check Sheet for Motor Replacement**

☑ What power is available? AC or DC? Single- or three-phase? What voltage?

☑ What speed is required for the load?

☑ What are the ambient conditions: abnormal temperature, excessive dirt, high humidity?

☑ What is the capacity required to operate the load?

☑ Will it require high starting torque, high running torque, or both? NEMA rating?

☑ What kind of bearing is required?

☑ What frame size?

Table 28-5 *Several Motor Problems and Procedures to Correct Them*

Failure	Cause	Correction
		ELECTRICAL
Will not run	No power	Check power supply to motor. Check all control devices to power.
	Winding shorted, open, or grounded	Replace the motor if the electrical problem is identified as internal. If a motor becomes flooded, the water may be baked out in an oven by a motor repair specialist.
	Over-heated, overload tripped	Allow time to cool and reset overload if manual, but check ambient conditions, current, and voltage levels before putting back in service.
Motor hums but does not turn	Open capacitor, start relay, or start winding	If voltage to motor is correct, then with power off, turn shaft by hand to check bearings.
		If shaft turns freely, check for open centrifugal switch, bad capacitor, or open start winding.
Wiring terminal housing hot or blackened	Loose electrical terminals	Tighten all electrical connections.
		MECHANICAL
Will not run	Worn centrifugal switch	Replace switch or motor.
Noisy	Worn bearings, end play or insufficient support	Check support bolts. If okay, have motor checked by repair specialist or replace.
	Requires lubrication	Bronze bushings may have a wick (small motors) or oil ring design (larger motors). Lubricate with proper viscosity oil. Follow routine maintenance schedule. (Too much lubrication can damage motor windings.)
	Hot or worn bearings	When worn, bearings transfer great amounts of heat to the housing. Have bearing replaced by specialist.
		Hot bearing may be caused by: oil too thin; oil too thick; dirty oil; excessive load on bearings; pulley rubbing against housing; improper alignment of the motor with respect to equipment; worn bearings.

MOTOR PROBLEMS

NOTE: Do not tamper with a motor if you are not appropriately trained in proper servicing and repair techniques. Table 28-5 has been prepared as a primer for basic knowledge about motor maintenance only. Consult a qualified electrician or motor specialist for assistance.

NOTE: Disconnect power before servicing. Be careful not to catch clothing, fingers, hair, or tools in a rotating piece of equipment.

REVIEW QUESTIONS

❶ What is another name for voltage?

❷ If the power is 75 VA and the voltage is 120 V, what is the amperage?

❸ How many energized wires (hots) are connected to a single-phase low voltage appliance?

❹ If a single-phase motor hums but does not turn, what could be the cause?

❺ Which type of bearing requires periodic re-lubrication?

CHAPTER

29

Electric Circuits, Circuit Protection, and Electrical Safety

LEARNING OBJECTIVES

The student will:

- ⚙ Explain how electricity flows through a circuit.
- ⚙ Calculate unknown values using Ohm's law and the power equation.
- ⚙ Describe different types of circuit protection.
- ⚙ Describe safe practices and procedures to follow on the job.

ELECTRIC CIRCUITS

Electric circuits are designed to accomplish a certain task. It might be to turn on a heater to maintain a certain temperature or to turn on a pump to control the pressure and flow in a water system. These circuits are usually drawn either as *schematics* or *ladder logic diagrams*. Schematics arrange the components of the circuit based on where they are physically located and also shows all the interconnected wiring. Ladder logic diagrams show the electrical components based on how the circuit logic works.

Once the designer knows what is to be controlled and how it will be controlled, he will usually break the circuit into three sections: *inputs, control logic,* and *outputs.* Inputs are devices that monitor conditions in the system. Good examples would be pressure switches, flow switches, and thermostats. Outputs are load devices such as pumps, valves, and heaters. The control logic section deals with making decisions based on the input information and using the outputs to keep the system running as intended. Figure 29-1 shows a conceptual view of ladder logic.

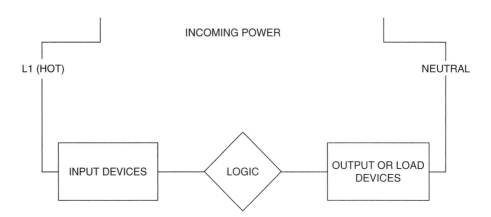

Figure 29-1 Concept drawing of ladder logic.

For example, some commercial water heaters have a water level sensor as well as a thermostat. If the thermostat senses a cool temperature, it would close and turn on the burner. But what if the cool temperature was because the tank was empty? The control logic section should prevent the water heater from starting. Using the wiring diagram, the technician should be able to identify what must happen in order for the system to work. In order to do this, you must learn the basic symbols and conventions used in these circuit drawings.

Originally, the elements in these diagrams were pictorial, so the drawing element looked like the actual device. Many diagrams attempt to follow this concept, but as electrical devices have been developed and modified, the symbols used are not as *picture perfect* as we might wish. Now, we use simplified symbols to represent the actual device. Figure 29-2 shows common symbols used for devices and the wiring symbol conventions that are most often seen. CAUTION: Because not all designers use the same symbols, always check the legend of the circuit to be sure.

These symbols or similar ones will be used in the diagrams to illustrate the points that follow. Lines and sometimes different line types will be used to identify how devices are connected. Remember, the lines are the wires that connect the power to the device.

Color Coding

In electrical work, neutral conductors are white or light gray, equipment ground conductors are bare, green, or green with a yellow stripe, and hot conductors are permitted to be any other color. In assembled cables, hot conductors are usually black or red.

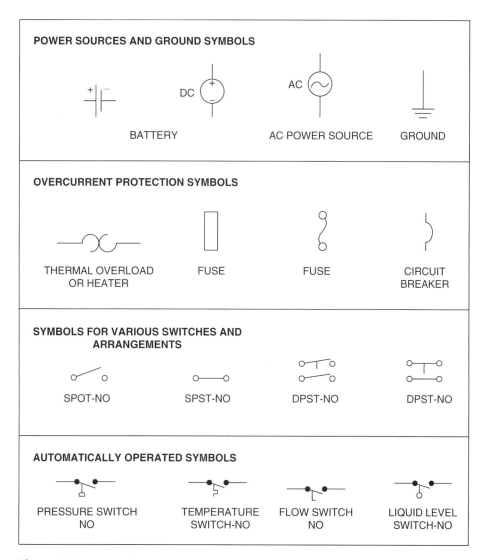

Figure 29-2 Conventional electrical symbols.

Series Circuits

A *power source* could be a battery, a circuit breaker panel, or a generator. The *load* is any device that is going to use that power such as a light bulb, an electric heater, or a motor. There can be any number of loads in a circuit as long as the power supply has enough capacity to supply them all with the correct voltage and the correct amperage. In a *series circuit*, there is only one path from the power source to the load. If that path is broken at any point, no device in that circuit will run. Figure 29-3 shows two series circuits: (a) one with only one load and (b) the other with several loads.

In Chapter 28, we described *current* as being similar to flow rate (gallons per minute) of a pump and voltage as being similar to the pressure (psi) developed by the pump. If you compare Figure 29-4 to Figure 29-3, you will see that the same current (or flow rate) goes through all of the devices while the voltage (or pressure) drops as it goes through the circuit. It should be pointed out that when a plumber says a valve is closed, there is no flow of water. But when an electrician says that a switch is closed, current will still be flowing.

The amount of voltage dropped across each lamp is determined by Ohm's Law, discussed later in this chapter. If all the lamps in Figure 29-3 have the same resistance and the circuit is on, the voltage of the source will be divided equally between the lamps and all will be equally bright. The more lights we add, the

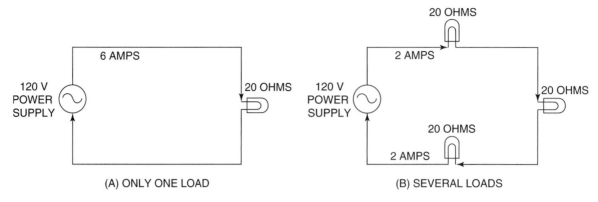

Figure 29-3 Electrical flow through series circuits with various loads.

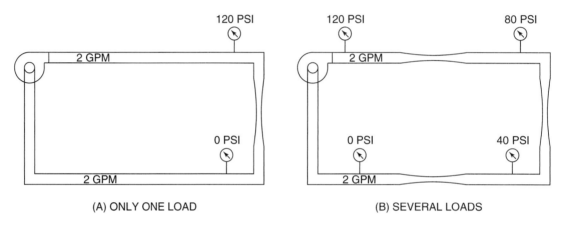

Figure 29-4 Water flow through series circuits with various loads.

more dim each light will glow. In order to determine the total resistance of resistors in series, each individual value is simply added. Using Equation 1, it was found that the three lamps in Figure 29-3(b) have a total resistance of 60 ohms.

$$R(total) = R1 + R2 + R3$$

Parallel Circuits

If we were to take the same lights and power source and rewire them as in Figure 29-5, there would be more than one path for the electricity to flow.

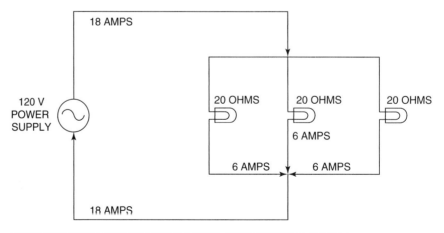

NOTE: THE THREE LIGHTS NOW HAVE A TOTAL RESISTANCE OF 6.67 OHMS

Figure 29-5 Current flow through a parallel circuit.

Now, the three lights are in parallel with each other, and since there are more paths for the electricity to flow, the resistance would be lower. This would be similar to having a one lane road feed into a four lane highway; the highway would allow the traffic to flow more easily, so we would say there is less restriction to flow. The total resistance for a parallel circuit can be found by using Equation 2.

$$\frac{1}{R(total)} = \frac{1}{R1} + \frac{1}{R2} + \frac{1}{R3} + \cdots$$

Sometimes electrical circuits are a combination of both series and parallel (see Figure 29-6). All of the current will flow through the series portion of the circuit, but will be divided when it gets to the parallel portion. The voltage will be divided between the series portion and the total resistance of the paralleled portion.

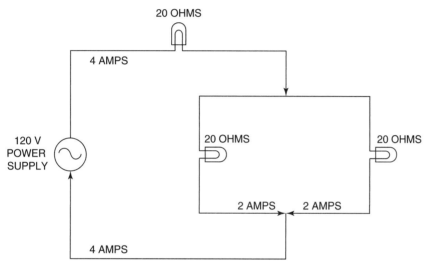

NOTE: THE TWO LIGHTS NOW HAVE A TOTAL RESISTANCE OF 10 OHMS

Figure 29-6 Current flow through a series-parallel circuit.

Series-Parallel Circuit

If the three lamps are the same, then lamp 1 will have the total current flowing through it and half the current will flow through lamps 2 and 3. Lamp 1 will therefore be brighter than lamps 2 and 3. If lamp 1 burns out, the total circuit is open and no current will flow. If lamp 2 or 3 burns out, the circuit would function with lamp 1 becoming dimmer and the remaining parallel lamp becoming brighter.

While most plumbing technicians will do very little wiring, they may be expected to replace defective components or install optional components to newly installed equipment. For these reasons, make sure you have a thorough understanding of what parallel and series mean. For example, if an additional safety stop button were supposed to be installed in series with the rest of the circuit and the technician places it in parallel, it would not serve its intended purpose.

Schematics

When an appliance or wiring circuit experiences a failure, the wiring diagram and knowledge of instrument techniques will help you to determine the problem and get it repaired. All of the drawings shown above are schematic diagrams. They are useful because they help us visualize the sequence and paths of current flow.

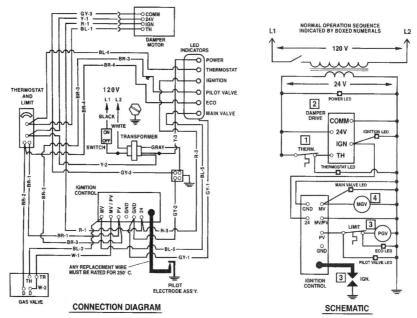

CONNECTION DIAGRAM

SCHEMATIC

NOTE: If any of the original wire as supplied with this appliance must be replaced, it MUST be replaced with 18 GA., 600V, 105°C wire or its equivalent, unless otherwise noted.

⚠ **CAUTION! Label all wires prior to disconnection when servicing controls. Wiring errors can cause improper and dangerous operation. VERIFY PROPER OPERATION AFTER SERVICING!**

Figure 29-7 Boiler-burner schematic diagram showing both the schematic and the ladder logic.

Figure 29-7 shows both the connection and schematic diagram of a commercial gas-fired water heater. The connection diagram shows where and how the components are connected while the schematic diagram is arranged so it is easy to see the logic of the wiring. For example, it is easier to use the schematic diagram to see that power from the secondary (24 volts) side of the transformer goes to the normally closed thermostat (*THERM*) and then to the damper motor drive. (Remember, closed means that current can flow through it.) But the schematic does not tell you where the components are. The connection diagram shows the same information, but it gives you the color and number of the wire as well as the approximate location of the components.

We can get information for solving problems by using these types of diagrams and they also tells us how the devices are supposed to function. The symbols used on a diagram should be understood. A symbol legend is usually part of the control diagram, but if not, each control component should be labeled in the circuit.

OHM'S LAW

The relationship between current, voltage, and resistance is given by Ohm's Law (named after George Ohm, a German physicist). It is usually expressed in the following equation form.

$$V = I \times R$$

V (or sometimes E) = Voltage

I = Current in amperes

R = Resistance in ohms

Like any algebraic equation, it can be rewritten to solve for any of the three values. For example:

$$I = V \div R \quad \text{or} \quad R = V \div I$$

Example 1

If the voltage of the source in Figure 29-3 is 120 volts and the resistance of the lamp is 20 ohms, what is the current?

$$I = 120V \div 60\,\Omega$$

$$I = 2\,A$$

Ohm's Law can be used to analyze the more complex circuits encountered in practical circuit applications. It is important to realize the relationship between the three values: if the resistance goes down, the current will go up.

Power

Another important consideration is the calculation of power. Remember that *power* is the rate of doing or ability to do work. Thus, if we total the power input to a building minute-by-minute, we will obtain the total work (or energy) taken from the electric utility for the building. A watt-hour meter performs this totalizing so we can be billed for the energy consumed. It is also important to realize that the power service, either the circuit breaker panel in a building or the transformer on the equipment, was sized to supply a limited amount of power. If too great of a load is placed on incoming power service, then overheating can occur. This overheating can lead to fires and/or equipment failure. To prevent this dangerous condition, we must understand the following relationship. The equation to find power using current and voltage is:

$$P = I \times E$$

P = **Power in watts**

I = **Current in amperes**

E = **Voltage in volts**

Example 2

What size *VA (volt amperes)* transformer would be required to supply 24 volts and 3 amps to a control circuit?

$$P(VA) = 24\ volts \times 3\ amps$$

$$P = 72\ VA$$

The 72 VA represents the power requirements for the 24 volts and 3 amps. The transformer chosen should have at least this rating or higher.

CIRCUIT PROTECTION

Fuses and Circuit Breakers

Fuses and *circuit breakers* are designed to automatically protect the circuit from damage due to excessive current flow. The possible damage from excessive current is usually considered to be overheating, but very high mechanical forces are also possible. Note that circuit breakers or fuses do not provide shock protection in the event that you contact energized parts of a circuit.

Fuses

A *fuse* is a low cost, highly reliable device that melts at a current flow slightly above the rating of the fuse. When the current-carrying part melts, the circuit is interrupted and the circuit components are protected. The fuse must be replaced to restore service. You should re-energize the circuit only after attempting to determine the cause of fuse opening. Replacement fuses must have the appropriate rating of the circuit being served. An oversized fuse may create a fire by allowing excessive current to flow through the wiring.

Dual-Element Fuse

A variation of the fuse concept is the *dual-element fuse*. These fuses combine the rapid-trip characteristic of normal fuses on high overload and a delayed trip (delay measured in seconds) with currents just slightly above the fuse rating. These fuses are commonly called *time-delay fuses*. They are used in motor applications because motors will have a larger than normal amount of current draw when they are first started. This will be discussed in more detail in the Motor Protection section of this chapter. Figure 29-8 shows several different fuse types.

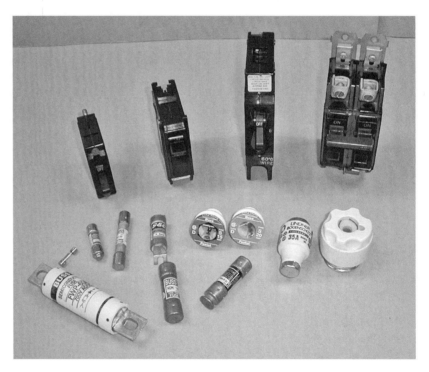

Figure 29-8 Different available fuses. Notice the current rating listed on each one.

Circuit Breakers

A *circuit breaker* trips when the circuit current becomes unusually high, thus protecting the system. The circuit breaker can be reset to restore service. Circuit breakers usually contain a magnetic trip mechanism and a thermal trip mechanism. Thus, they trip quickly on large over-current flows and after a short time on small over-currents. Fuses or circuit breakers for most circuits must be selected so that their operating set point does not exceed the rating of the conductors used in the circuit.

Ground Fault Circuit Interrupters

A *Ground Fault Circuit Interrupter (GFCI)* is a device that operates by comparing the current in the energized hot line to the current in the neutral line and opening the circuit if these currents differ by more than a 4 to 6 milliamperes

(0.004 amps–0.006 amps). Ground fault circuit interrupters are highly sensitive switching devices that open the circuit if there is any unbalance in the current out to the load versus the current returning from the load. The idea is that if current is going to the load through the hot wire and only some of the current is returning through the neutral, then the difference is traveling through a ground fault. Figure 29-9 shows how a situation such as a frayed wire can make contact with the metal housing of an appliance and allow some of the current to travel in an unexpected path.

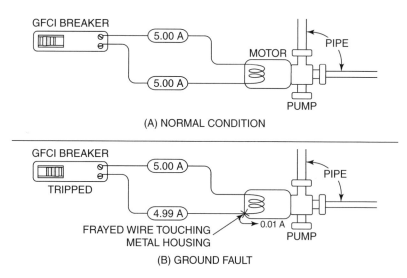

Figure 29-9 Ground faults occur whenever the current is allowed to flow to ground without going through the neutral conductor.

If the frayed wire makes a high resistance circuit with the metal appliance, only a small amount of current will flow (Ohm's Law: I = V/R Big R means little I). But even a small amount of current like the one in Figure 29-9, could kill a person when he/she is the connecting path.

GFCIs (either a circuit breaker in the panel or a special outlet) are installed to protect people from shock hazards. Article 210 of the *National Electric Code* (NEC) lists several places that require GFCI protection. Table 29-1 lists some of those locations.

Table 29-1   Locations Requiring GFCI Protection (Based on Article 210 of the NEC)

Location
Bathrooms
Commercial garages
Elevators, escalators, and moving walkways
Fountains
Health Care Facilities
Pools
Temporary installations (temporary power on a construction site)

Any outlet in or near a wet location should be protected. Note that GFCI receptacles are not over-current devices and therefore must be used with a normal circuit breaker. Figure 29-10 shows the GFCI receptacle.

Figure 29-10 GFCI receptacle.

MOTOR PROTECTION

When a motor first starts up, it will draw large amounts of current until it gets up to its rated speed. This current can easily be six to eight times the normal running current. Fuses, circuit breakers, and wiring for motors must be sized large enough to handle this short term but high current. A motor can become overloaded while it is running, but this increase in current may not be high enough to trip the fuses or circuit breakers. If left running for an extended time, these higher than normal currents could cause the motor to fail. Therefore, they must be guarded with temperature sensing devices that respond to over-currents in a similar way as the motor itself. Motors may be overloaded mechanically (trying to move a heavy load) or they may be overloaded electrically if the supplied power suffers a partial breakdown. Electrical problems can cause motor overloads. The two most common problems are: (1) low voltage and (2) loss of a phase on a three-phase motor. Either of these events produces excessive motor currents.

Motors can be protected with automatic reset devices or manual resets. Automatic reset restores service when the overload device has cooled, but repeated overloads can result in damage to the motor. The manual reset version means that service is not restored until a person recognizes a problem and resets the control. Typically, sump pumps will have automatic reset while motors for equipment like portable air compressors or drain cleaning equipment will have manual resets.

Since there is such a large variety and combination of motors, applications, and protective devices, check with an experienced individual before making the final determination of motor overload type and size.

Small motors typically use thermostatic switches that mount either inside or on the outside of the motor to sense high winding temperature and shut down the motor. The thermal overload device is in series with the windings of the motor. When the motor temperature is too high, the thermostat opens and the power to the motor is turned off. Figure 29-11 shows an externally mounted thermal disk-type overload commonly used on household refrigerator compressors. Upon cooling, it automatically resets itself. Note this same type of overload is used as a high limit for a water heater except it has to be manually reset.

In other cases, a thermostat or thermal disk is built into the motor. The button on the end of the motor, as shown in Figure 29-12, is for manual reset once the motor has cooled.

For larger motors, the motor starting component contains an overload device that responds to the current flow to the motor. The two most common types are: (1) thermal overloads (see Figure 29-13) and (2) magnetic overloads (see Figure 29-14).

Figure 29-11 Externally mounted thermal disk used to break power to the motor if it becomes too hot.

Figure 29-12 Motor with internal overload. Red Button is a manual reset after the motor has had time to cool.

Figure 29-13 Melting alloy type thermal overloads, sometimes called "heaters" (three-phase).

Thermal overloads are designed to heat up as the current to the motor increases. Once they reach their set temperature, a set of contacts opens and breaks the power going to the motor. This type of overload needs time to cool down before it can be reset. The overload device uses a melting alloy-type overload relay. The control switch is spring-loaded to open and a ratchet gear, which is held fixed by a solder-type material, restrains it. The overload coil heats the solder holder so that an over-current produces a temperature rise sufficient to melt the alloy, which allows the ratchet gear to turn, and the switch contact to open. After cooling, the solder solidifies and the switch can be reset manually. CAUTION: Sometimes if the ambient temperature is too high, the motor will overheat and the thermal overload will trip.

Magnetic overloads use a current coil to detect the amount of current flowing in each of the energized conductors to the motor. If the current becomes too high, the current coils will open a set of contacts in the control wiring circuit. Magnetic overloads are more adjustable and quicker to reset, but are usually more expensive than thermal overloads. It is important that all three lines in a three-phase motor be monitored. If only one of the three power lines were lost, this condition would be called *single phasing*. The motor would be trying to work with only two of its three windings. This would cause a loss of torque and an increase in the amount of current flowing through the remaining windings.

Article 430 of the NEC allows for the overload device settings 115% to 125% of motor full load current. This slight increase makes allowances for times that the motor is temporarily subjected to an increase in load or when a longer starting up time occurs because of a load.

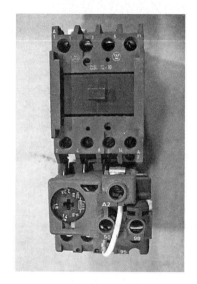

Figure 29-14 Magnetic overloads with adjustable setting (three-phase).

Electrical Safety

Electrical installations in buildings are capable of doing large amounts of work. However, if any part of the electrical system is out of control, serious injury or damage may occur. If any of the components that make up the circuit break down or fail, the electricity will flow through the path of least resistance, which could be through the metal frame of a machine or through the person operating the equipment. This unexpected event is known as a *ground fault*. Therefore, we must be sure to use materials and methods that minimize the chance of electrical failure.

A typical 120 volt alternating current circuit found in a home or office consists of an *ungrounded conductor*, commonly referred to as the *hot conductor* (usually black), and a *neutral conductor* (usually white). The black and white wires complete the path of flow for the electricity similar to the supply and return line in a hydraulic system. When this path is complete, the electrons can flow and the equipment attached to the circuit can work. Normally, the neutral conductor is attached to a grounding bar where the electrical service enters the building (see Figure 29-15). This is referred to as a *polarized circuit*, which simply means that one side of the power source is referenced to ground.

Most computer controlled equipment or appliances with circuit boards, such as furnaces with electronic ignition, would not function properly if the circuit were not polarized. In the case of a furnace, the burner controller uses flame rectification to determine the presence of the pilot flame. This process relies on the fact that the burner is grounded and a very small current will only pass one way through the circuit board. If a good ground is not present or the incoming power (energized conductor and neutral conductor) is reversed from the way the manufacturer has labeled the equipment, the appliance will not function. Figure 29-16 shows how to tell whether a typical receptacle is polarized.

Once the incoming service has been properly grounded, as shown in Figure 29-15, the rest of the system must be grounded as well. An additional conductor, which

IN THE FIELD

Circuit analyzers are low cost, simple to use, and available to detect many of the common electric problems (see Figure 29-17).

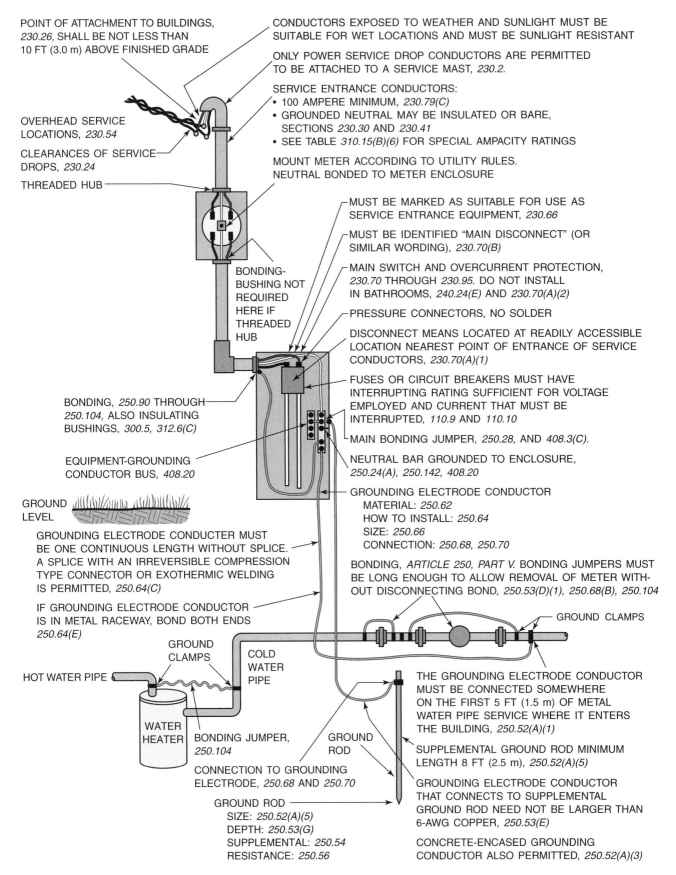

POINT OF ATTACHMENT TO BUILDINGS, *230.26,* SHALL BE NOT LESS THAN 10 FT (3.0 m) ABOVE FINISHED GRADE

CONDUCTORS EXPOSED TO WEATHER AND SUNLIGHT MUST BE SUITABLE FOR WET LOCATIONS AND MUST BE SUNLIGHT RESISTANT

ONLY POWER SERVICE DROP CONDUCTORS ARE PERMITTED TO BE ATTACHED TO A SERVICE MAST, *230.2.*

SERVICE ENTRANCE CONDUCTORS:
• 100 AMPERE MINIMUM, *230.79(C)*
• GROUNDED NEUTRAL MAY BE INSULATED OR BARE, SECTIONS *230.30* AND *230.41*
• SEE TABLE *310.15(B)(6)* FOR SPECIAL AMPACITY RATINGS

OVERHEAD SERVICE LOCATIONS, *230.54*

CLEARANCES OF SERVICE DROPS, *230.24*

THREADED HUB

MOUNT METER ACCORDING TO UTILITY RULES. NEUTRAL BONDED TO METER ENCLOSURE

MUST BE MARKED AS SUITABLE FOR USE AS SERVICE ENTRANCE EQUIPMENT, *230.66*

MUST BE IDENTIFIED "MAIN DISCONNECT" (OR SIMILAR WORDING), *230.70(B)*

MAIN SWITCH AND OVERCURRENT PROTECTION, *230.70* THROUGH *230.95.* DO NOT INSTALL IN BATHROOMS, *240.24(E)* AND *230.70(A)(2)*

BONDING-BUSHING NOT REQUIRED HERE IF THREADED HUB

PRESSURE CONNECTORS, NO SOLDER

DISCONNECT MEANS LOCATED AT READILY ACCESSIBLE LOCATION NEAREST POINT OF ENTRANCE OF SERVICE CONDUCTORS, *230.70(A)(1)*

FUSES OR CIRCUIT BREAKERS MUST HAVE INTERRUPTING RATING SUFFICIENT FOR VOLTAGE EMPLOYED AND CURRENT THAT MUST BE INTERRUPTED, *110.9* AND *110.10*

BONDING, *250.90* THROUGH *250.104,* ALSO INSULATING BUSHINGS, *300.5, 312.6(C)*

MAIN BONDING JUMPER, *250.28,* AND *408.3(C).*

NEUTRAL BAR GROUNDED TO ENCLOSURE, *250.24(A), 250.142, 408.20*

EQUIPMENT-GROUNDING CONDUCTOR BUS, *408.20*

GROUNDING ELECTRODE CONDUCTOR
    MATERIAL: *250.62*
    HOW TO INSTALL: *250.64*
    SIZE: *250.66*
    CONNECTION: *250.68, 250.70*

GROUND LEVEL

GROUNDING ELECTRODE CONDUCTOR MUST BE ONE CONTINUOUS LENGTH WITHOUT SPLICE. A SPLICE WITH AN IRREVERSIBLE COMPRESSION TYPE CONNECTOR OR EXOTHERMIC WELDING IS PERMITTED, *250.64(C)*

BONDING, *ARTICLE 250, PART V.* BONDING JUMPERS MUST BE LONG ENOUGH TO ALLOW REMOVAL OF METER WITH-OUT DISCONNECTING BOND, *250.53(D)(1), 250.68(B), 250.104*

IF GROUNDING ELECTRODE CONDUCTOR IS IN METAL RACEWAY, BOND BOTH ENDS *250.64(E)*

GROUND CLAMPS

GROUND CLAMPS

COLD WATER PIPE

HOT WATER PIPE

WATER HEATER

BONDING JUMPER, *250.104*

CONNECTION TO GROUNDING ELECTRODE, *250.68* AND *250.70*

GROUND ROD

GROUND ROD
    SIZE: *250.52(A)(5)*
    DEPTH: *250.53(G)*
    SUPPLEMENTAL: *250.54*
    RESISTANCE: *250.56*

THE GROUNDING ELECTRODE CONDUCTOR MUST BE CONNECTED SOMEWHERE ON THE FIRST 5 FT (1.5 m) OF METAL WATER PIPE SERVICE WHERE IT ENTERS THE BUILDING, *250.52(A)(1)*

SUPPLEMENTAL GROUND ROD MINIMUM LENGTH 8 FT (2.5 m), *250.52(A)(5)*

GROUNDING ELECTRODE CONDUCTOR THAT CONNECTS TO SUPPLEMENTAL GROUND ROD NEED NOT BE LARGER THAN 6-AWG COPPER, *250.53(E)*

CONCRETE-ENCASED GROUNDING CONDUCTOR ALSO PERMITTED, *250.52(A)(3)*

Figure 29-15 Residential circuit breaker panel. Notice how the grounding rod is connected to the equipment ground bus bar and the neutral bus bar.

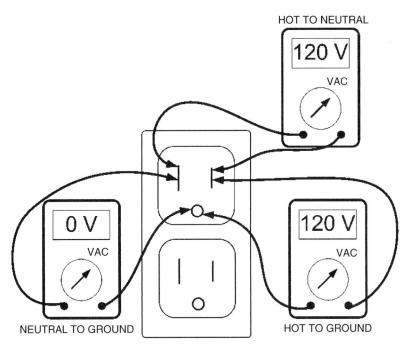

HOT TO NEUTRAL

120 V
VAC

0 V
VAC

120 V
VAC

NEUTRAL TO GROUND   HOT TO GROUND

COMMON POLARIZED 120 VAC RECEPTACLE

Figure 29-16 Voltage relations between posts of a polarized circuit.

Figure 29-17 A Circuit analyzer used to test common 120 volt receptacles.

is either a bare wire or a wire with green insulation, is added to the circuit. This additional wire is known as an *equipment ground*. The equipment ground conductor is provided to connect to the metal frames of all electrical equipment. This concept is shown in Figure 29-18.

Connecting all these metal components together is called *bonding*. Thus, all equipment frames in the building will be bonded together at ground potential

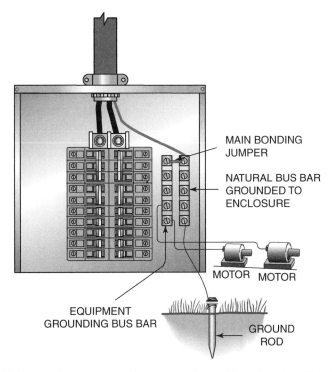

MAIN BONDING
JUMPER

NATURAL BUS BAR
GROUNDED TO
ENCLOSURE

MOTOR   MOTOR

EQUIPMENT
GROUNDING BUS BAR

GROUND
ROD

Figure 29-18 The equipment ground connects all metal housings together to help prevent electrocution.

and when a person touches one of these devices, it will be less likely to transmit an electrical shock. If any device develops an electrical short to the frame, the equipment ground will usually provide a low resistance path for fault currents so that the circuit breaker or fuse protecting the device will open, clearing the ground fault. It is possible, however, if the internal failure is of high resistance, the circuit breaker or fuse will not trip.

The grounding path for the electrical service must be considered when replacing or repairing water-service piping. If any appliance or section of the electrical system depends on the water-service piping consult with a qualified electrician before removing any of that piping, and never use any portion of a gas system as a grounding conductor or electrode. (This is covered in Article 7.13 of the 2006 *National Fuel Gas Code*.)

ELECTRICAL SHOCK

Electric shock may cause instant death, unconsciousness, cessation of breathing, and burns of all kinds. If the current path is through your body, from one hand to the other hand, or to the foot, the following conditions may exist:

- A current of one milliampere (0.001 amp) can be felt.
- A current of ten milliamperes (0.010 amp) will paralyze muscles, so you cannot release your grip on the conductor.
- A current of 100 milliamperes (0.10 amps) is fatal if it persists for one second.

The electrical resistance of the human body can be quite low (even more so when the skin is wet) so lethal currents can be received even when the voltage is low. (Remember, current is what kills.)

Shock is prevented primarily by grounding the frames of all equipment, grounding the frames of portable tools, or by using portable or fixed ground-fault interrupting devices. Avoid any situation that allows your body to provide a path from an electrical source to ground. When working on electrical devices, follow this advice:

- Avoid standing on damp surfaces.
- Do not use metal ladders.
- Do not hold onto a grounded surface when using, testing, or otherwise inspecting electrical equipment.
- Use extension cords in good repair and of adequate size for the load being served. Use the minimum length cord that will reach the work site.
- Be sure circuits are off before you work on them or the equipment they serve. If necessary, padlock off and tag the safety switch serving the circuit you are repairing or checking.

Figure 29-19 shows a lockout for an electrical disconnect being used while Figure 29-20 shows a lockout for a circuit breaker being used.

Many portable tools are *double insulated*. That is, besides the winding insulation, the case of the tool is made of insulating material so the user cannot be shocked by an internal failure.

What to Do if an Electrical Shock Occurs

If you are near someone who is being shocked by an electrical circuit: **DO NOT TOUCH THE PERSON UNTIL HE/SHE IS FREE OF THE POWER SOURCE! Do not touch the person, unless you are 100% sure the power is turned off.** If you are not sure that the power is turned off, then do the following:

1. De-energize the circuit.
2. After removing the person from the electrical source, examine the victim. Call 911 or other emergency professionals.

Figure 29-19 Lockout for disconnect. Notice that there is room for each person to install their own lockout device.

Figure 29-20 Lockout for circuit breaker.

3. If the heart has stopped **and you are trained,** begin CPR reviving methods. CPR training is a skill worth having. Take the time to become certified.
4. If the person is not breathing **and you are trained,** initiate mouth-to-mouth resuscitation at once.
5. Keep the person warm.
6. Get help as soon as possible.

REVIEW QUESTIONS

1 What colors are normally used to identify the neutral conductor?

2 What would be the maximum current available on a 150 VA transformer with a voltage of 24 volts?

3 According to Ohm's Law, what would happen to the current if the voltage stayed the same and the resistance increased?

4 At what amount of current would the human body lose muscle control?

5 When someone is being electrocuted, what is the first thing to do?

30

Electric Circuits Troubleshooting

The student will:

- Employ proper techniques when using basic test equipment.
- Demonstrate the use of a schematic or wiring diagram for testing electrical circuits.
- Demonstrate the use of systematic approach for finding a fault in an electrical circuit.

TEST INSTRUMENTS

To locate the precise area and nature of component or wiring failures in a piece of equipment, we need the schematic diagram and the knowledge of how to use various handheld test instruments. The most commonly used instruments include the following:

- Voltmeter to measure voltage
- Ammeter to measure current
- Ohmmeter to measure resistance
- Test lamps or voltage detectors
- Multimeter (a device that can be used as a voltmeter, ammeter, and ohmmeter)

All of the above instruments can be found in either of two forms: (1) a *digital meter* with a *Light Emitting Diode (LED)* display or (2) *an analog meter* which uses a scale and a needle. In either form, the meter needs to be set to the correct type of measurement (ohms, volts, etc.) and the correct scale (0-6A, 0-15A). Scales are set to the highest value that you expect to find when taking a measurement. CAUTION: If you are unsure of the expected value, always start at the highest scale and work your way down. Even though most meters are fuse-protected, setting it on a low scale while attempting to read a much higher value could cause damage to the meter.

Picking the correct scale is relatively simple. If the highest scale on the meter is 600 VAC, this means it can read any value from 0 to 600 volts AC. Using it to measure the voltage of a common household receptacle, the display may read 120, and if the next lower scale setting on the meter is 200 VAC, then it can be used. Moving to the 200 VAC scale will help obtain a more accurate reading; the same receptacle may now show a reading of 118.5 VAC. But if the next lower scale is 60 VAC, it should not be used because the 118.5 VAC is well above the 60 VAC maximum of this setting. The same is true for any of the instruments covered in this chapter. Test lamps or voltage detectors are a very quick and convenient way to determine whether power is being applied to a circuit, but they cannot distinguish between 24, 120, 208, or 240 volts. A voltmeter must be used to determine if the correct voltage is being applied.

Voltmeters

Voltmeters are very versatile tools. They can be used to find the presence of power, open switches, defective capacitors, broken wires, or even blown fuses. Voltmeters have maximum values up the which they can test. If this value is exceeded, the user and the meter could be harmed. If the insulation on the test leads is cracked or worn thin, it is possible to be shocked even when holding the insulated leads. Always test the voltmeter with a known working power supply before using it to troubleshoot a circuit. If the meter or the test leads are broken, an energized circuit would appear to be off, but the user could become electrocuted. Voltmeters are always attached parallel to the part of the circuit they are testing. This is convenient because no wires have to be removed or broken. Figure 30-1 shows a fuse being tested using a voltmeter on an energized circuit. If there is no voltage drop across a fuse, then that fuse is good.

Figure 30-2 shows a voltmeter using a special adapter to measure the output of a water heater thermocouple.

Ammeter

Ammeters are used to measure the current flowing through the circuit to help determine the approximate load condition of a motor or the operating load of

Figure 30-1 Checking for a voltage drop across a fuse. If there is a voltage drop, the fuse is bad.

Figure 30-2 Voltmeter and special adapter for measuring thermocouples. Caution: Do not over-tighten adapter or thermocouple.

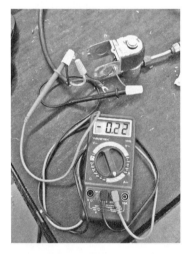

Figure 30-3 Ammeters used in series. The circuit had to be disassembled to insert the ammeter.

Figure 30-5 Ohmmeter used to check water heater element. This element has infinite resistance, so it is bad.

a circuit. Normally, in order to use the ammeter function of a multimeter, the meter must be installed in series with the circuit. This can be time consuming, because the circuit must first be de-energized, separated, and then connected to the leads of the ammeter (see Figure 30-3). The *amp clamp* avoids this problem because it can be clamped around a single conductor to read the current flow without breaking the circuit. Amp clamps are commonly referred to as an *Amprobe*, but this is a brand name, not an instrument. Placing an amp clamp around an extension cord or appliance cord will not work because there are two current-carrying conductors (hot and neutral) in these cords and their magnetic fields cancel each other out.

An amp clamp can be used to quickly determine if a solenoid coil is being energized or not. If the jaws are held open and placed near the coil, they will pick up the magnetic field and let you know that power is being applied to the coil (see Figure 30-4).

Figure 30-4 Amp clamp picking up the magnetic field of a solenoid on a control valve.

Ohmmeter

Ohmmeters measure the amount of resistance in a circuit and are used most often as a simple continuity tester. *Continuity* means whether the circuit is complete or not. The ohmmeter uses the battery power supply in the meter to energize the circuit and then reads the resistance to the applied DC voltage. This is important because if an ohmmeter were connected to an energized circuit, the meter could become damaged. A common mistake made by technicians is to use the diode setting (or beeper) to check the coil of a valve or relay or to check the windings of a transformer. The *diode setting* is for measuring low resistance values. When connected to a higher resistance coil or winding, it will give the user the impression that the coil or winding is bad when it is not. Always check loads with the highest ohms setting and work down, just like the steps used with the voltmeter in the beginning of the chapter. The diode setting can be used to check fuses and switches because they should have very low resistance values. Figure 30-5 shows an ohmmeter being used to check the resistance of a water heater element.

TESTING PORTABLE TOOLS

A quick and easy check of power tools can be performed using an ohmmeter. If the tool is double insulated, it will only have two prongs on its cord. With the trigger squeezed, there should be a resistance value present between the two prongs. When the trigger is released, this resistance value should go to infinity (gives the same reading as when the test leads are not touching anything).

If the tool has an equipment ground, it will have three prongs on its cord. The two flat prongs will test as described above, but there should never be a circuit between either of the two flat prongs and the round grounding prong.

TEST PROCEDURES

In order to protect yourself and your equipment, it is necessary to develop safe work habits. The following is a list of procedures that should help develop a safe troubleshooting routine for testing equipment.

1. Think about and determine what you need to measure, and then select the proper instrument.
2. Select the highest range on the meter and make the test. Repeat on reduced range settings until a satisfactory reading is obtained.
3. Do not use an ohmmeter on an energized circuit.
4. Use AC ranges for AC, and DC ranges for DC. An AC range will read on a DC application, but a DC range will not read on AC voltage. The problem with using the AC scales for DC readings is that the exact value will not register properly on the typical portable meter.
5. Use the proper terminals on the test meter. If checking DC, be sure to use the red lead on positive polarity and the black lead on negative.
6. A voltmeter is always connected parallel to the device being checked.
7. An ammeter is always placed in series with the device being checked. If an amp clamp is being used, only clamp the jaws around one conductor at a time.
8. Be especially careful not to become part of the circuit. Use extra caution if you are on a ladder or otherwise elevated. Remember that a minor shock on a ladder may result in a serious fall.

In addition to knowing the test instruments and developing safe work habits, always make sure that you have a good understanding of the operation of the equipment. The following example describes how to troubleshoot a thermocouple on a residential water heater, but before beginning this test always read the startup procedure, listed on the piece of equipment.

RESIDENTIAL GAS WATER HEATER THERMOCOUPLE

Testing Procedure

A thermocouple testing procedure includes the following operations (see Table 30-1 for safety rules):

1. Unscrew the thermocouple from the gas valve and insert the special adapter for testing thermocouples (refer back to Figure 30-2).
   CAUTION: Over tightening the thermocouple will damage the *energy cut off (ECO)* device of the gas valve. Screw the thermocouple in hand tightly and then 1/4 turn with a wrench.
2. Light the pilot burner and check the flame. It should be blue with a sharp cone in the center. If the flame is large and yellow, clean the pilot burner and check the orifice size.
3. After the flame has been applied to the thermocouple for approximately three minutes, check the millivolt reading on the thermocouple when it is connected to its load. If the reading is less than 13 millivolt (mV) (*1 millivolt = .001 volts*), reposition the thermocouple so that the pilot flame contacts more of the thermocouple or replace the thermocouple.

4. If the thermocouple checks out okay, check the magnetic safety valve in the gas valve. Turn off the pilot and watch the millivolt reading drop. The magnet should stay energized for a 5 mV drop, meaning it should click at approximately 8 mV (13 mV–5 mV). If it clicks sooner, then the magnet is weak and the gas valve should be replaced.

Table 30-1   Safety Rules When Testing an Electric Water Heater

☑ When servicing an electric appliance, turn the power off.

☑ Use a voltage meter or sensing tool to locate the voltage. Obviously, if voltage readings are needed, the power must be on, so use proper personal protection equipment and proceed with caution.

☑ Never touch exposed leads, which may be energized.

☑ Do not work on equipment with wet hands or when standing in water. Insulate your body from wet floors by standing on dry multiple layers of wood, canvas, approved rubber insulating mat, or other non-conducting material.

☑ Use insulated testing tools.

☑ Wear safety glasses when operating electrical testing tools.

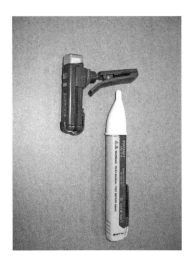

Figure 30-6 Voltage detector pens are excellent for quick checks to detect if a circuit is energized.

In order to properly troubleshoot an electrical circuit, a good quality voltmeter, ammeter, and ohmmeter are required. Voltage detector pens and test lights could be used, but they can lead to false readings or assumptions (see Figure 30-6). For instance, the common non-simultaneous water heater that we are testing will have a constant 120 volts applied to both elements at all times but it requires a complete flow path and the proper 208 or 240 volts in order to generate heat. The voltage detectors will start to glow at around 24 volts, so the user would have no idea what level of power is being applied. Your life could depend on the tools you use, so purchase a good quality meter and take care of it.

Testing with Water Heater Energized

Some of the most useful information concerning the circuit and its components is obtained by checking the circuit with the power on. Always assume the power is on until verified by a voltmeter. It is usually more convenient and quicker to check and see if the circuit breaker is tripped or if the fuse is blown before removing water heater access panels or climbing into a crawlspace or an attic. If either the fuse or the circuit breaker is open, the real problem must be discovered before the power can be restored. A blown fuse or a tripped circuit breaker is the symptom, not the problem. Check all of the wire connections between the power supply and the final element—a loose wire connection will cause high current and overheat.

Figure 30-7 shows a voltage meter being used to check fuses for incoming power. Incoming power generally comes into the top of the fuse and leaves out the bottom of the fuse. The first logical step would be to check both current carrying conductors going to the water heater (the load side). If the correct voltage is present, then the fuses are good and there is no need to check them. If the voltage differs from what is expected (208 V–240 V) by more than 10%, check each fuse individually. There should not be a voltage drop across a fuse. Be sure to test across the lines because a 240 volt supply on a 120/240 volt system will probably show 120 volts to ground on each line (wire) even if one line is open (fuse is bad). This is because the load side of the bad fuse is connected to the rest of the circuit through the heating element and since there is no current, there is no voltage drop across the circuit (see Figure 30-11 to trace the path).

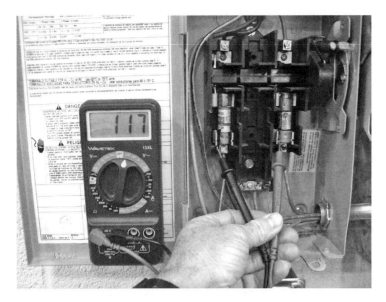

Figure 30-7 The voltage coming into the fuses (at top) was 240 volts, but because the left–side fuse is bad, the load side (at bottom) voltage is lower than expected.

If the power supply is intact, proceed to the heater. Before troubleshooting can continue, the order of operation and the function of each component must first be understood. Figure 30-8 shows what a typical dual element, non-simultaneous water heater electrical connection looks like.

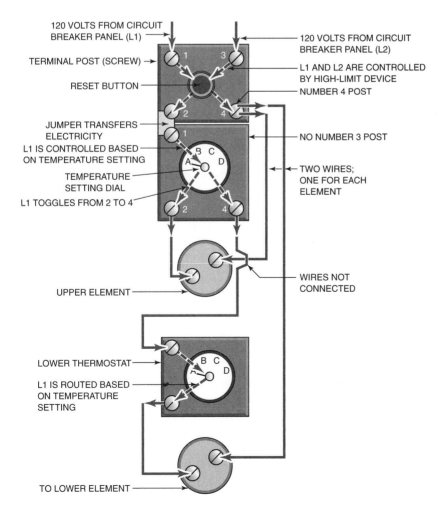

Figure 30-8 Wiring connections for a non-simultaneous, dual element residential water heater.

The first component to receive power is the ECO, or the *high-limit switch* (see Figure 30-9). It is a manual reset thermal disk that monitors the temperature of the tank. By placing it firmly against the outside of the metal tank, high water temperatures can be sensed and power from both L1 and L2 can be interrupted.

Figure 30-9 High-limit device and upper thermostat assembly for the top heating element.

Figure 30-10 Common submersible electric water heater element. Note the wattage and voltage ratings.

The second component is the *adjustable thermostat* that controls the temperature of the water in the upper 1/3 of the tank. This device is never off. It either sends power to the top heating element or the lower thermostat. The ECO and upper thermostat are bonded together with a copper *busbar* and mounted as one unit, which means if either component fails, both will have to be replaced.

The third and fourth components are the *heating elements* which are submersed in the water (see Figure 30-10). They either screw in or are flange mounted to the side of the tank. What's most important is that they have the same voltage and wattage rating that is listed on the manufacturer's label.

The fifth component is another *adjustable surface-mounted thermostat*. This one controls the temperature of the bottom 2/3 of the tank. If power is supplied to it from the upper thermostat and the tank temperature is below its setpoint, it will connect L1 power to the lower heating element. It differs from the top thermostat in that it is either on or off (single pole, single throw) whereas the top thermostat simply diverts power from one device to another.

Figure 30-11 shows a ladder logic diagram, which gives a better idea of the operating logic and will be used to get a better picture of how the components work together. A common misconception is that if voltage is present, water is being heated. Before heat can be generated, there must be a completed path for the electrons to flow from L1 to L2. If a heating element is blown (open) then the proper voltage will be present, but there will be no current flow and no hot water.

Power is first applied to the top terminals (1 and 3) of the ECO. Under normal operating conditions, a circuit will be made from terminal 1 to terminal 2 of the ECO, which will also be connected to the common terminal (1) of the upper thermostat. At the same time, a circuit will be made from terminal 3 to terminal 4 of the ECO. Terminal 4 is connected to one side of both the upper and lower heating elements.

If the water in the upper portion of the tank is cooler than the setpoint of the upper thermostat, L1 will be connected to the upper heating element. Now the upper heating element will have the proper voltage and there will be a complete path back to L2. This means that current will flow and heat will be generated. Notice that there is no complete path to the lower thermostat, so it will not be producing heat.

IN THE FIELD

Warning: Both heating elements will have 120 volts applied to them at all times.

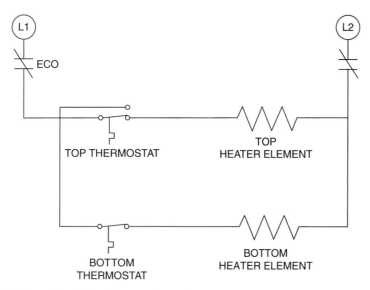

Figure 30-11 Ladder logic diagram shows how components are controlled. It does not show their physical location with respect to each other.

Once the setpoint of the upper thermostat is reached, then L1 power will be routed directly to the lower thermostat, which will control the flow to the lower heating element, if needed.

Now that the order of operation is understood, let's look at some procedures for testing the water heater: (1) with power applied and (2) without power.

Testing with Heater Energized

Using Figure 30-11, step through the circuit, checking voltages between terminals.

1. Check the voltage coming into the top of the high-limit device at terminals 1 and 3. Having already checked the circuit breakers for proper voltage, a low voltage here would mean a loose connection, possibly where the power comes into the top of the unit. If the proper voltage is applied to 1 and 3, then check for voltage at the output of the high-limit device between terminals 2 and 4.

2. Check for voltage between terminals 2 on the top thermostat and terminal 4 of the high limit. If the tank is cooler than the temperature setting of the thermostat, voltage should be present. If the tank is hotter than the thermostat setting, then the proper voltage should be present between terminal 4 on the top thermostat and terminal 4 on the high-limit device. In either case, voltage should be present between terminals 2 and 4, or terminals 4 and 4. If it is not, then the top thermostat is defective and should be replaced.

3. If the correct voltage is present between terminals 2 (thermostat) and terminal 4 (high-limit device), then check for voltage at the top-heating element. If the correct voltage is present between terminals 4 (thermostat) and terminal 4 (high-limit device), then check for voltage at the lower thermostat.

4. If the proper voltage is being applied to the heating element, then use an ammeter to check the current to that element. The current should be within 10% of the calculated value. Use the power equation to calculate what the power should be:

$$P = I \times E$$

$$3500 \ w = I \times 208 \ v$$

$$I = \frac{3500 \ w}{208 \ v} = 16.8 \ A$$

If the ammeter reading does not show this current (+/− 10%), then the heating element is defective. As an alternate method, an ohmmeter can be used to check the continuity of the element, which is described in the next section.

Testing with Heater De-Energized

In this test, each device will be checked with an ohmmeter. Always disconnect the power from the heater before using an ohmmeter. The component being tested should be unwired from the circuit because it is simple to get false readings through other devices.

Figure 30-12 Before using an ohmmeter to check the resistance of the element, be sure to disconnect the power and remove any wires from the element.

1. Check for continuity between terminals 1 and 2 on the high-limit device. If near zero resistance, check between terminals 3 and 4—both sets should read about the same. If not, push the red reset button and check again. If the device will not reset, it should be replaced. There should not be continuity between 1 and 4, or between 2 and 3.
2. If the high-limit device is good, then check for continuity between terminals 1 and 2, or between terminals 1 and 4 on the top thermostat. There should be continuity between one set of terminals or the other. If not, the thermostat is defective and should be replaced.
3. If both the high limit and the thermostat are good, then the element should be checked. The *element* is a load device, so there should be more resistance than compared to a switch. A normal heating-element will have approximately 13 ohms across it. If it reads zero ohms, it is shorted closed and if it has infinite resistance (the meter will display the same reading as when the leads are apart and not touching anything), then the element is shorted open (see Figure 30-12). In either case, the element should be replaced.
4. If the water heater is producing some hot water but runs out quickly, it should have passed all of the above tests and the problem is with the bottom circuit. If the water in the lower portion is cooler than the lower thermostat's temperature setting, then the contacts should be closed and there would be a near-zero resistance reading between the two screw terminals on the thermostat. Usually, turning the thermostat to its maximum setting will cause the thermostat to close, which should make the resistance go to zero. If not, the thermostat is bad and should be replaced.
5. If the lower thermostat is good, then the lower heating element is the last to be checked. It should be tested in the same manner as the upper heating element.

It is important to remember that *switches* should have zero ohms resistance when closed and infinite resistance when open. Loads, such as motors and heating elements, should have some resistance, never zero or infinite.

LOW VOLTAGE CONTROLS FOR POWER VENT WATER HEATERS

As equipment manufacturers make units with higher efficiencies and more safeties, the control circuits become more complex. Earlier, the testing of a standing pilot gas water heater was discussed. The controls consisted mostly of a thermocouple and high-limit controls. As the cost of control modules has decreased, some modern residential water heaters are using an electronic ignition control to eliminate the constant burning of a pilot flame. This will decrease the amount of fuel being burned and save the customer money over the life of the unit. Manufacturers also add power ventilators to allow more options as to where the units can be installed. Figure 30-13 shows the connection diagram and the schematic diagram for an electronic ignition burner with a power vent-type water heater.

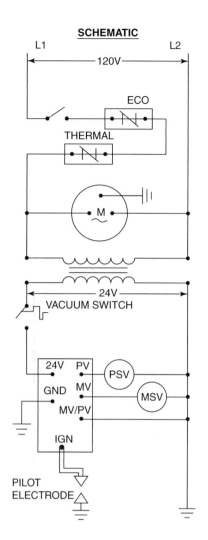

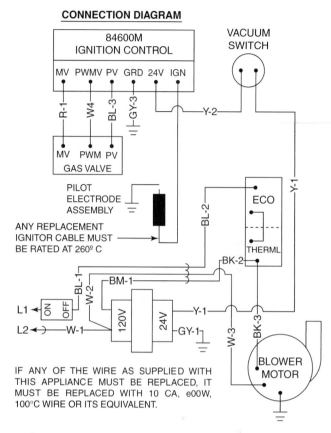

Figure 30-13 Schematic and connection diagram for a power vent water heater. Courtesy of Rheem Manufacturing Company, Water Heater Division.

As control systems become more complex, the order of operations becomes increasingly more important. A good procedure to use when troubleshooting is to ask yourself, "When the unit starts, what must happen first?" Use the schematic diagram in Figure 30-13 to follow the sequence of operations.

SEQUENCE OF OPERATIONS

Power flows from L1, through the on/off switch, through the ECO switch, through the thermostat and then it is connected to both the transformer and the vent blower motor. So, the first event on a call for heat is that the blower should start and the transformer should be energized. (See Chapters 28 and 29 for review of transformers.) The transformer will supply the 24 volt power necessary for the remainder of the circuit. Notice how the secondary side of the transformer is grounded. This is important because some electronic ignition modules won't work if not properly grounded.

Once the vent motor has started, the draft (vacuum) switch should close and supply power to the 24-volt terminal on the ignition control module. The vacuum switch proves that not only did the vent motor start, but it also is generating enough draft for the burner to function safely. The vacuum switch requires approximately .75 inches of water column pressure in order to close. Figure 30-14 shows checking for a voltage drop across the draft switch.

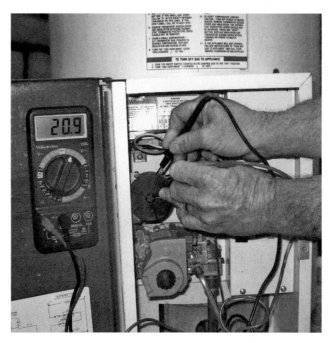

Figure 30-14 Voltage drop across the draft switch means that the switch is open; either there is not enough flow of combustion air or the switch is defective.

So, the second event on a call for heat is that the vacuum switch should close and send 24 volts to the ignition control module. Once energized, the ignition control module will send power to the pilot valve solenoid and a high voltage to the pilot electrode assembly. **CAUTION: This is usually very high voltage. Do not attempt to measure this voltage.** The control module will now look for verification of a pilot flame for 90 seconds. If a pilot flame is sensed, it will then send power to the main gas valve solenoid and the main burner should light. The burner will run for as long as the thermostat calls for heat. Once the thermostat is satisfied, the blower will stop, causing the vacuum switch to open and remove power from the ignition control module. The ignition control module will then remove power to the gas valve and the flame should go out.

Now that the sequence of operations is understood, let's look at a possible scenario and see how to troubleshoot it.

Example 1

On a call for heat, the vent blower runs, but there is no flame. What is the first item to check? We know the power is on and the blower works, so there is no need to check the ECO or the thermostat. The next condition to check is to see if the gas is turned on. Remember to check the easiest items first. Next, check to see if 24 volts is being applied to the ignition control module. If 24 volts is being applied, but there is no power to the pilot valve solenoid, then the ignition module could be faulty. Before replacing any ignition module, turn off all power and allow everything time to reset (usually 1–3 minutes). Then, reapply the power and check the output from the module again. Sometimes if a control module tries to start multiple times and fails, it will enter a *hard lockout state*. This means that the power has to be manually reset before it will try to restart again. This is a safety feature that prevents gas build up or damage to an igniter. After being reset, if the burner still does not try to ignite, then the control module is faulty and should be replaced.

Placing the common (black) lead of the voltmeter on the transformer secondary side ground, and moving through the circuit will reveal a broken wire as well as a defective control. Figure 30-15 illustrates this procedure.

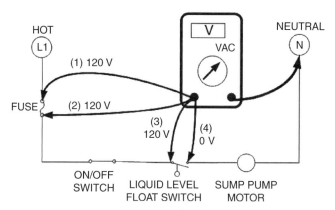

Figure 30-15 Using a voltmeter to step through a circuit until the voltage is not detected is a good way to help determine where a break in the circuit is. The fuse is good, the switch is good, but the float switch is open. The left side (3) has 120 volts, but the right side (4) does not.

After some experience at troubleshooting, you will be surprised to see how many times the problem will be caused by a broken wire or a loose connection.

Once again, establish a procedure when troubleshooting. Proceed from the known to the unknown. Learn how to use the handheld test meters properly, and memorize the rules for safe testing. Study the schematic diagram to enable yourself to check out the unit. With these methods, you will be able to check out the electrical circuits in your equipment.

Table 30–2   Troubleshooting Table
    Courtesy of Rheem Manufacturing Company, Water Heater Division

Nature of Trouble	Possible Cause	Service
No Hot Water	1. Manual disconnect switch turned off.	Turn to "ON."
	2. Improper wiring.	Rewire per wiring diagram.
	3. No power-overcurrent protective device open.	
	a. Shorted wiring.	Replace or repair.
	b. Circuit overload.	Provide adequate circuit or reduce load.
	c. Improper wiring.	Rewire per wiring diagram.
	d. Grounded transfer or thermostat.	Replace.
	4. Manual reset limit open.	Refer to "OPERATION": Section 4.
	a. Thermostat breakdown.	Replace thermostat.
	b. Thermostat out of calibration.	Replace thermostat.
	c. Limit breakdown.	Replace thermostat.
	5. Blower inoperative.	
	a. Blower not plugged in.	Plug in blower.
	b. Blower breakdown.	Replace.
	6. Ignitor inoperative.	
	a. Vacuum switch breakdown.	Replace or check tubing connections.
	b. Improper wing.	Rewire per wiring diagram.
	c. Intermittent pilot ignitor gap too large.	Replace, or adjust to 1/8'.
	d. Broken Intermittent pilot ignitor.	Replace.
	e. Improperly vented.	Refer to "Installation" Section 7.
	f. Ignition Control breakdown.	Replace
Yellow Flame Sooting	1. Scale on top of burner.	Shut off heater and remove scale.
	2. Combustion air inlets of flue ways restricted.	Remove lint or debris and inspect air inlet.
	3. Not enough combustion or ventilation air supplied to the room.	Refer to Introduction, Section D.
Rumbling Noise	1. Scale or sediment in tank.	Clean tank—See Maintenance, Section 1H.

CAUTION: Make certain power to water heater is "OFF" before removing access panel FOR ANY REASON.

REVIEW QUESTIONS

❶ When using a voltmeter, what scale is used first?

❷ A voltmeter is placed so that one lead is connected to the top of a fuse, while the second lead is connected to the bottom of the fuse. If a voltage is read, is the fuse good or bad?

❸ What is the first step before checking a heating element with an ohmmeter?

❹ According to Table 30-2, what is a cause of a humming noise in a water heater?

❺ On a non-simultaneous, dual element water heater, which thermostat must be satisfied first?

CHAPTER 31

Control Wiring

SWITCHES

A *switch* is a device that is capable of interrupting the flow of electricity in a circuit. Switches can be placed in series with a load or with other switches (see Figure 31-1).

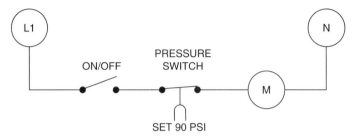

Figure 31-1 Switches in series for an air compressor motor. Notice both switches must be on for the motor to run.

Switches can also be installed in parallel with other switches, but not in parallel with a load. Putting a switch in parallel with a load would bypass the load and create a dead short (see Figure 31-2).

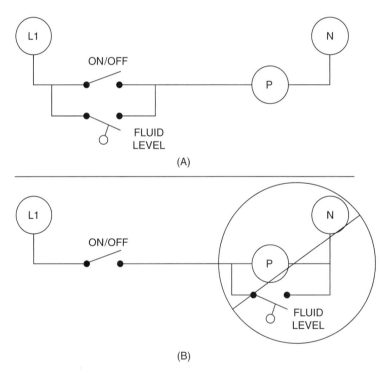

Figure 31-2 (a) Either the on/off switch or the liquid level switch will turn on the pump. (b) The liquid level switch is parallel to the pump motor. If the liquid level switch and the on/off switch close, the current will be very high (because resistance is low).

A *dead short* is when there is no load in the circuit. Since switches have very low resistance when closed, the circuit resistance would be low and the current would go very high. (Ohm's Law: $I = V/R$)

Switches can be divided into two main categories: (1) manually operated and (2) system-operated. A *manually operated* switch is one that changes its state (on/off) based on the actions of an individual. A *system-operated* switch is one that changes its state based on operating conditions of the system such as pressure, temperature, flow, or level. Switches will have voltage and amperage ratings that should always be checked. For example, a common residential light switch would have a voltage rating of 120 volts and a current rating of 15 amps.

Switches can also be identified by the number of poles that they have. A *pole* refers to a single contact that will open or close a path that connects two wires. A double-pole switch has two separate contacts that will open and close simultaneously, but are still considered to be separate. Figure 31-3 shows this concept.

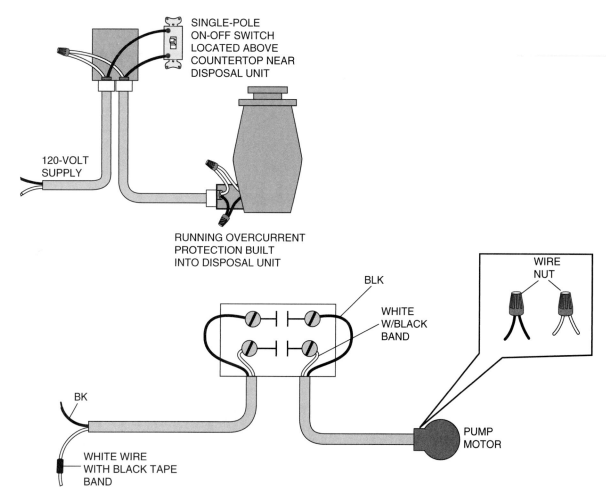

Figure 31-3 Single pole switch on a 120 V garbage disposal and double pole pressure switch on a 220 V well pump.

Manually Operated Switches

The most common type of switch that everyone is familiar with is the manually operated single pole toggle switch that controls the lights in a residential building. Figure 31-4 shows what the wiring connections would look like for a standard light fixture with a plastic housing. Notice that with plastic housings, the ground wire is not attached. Never cut the ground wire short; just leave it, in case any replacement fixture requires it.

A slightly different version of this switch can be found in the pushbutton control of a hydro massage bathtub or hot tub. These are usually mounted flush with the deck of the tub and operated by pushing "in" on the button, instead of flipping a lever (see Figure 31-5).

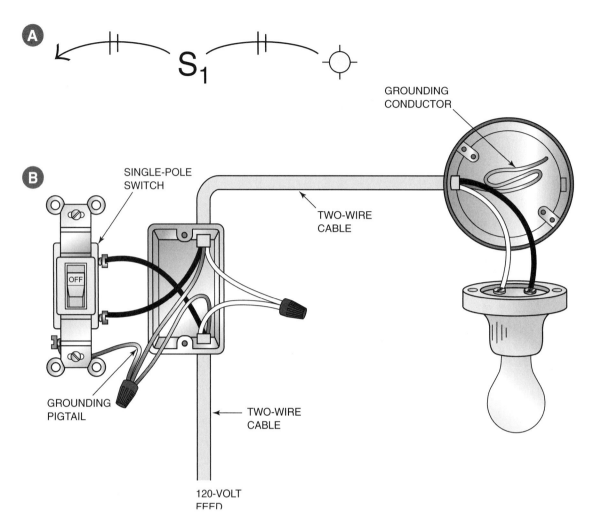

Figure 31-4 Switch being used to control the power feed to a standard light with a plastic housing. This would be very similar to the power connect to a dishwasher except the grounding wire would be attached to the grounding screw. (Never cut the ground wire short, even if not in use.)

Figure 31-5 Pushbutton control switch for a hot tub.

Automatically Controlled Switches

Automatic switches operate on various inputs such as time, pressure, liquid level, or temperature. These switches are designed to control a circuit based on some external condition and function in many ways; some will open when the signal is increased while others will close when the signal is increased.

Automatic control switches will also be identified as either *normally open* or *normally closed*. The normal state is the position the contacts would be in if there were no external signal being applied, or straight out of the box. For example, a normally open pressure switch would have the contacts open when there is 0 psi applied to it. A normally closed one would have its contacts closed. It should be pointed out that in electrical terms *closed* means that current is flowing while *open* means that current has stopped.

One of the most common types of automatically controlled switches that a plumber will encounter will be the *pressure switch*. They can be found on well pumps and air compressors. Figure 31-6 shows a pressure switch that is adjustable.

This particular pressure switch has two pressure adjustments: (1) cut-out or range and (2) differential. *Cut-out* is the pressure at which the contacts will open and the piece of equipment will stop running. *Cut-in* is the pressure at which the contacts will close and the piece of equipment will start running again. *Differential* is the difference between the cut-in and the cut-out. Figure 31-7 shows the relationship between these variables using pressure settings that would be common for a residential well pump. Many pressure switches come with a fixed differential.

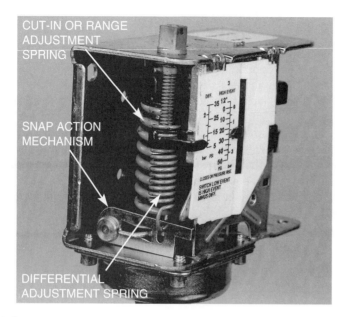

Figure 31-6 Pressure switch with both range and differential adjustments. Compressing the spring more will increase the pressure setpoint or increase the pressure differential.

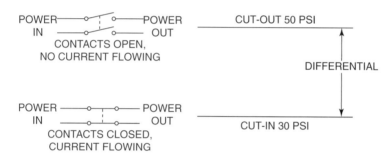

Figure 31-7 Cut-in, cut-out, and differential. Notice how the contacts open and close.

The pressure settings can be adjusted by turning the adjustment screw or nut either clockwise (increase pressure) or counterclockwise (lower pressure). In the application of the well pump, the switch would have a normally closed set of contacts that would open when the system pressure increases to the cut-out setpoint. Other pressure switches may be purchased that have normally closed contacts which open when the system pressure drops. A low pressure cut off for a pump or a refrigeration compressor would be an example of this. Figure 31-8 shows the symbols used to describe these different actions.

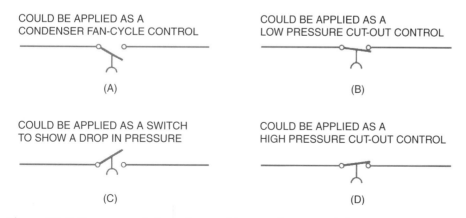

Figure 31-8 Pressure symbols as they would appear in a control diagram. Symbols A and B indicate that the switch will close when the pressure increases. Symbols C and D will open when the pressure increases.

Wire Sizing and Connections

Plumbing technicians should not size or install wire unless they are licensed to do so, but they should be familiar with the concept so that they can recognize the problems associated with an *undersized conductor*. These problems include: drop in voltage at the appliance, circuit breakers or fuses tripping, or conductors overheating. When trying to determine a voltage drop at an appliance, compare the voltage readings at the appliance with the appliance turned off to the voltage readings with the appliance turned on. This is just like measuring pressure in a hydraulic system. When there is no flow of fluid, there is no pressure drop; when the fluid starts to flow through small pipes, the pressure will drop.

The five factors that determine the *current carrying capacity (ampacity)* of a wire are the following:

1. Material
2. Diameter
3. Length
4. Insulation
5. Temperature

In modern construction, copper is the most widely used material. The table most often used by electricians is Table 310-16 of the *National Electric Code* (NEC). The values in this table can be safely used for runs up to 1000' and temperatures up to 86° F. The insulation type is chosen based on the temperature and environmental conditions it will be exposed to. This only leaves diameter.

The sizes of wire that most plumbers will encounter will be measured in the American Wire Gauge number system. It is important to know that as the wire number increases, the diameter of the wire decreases. For example, a #14 wire size is smaller than a #12. Note that as the diameter of the wire decreases, its ability to carry current decreases. This is just like in piping; the smaller diameter pipe cannot transfer as much water (current) as a larger pipe, without a bigger pressure drop (voltage drop).

When a plumber suspects undersized wiring for a reduced voltage, the manufacturer's literature is the best source for the proper sizing. Equipment manufacturers will list the size of the wire, the minimum temperature rating of the insulation, and the correct circuit breaker sizing. If this information is not available, the *National Electric Code* (NEC) has tables that will list the ampacity of the wire. Table 31-1 shows examples of sizes from Table 310-16 of the NEC.

Connections

Another problem that can cause voltage drops, tripped circuit breakers, or blown fuses, and overheating is a loose connection. For this reason, the following section

Table 31-1 Wire Ampacity

Size, AWG	60°C (140° F) Insulation Types: TW, UF	90°C (194° F) Insulation Types: THHN, THHW, RHH, TBS, SA . . .
	COPPER	
18	Not rated	14
14*	20	25
12*	25	30
10	30	40
8	40	55

*For these conductor sizes, the NEC has set limits on the overcurrent protection as follows:
15 amperes for #14
20 amperes for #12
30 amperes for #30
(For a complete listing, refer to Table 310-16 of the NEC)

will discuss the proper procedures for installation of *wirenuts* and *terminal loops* on screw terminals. Remember to always de-energize and lock out the power before attempting to work on electrical appliances. Figure 31-9 shows how to properly install a wirenut. This would be a common practice when installing a garbage disposal. Some technicians will also secure the wirenut with a couple wraps of electrical tape, which may help prevent corrosion in high humidity areas.

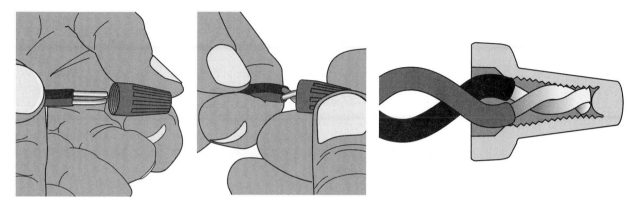

Figure 31-9 **Connecting wires together with a wirenut.**

When the wire is being installed on a screw terminal, such as a pressure switch on a well pump, it is best to form a loop in the end of the wire and then wrap it in the direction the screw will be tightened. Figure 31.10 shows how to properly make this connection.

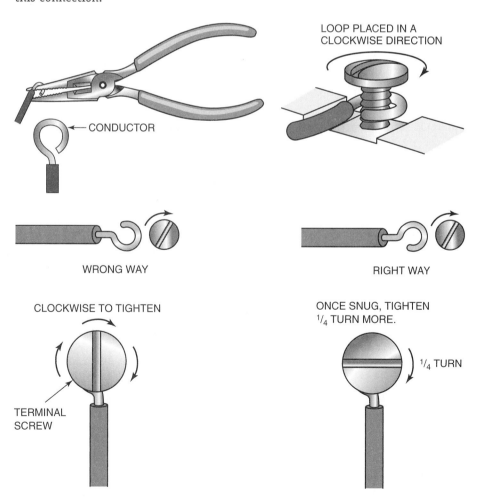

Figure 31-10 **Connecting wires to screw terminals. Always wrap the wire in the direction the screw will be tightened.**

If after checking all the connections and low voltage is still present, the incoming power should be checked. This may be supplied through the circuit breaker panel or through a transformer.

TRANSFORMERS

A *transformer* is a highly efficient device that can either raise or lower the voltage through the use of magnetic fields produced by coils of wire. A transformer is commonly used on equipment to lower the control voltage so that smaller devices and wiring may be used for controls. A plumbing technician will not normally install transformers, but the following information may help in diagnosing problems when troubleshooting equipment.

A transformer consists of a coil of wire wrapped around an iron core. The coil of wire that will receive the incoming power is called the *primary winding*. The primary winding will generate a magnetic field that will generate a voltage in the other separate winding. This second coil of wire is referred to as the *secondary winding* of the transformer. Figure 31-11 shows a conceptual drawing of a transformer.

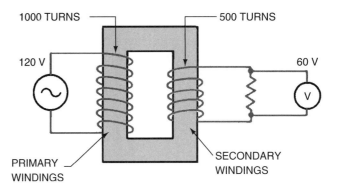

Figure 31-11 Conceptual drawing of a control transformer.

If the secondary windings have fewer wraps (or coils) than the primary windings the generated voltage in it will be smaller than that in the primary windings. This is known as a *step down transformer* because the voltage leaving will be lower than that coming in. If the secondary winding has more coils than the primary winding, then it would increase the voltage and be considered a *step up transformer*.

Single-phase transformers are typically sized by the amount of power they are capable of handling. This is commonly referred to as *VA*, which stands for *volt-amperes*. Since a transformer cannot create power, what comes in must equal what goes out.

> *Power (P) in = Power(P) out*
>
> *Voltage (V) in × Amperage(A) in = Voltage (V) out × Amperage(A) out*

If the voltage is reduced by half, say 120 V (primary) is reduced to 60 V (secondary), then the current capability would be doubled. Example 1 shows this concept.

Example 1

A control transformer has a rating of 100 VA with a primary voltage rating of 120 volts and a secondary voltage of 24 volts. What are the current ratings for both the primary and the secondary sides?

$$\text{Primary Side Power} = \text{Voltage} \times \text{Current}$$

$$100 \text{ VA} = 120 \text{ V} \times I$$

$$I = \frac{100 \text{ VA}}{120 \text{ V}}$$

$$I = .833 \text{ A}$$

$$\text{Secondary Side Power} = \text{Voltage} \times \text{Current}$$

$$I = \frac{100 \text{ VA}}{24 \text{ V}}$$

$$I = 4.17 \text{ A}$$

The transformer selected for the control circuit power supply must have enough capacity to supply the necessary volt-amperes, but it should not be significantly oversized. The primary voltage rating must be equal to the line voltage applied to the equipment. If the line voltage is more than 110% of the transformer rating, the transformer life will be reduced. If the primary voltage is less than 90% of the transformer rating, the secondary output voltage may be too low for reliable operation of the controls. Table 31-2 lists some of the more common voltages found in industry.

Table 31-2 **Transformer Voltages**

Primary Voltages	Secondary Voltages
480 V	120 V
277 V	24 V
240 V	
220 V	
208 V	
120 V	

Depending on the turns ratio, any primary voltage could produce any of the secondary voltages.

REVIEW QUESTIONS

1 What does "normally open" mean when referring to a switch?

2 What does a "step down transformer" mean?

3 If a control transformer has a rating of 40 VA and the secondary voltage is 24 volts, what is the maximum current available on the secondary side?

4 Which wire is smaller, #14 or #10?

5 Why does a loose connection reduce the available voltage?

APPENDIX

A

MasterFormat 2004™ Edition Numbers and Titles

DIVISION NUMBERS AND TITLES

PROCUREMENT AND CONTRACTING REQUIREMENTS GROUP

DIVISION 00—PROCUREMENT AND CONTRACTING REQUIREMENTS

INTRODUCTORY INFORMATION

PROCUREMENT REQUIREMENTS
00 10 00 SOLICITATION
00 11 00 Advertisements and Invitations
00 20 00 INSTRUCTIONS FOR PROCUREMENT
00 21 00 Instructions
00 22 00 Supplementary Instructions
00 23 00 Procurement Definitions
00 24 00 Procurement Scopes
00 25 00 Procurement Meetings
00 26 00 Procurement Substitution Procedures
00 30 00 AVAILABLE INFORMATION
00 31 00 Available Project Information
00 40 00 PROCUREMENT FORMS AND SUPPLEMENTS
00 41 00 Bid Forms
00 42 00 Proposal Forms
00 43 00 Procurement Form Supplements
00 45 00 Representations and Certifications

CONTRACTING REQUIREMENTS
00 50 00 CONTRACTING FORMS AND SUPPLEMENTS
00 51 00 Notice of Award
00 52 00 Agreement Forms
00 54 00 Agreement Form Supplements
00 55 00 Notice to Proceed
00 60 00 PROJECT FORMS
00 61 00 Bond Forms
00 62 00 Certificates and Other Forms
00 63 00 Clarification and Modification Forms
00 65 00 Closeout Forms
00 70 00 CONDITIONS OF THE CONTRACT
00 71 00 Contracting Definitions
00 72 00 General Conditions
00 73 00 Supplementary Conditions
00 80 00 Unassigned
00 90 00 REVISIONS, CLARIFICATIONS, AND MODIFICATIONS
00 91 00 Precontract Revisions
00 93 00 Record Clarifications and Proposals
00 94 00 Record Modifications

DIVISION 01—GENERAL REQUIREMENTS

01 00 00 GENERAL REQUIREMENTS

01 10 00 SUMMARY
01 11 00 Summary of Work
01 12 00 Multiple Contract Summary
01 14 00 Work Restrictions
01 18 00 Project Utility Sources

01 20 00 PRICE AND PAYMENT PROCEDURES
01 21 00   Allowances
01 22 00   Unit Prices
01 23 00   Alternates
01 24 00   Value Analysis
01 25 00   Substitution Procedures
01 26 00   Contract Modification Procedures
01 29 00   Payment Procedures
01 30 00 ADMINISTRATIVE REQUIREMENTS
01 31 00   Project Management and Coordination
01 32 00   Construction Progress Documentation
01 33 00   Submittal Procedures
01 35 00   Special Procedures
01 40 00 QUALITY REQUIREMENTS
01 41 00   Regulatory Requirements
01 42 00   References
01 43 00   Quality Assurance
01 45 00   Quality Control
01 50 00 TEMPORARY FACILITIES AND CONTROLS
01 51 00   Temporary Utilities
01 52 00   Construction Facilities
01 53 00   Temporary Constructio
01 54 00   Construction Aids
01 55 00   Vehicular Access and Parking
01 56 00   Temporary Barriers and Enclosures
01 57 00   Temporary Controls
01 58 00   Project Identification
01 58 13   Temporary Project Signage
01 58 16   Temporary Interior Signage
01 60 00 PRODUCT REQUIREMENTS
01 61 00   Common Product Requirements
01 62 00   Product Options
01 64 00   Owner-Furnished Products
01 65 00   Product Delivery Requirements
01 66 00   Product Storage and Handling Requirements
01 70 00 EXECUTION AND CLOSEOUT REQUIREMENTS
01 71 00   Examination and Preparation
01 73 00   Execution
01 74 00   Cleaning and Waste Management
01 75 00   Starting and Adjusting
01 76 00   Protecting Installed Construction
01 77 00   Closeout Procedures
01 78 00   Closeout Submittals
01 79 00   Demonstration and Training
01 80 00 PERFORMANCE REQUIREMENTS
01 81 00   Facility Performance Requirements
01 82 00   Facility Substructure Performance Requirements
01 83 00   Facility Shell Performance Requirements
01 84 00   Interiors Performance Requirements
01 85 00   Conveying Equipment Performance Requirements
01 86 00   Facility Services Performance Requirements
01 87 00   Equipment and Furnishings Performance Requirements
01 88 00   Other Facility Construction Performance Requirements
01 89 00   Site Construction Performance Requirements
01 90 00 LIFE CYCLE ACTIVITIES
01 91 00   Commissioning
01 92 00   Facility Operation

01 93 00 Facility Maintenance
01 94 00 Facility Decommissioning

DIVISION 02—EXISTING CONDITIONS

02 00 00 EXISTING CONDITIONS
02 01 00 Maintenance of Existing Conditions
02 05 00 Common Work Results for Existing Conditions
02 06 00 Schedules for Existing Conditions
02 08 00 Commissioning of Existing Conditions
02 10 00 Unassigned
02 20 00 ASSESSMENT
02 21 00 Surveys
02 22 00 Existing Conditions Assessment
02 24 00 Environmental Assessment
02 25 00 Existing Material Assessment
02 26 00 Hazardous Material Assessment
02 30 00 SUBSURFACE INVESTIGATION
02 31 00 Geophysical Investigations
02 32 00 Geotechnical Investigations
02 40 00 DEMOLITION AND STRUCTURE MOVING
02 41 00 Demolition
02 42 00 Removal and Salvage of Construction Materials
02 43 00 Structure Moving
02 50 00 SITE REMEDIATION
02 51 00 Physical Decontamination
02 52 00 Chemical Decontamination
02 53 00 Thermal Decontamination
02 54 00 Biological Decontamination
02 55 00 Remediation Soil Stabilization
02 56 00 Site Containment
02 57 00 Sinkhole Remediation
02 58 00 Snow Control
02 60 00 CONTAMINATED SITE MATERIAL REMOVAL
02 61 00 Removal and Disposal of Contaminated Soils
02 62 00 Hazardous Waste Recovery Processes
02 65 00 Underground Storage Tank Removal
02 66 00 Landfill Construction and Storage
02 70 00 WATER REMEDIATION
02 71 00 Groundwater Treatment
02 72 00 Water Decontamination
02 80 00 FACILITY REMEDIATION
02 81 00 Transportation and Disposal of Hazardous Materials
02 82 00 Asbestos Remediation
02 83 00 Lead Remediation
02 84 00 Polychlorinate Biphenyl Remediation
02 85 00 Mold Remediation
02 86 00 Hazardous Waste Drum Handling
02 90 00 Unassigned

DIVISION 03—CONCRETE

03 00 00 CONCRETE
03 01 00 Maintenance of Concrete
03 05 00 Common Work Results for Concrete

04 27 00 Multiple-Wythe Unit Masonry
04 28 00 Concrete Form Masonry Units
04 30 00 Unassigned
04 40 00 STONE ASSEMBLIES
04 41 00 Dry-Placed Stone
04 42 00 Exterior Stone Cladding
04 43 00 Stone Masonry
04 50 00 REFRACTORY MASONRY
04 51 00 Flue Liner Masonry
04 52 00 Combustion Chamber Masonry
04 53 00 Castable Refractory Masonry
04 54 00 Refractory Brick Masonry
04 57 00 Masonry Fireplaces
04 60 00 CORROSION-RESISTANT MASONRY
04 61 00 Chemical-Resistant Brick Masonry
04 62 00 Vitrified Clay Liner Plate
04 70 00 MANUFACTURED MASONRY
04 71 00 Manufactured Brick Masonry
04 72 00 Cast Stone Masonry
04 73 00 Manufactured Stone Masonry
04 80 00 Unassigned
04 90 00 Unassigned

DIVISION 05—METALS

05 00 00 METALS
05 01 00 Maintenance of Metals
05 05 00 Common Work Results for Metals
05 06 00 Schedules for Metals
05 08 00 Commissioning of Metals
05 10 00 STRUCTURAL METAL FRAMING
05 12 00 Structural Steel Framing
05 13 00 Structural Stainless-Steel Framing
05 14 00 Structural Aluminum Framing
05 15 00 Wire Rope Assemblies
05 16 00 Structural Cabling
05 20 00 METAL JOISTS
05 21 00 Steel Joist Framing
05 25 00 Aluminum Joist Framing
05 30 00 METAL DECKING
05 31 00 Steel Decking
05 33 00 Aluminum Decking
05 34 00 Acoustical Metal Decking
05 35 00 Raceway Decking Assemblies
05 36 00 Composite Metal Decking
05 40 00 COLD-FORMED METAL FRAMING
05 41 00 Structural Metal Stud Framing
05 42 00 Cold-Formed Metal Joist Framing
05 43 00 Slotted Channel Framing
05 44 00 Cold-Formed Metal Trusses
05 45 00 Metal Support Assemblies
05 50 00 METAL FABRICATIONS
05 51 00 Metal Stairs
05 52 00 Metal Railings
05 53 00 Metal Gratings
05 54 00 Metal Floor Plates

05 55 00 Metal Stair Treads and Nosings
05 56 00 Metal Castings
05 58 00 Formed Metal Fabrications
05 59 00 Metal Specialties
05 60 00 Unassigned
05 70 00 DECORATIVE METAL
05 71 00 Decorative Metal Stairs
05 73 00 Decorative Metal Railings
05 74 00 Decorative Metal Castings
05 75 00 Decorative Formed Metal
05 76 00 Decorative Forged Metal
05 80 00 Unassigned
05 90 00 Unassigned

DIVISION 06—WOOD, PLASTICS, AND COMPOSITES

06 00 00 WOOD, PLASTICS, AND COMPOSITES
06 01 00 Maintenance of Wood, Plastics, and Composites
06 01 10 Maintenance of Rough Carpentry
06 05 00 Common Work Results for Wood, Plastics, and Composites
06 06 00 Schedules for Wood, Plastics, and Composites
06 10 00 ROUGH CARPENTRY
06 11 00 Wood Framing
06 12 00 Structural Panels
06 13 00 Heavy Timber
06 14 00 Treated Wood Foundations
06 15 00 Wood Decking
06 16 00 Sheathing
06 17 00 Shop-Fabricated Structural Wood
06 18 00 Glued-Laminated Construction
06 20 00 FINISH CARPENTRY
06 22 00 Millwork
06 25 00 Prefinished Paneling
06 26 00 Board Paneling
06 30 00 Unassigned
06 40 00 ARCHITECTURAL WOODWORK
06 41 00 Architectural Wood Casework
06 42 00 Wood Paneling
06 43 00 Wood Stairs and Railings
06 44 00 Ornamental Woodwork
06 46 00 Wood Trim
06 48 00 Wood Frames
06 49 00 Wood Screens and Exterior Wood Shutters
06 50 00 STRUCTURAL PLASTICS
06 51 00 Structural Plastic Shapes and Plates
06 52 00 Plastic Structural Assemblies
06 53 00 Plastic Decking
06 60 00 PLASTIC FABRICATIONS
06 61 00 Simulated Stone Fabrications
06 63 00 Plastic Railings
06 64 00 Plastic Paneling
06 65 00 Plastic Simulated Wood Trim
06 66 00 Custom Ornamental Simulated Woodwork
06 70 00 STRUCTURAL COMPOSITES
06 71 00 Structural Composite Shapes and Plates
06 72 00 Composite Structural Assemblies

06 73 00 Composite Decking
06 80 00 COMPOSITE FABRICATIONS
06 81 00 Composite Railings
06 82 00 Glass-Fiber-Reinforced Plastic
06 90 00 Unassigned

DIVISION 07—THERMAL AND MOISTURE PROTECTION

07 00 00 THERMAL AND MOISTURE PROTECTION
07 01 00 Operation and Maintenance of Thermal and Moisture Protection
07 05 00 Common Work Results for Thermal and Moisture Protection
07 06 00 Schedules for Thermal and Moisture Protection
07 08 00 Commissioning of Thermal and Moisture Protection
07 10 00 DAMPPROOFING AND WATERPROOFING
07 11 00 Dampproofing
07 12 00 Built-Up Bituminous Waterproofing
07 13 00 Sheet Waterproofing
07 14 00 Fluid-Applied Waterproofing
07 15 00 Sheet Metal Waterproofing
07 16 00 Cementitious and Reactive Waterproofing
07 17 00 Bentonite Waterproofing
07 18 00 Traffic Coatings
07 19 00 Water Repellents
07 20 00 THERMAL PROTECTION
07 21 00 Thermal Insulation
07 22 00 Roof and Deck Insulation
07 24 00 Exterior Insulation and Finish Systems
07 25 00 WEATHER BARRIERS
07 26 00 Vapor Retarders
07 27 00 Air Barriers
07 30 00 STEEP SLOPE ROOFING
07 31 00 Shingles and Shakes
07 32 00 Roof Tiles
07 33 00 Natural Roof Coverings
07 40 00 ROOFING AND SIDING PANELS
07 41 00 Roof Panels
07 42 00 Wall Panels
07 44 00 Faced Panels
07 46 00 Siding
07 50 00 MEMBRANE ROOFING
07 51 00 Built-Up Bituminous Roofing
07 52 00 Modified Bituminous Membrane Roofing
07 53 00 Elastomeric Membrane Roofing
07 54 00 Thermoplastic Membrane Roofing
07 55 00 Protected Membrane Roofing
07 56 00 Fluid-Applied Roofing
07 57 00 Coated Foamed Roofing
07 58 00 Roll Roofing
07 60 00 FLASHING AND SHEET METAL
07 61 00 Sheet Metal Roofing
07 62 00 Sheet Metal Flashing and Trim
07 63 00 Sheet Metal Roofing Specialties
07 65 00 Flexible Flashing
07 70 00 ROOF AND WALL SPECIALTIES AND ACCESSORIES
07 71 00 Roof Specialties
07 72 00 Roof Accessories

07 76 00   Roof Pavers
07 77 00   Wall Specialties
07 80 00 FIRE AND SMOKE PROTECTION
07 81 00   Applied Fireproofing
07 82 00   Board Fireproofing
07 84 00   Firestopping
07 86 00   Smoke Seals
07 87 00   Smoke Containment Barriers
07 90 00 JOINT PROTECTION
07 91 00   Preformed Joint Seals
07 92 00   Joint Sealants
07 95 00   Expansion Control

DIVISION 08—OPENINGS

08 00 00 OPENINGS
08 01 00   Operation and Maintenance of Openings
08 05 00   Common Work Results for Openings
08 06 00   Schedules for Openings
08 08 00   Commissioning of Openings
08 10 00 DOORS AND FRAMES
08 11 00   Metal Doors and Frames
08 12 00   Metal Frames
08 13 00   Metal Doors
08 14 00   Wood Doors
08 15 00   Plastic Doors
08 16 00   Composite Doors
08 17 00   Integrated Door Opening Assemblies
08 20 00   Unassigned
08 30 00 SPECIALTY DOORS AND FRAMES
08 31 00   Access Doors and Panels
08 32 00   Sliding Glass Doors
08 33 00   Coiling Doors and Grilles
08 34 00   Special Function Doors
08 35 00   Folding Doors and Grilles
08 36 00   Panel Doors
08 38 00   Traffic Doors
08 39 00   Pressure-Resistant Doors
08 40 00 ENTRANCES, STOREFRONTS, AND CURTAIN WALLS
08 41 00   Entrances and Storefronts
08 42 00   Entrances
08 43 00   Storefronts
08 44 00   Curtain Wall and Glazed Assemblies
08 45 00   Translucent Wall and Roof Assemblies
08 50 00 WINDOWS
08 51 00   Metal Windows
08 52 00   Wood Windows
08 53 00   Plastic Windows
08 54 00   Composite Windows
08 55 00   Pressure-Resistant Windows
08 56 00   Special Function Windows
08 60 00 ROOF WINDOWS AND SKYLIGHTS
08 61 00   Roof Windows
08 62 00   Unit Skylights
08 63 00   Metal-Framed Skylights

08 64 00 Plastic-Framed Skylights
08 67 00 Skylight Protection and Screens
08 70 00 HARDWARE
08 71 00 Door Hardware
08 74 00 Access Control Hardware
08 75 00 Window Hardware
08 78 00 Special Function Hardware
08 79 00 Hardware Accessories
08 80 00 GLAZING
08 81 00 Glass Glazing
08 83 00 Mirrors
08 84 00 Plastic Glazing
08 85 00 Glazing Accessories
08 87 00 Glazing Surface Films
08 88 00 Special Function Glazing
08 90 00 LOUVERS AND VENTS
08 91 00 Louvers
08 92 00 Louvered Equipment Enclosures
08 95 00 Vents

DIVISION 09—FINISHES

09 00 00 FINISHES
09 01 00 Maintenance of Finishes
09 05 00 Common Work Results for Finishes
09 06 00 Schedules for Finishes
09 08 00 Commissioning of Finishes
09 10 00 Unassigned
09 20 00 PLASTER AND GYPSUM BOARD
09 21 00 Plaster and Gypsum Board Assemblies
09 22 00 Supports for Plaster and Gypsum Board
09 23 00 Gypsum Plastering
09 24 00 Portland Cement Plastering
09 25 00 Other Plastering
09 26 00 Veneer Plastering
09 27 00 Plaster Fabrications
09 28 00 Backing Boards and Underlayments
09 29 00 Gypsum Board
09 30 00 TILING
09 31 00 Thin-Set Tiling
09 32 00 Mortar-Bed Tiling
09 33 00 Conductive Tiling
09 34 00 Waterproofing-Membrane Tiling
09 35 00 Chemical-Resistant Tiling
09 40 00 Unassigned
09 50 00 CEILINGS
09 51 00 Acoustical Ceilings
09 53 00 Acoustical Ceiling Suspension Assemblies
09 54 00 Specialty Ceilings
09 56 00 Textured Ceilings
09 57 00 Special Function Ceilings
09 58 00 Integrated Ceiling Assemblies
09 60 00 FLOORING
09 61 00 Flooring Treatment
09 62 00 Specialty Flooring

09 63 00 Masonry Flooring
09 64 00 Wood Flooring
09 65 00 Resilient Flooring
09 66 00 Terrazzo Flooring
09 67 00 Fluid-Applied Flooring
09 68 00 Carpeting
09 69 00 Access Flooring
09 70 00 WALL FINISHES
09 72 00 Wall Coverings
09 73 00 Wall Carpeting
09 74 00 Flexible Wood Sheets
09 75 00 Stone Facing
09 76 00 Plastic Blocks
09 77 00 Special Wall Surfacing
09 80 00 ACOUSTIC TREATMENT
09 81 00 Acoustic Insulation
09 83 00 Acoustic Finishes
09 84 00 Acoustic Room Components
09 90 00 PAINTING AND COATING
09 91 00 Painting
09 93 00 Staining and Transparent Finishing
09 94 00 Decorative Finishing
09 96 00 High-Performance Coatings
09 97 00 Special Coatings

DIVISION 10—SPECIALTIES

10 00 00 SPECIALTIES
10 01 00 Operation and Maintenance of Specialties
10 05 00 Common Work Results for Specialties
10 06 00 Schedules for Specialties
10 08 00 Commissioning of Specialties
10 10 00 INFORMATION SPECIALTIES
10 11 00 Visual Display Surfaces
10 12 00 Display Cases
10 13 00 Directories
10 14 00 Signage
10 17 00 Telephone Specialties
10 18 00 Informational Kiosks
10 20 00 INTERIOR SPECIALTIES
10 21 00 Compartments and Cubicles
10 22 00 Partitions
10 25 00 Service Walls
10 26 00 Wall and Door Protection
10 28 00 Toilet, Bath, and Laundry Accessories
10 30 00 FIREPLACES AND STOVES
10 31 00 Manufactured Fireplaces
10 32 00 Fireplace Specialties
10 35 00 Stoves
10 40 00 SAFETY SPECIALTIES
10 41 00 Emergency Access and Information Cabinets
10 43 00 Emergency Aid Specialties
10 44 00 Fire Protection Specialties
10 50 00 STORAGE SPECIALTIES
10 51 00 Lockers
10 55 00 Postal Specialties
10 56 00 Storage Assemblies
10 57 00 Wardrobe and Closet Specialties

10 60 00 Unassigned
10 70 00 EXTERIOR SPECIALTIES
10 71 00 Exterior Protection
10 73 00 Protective Covers
10 74 00 Manufactured Exterior Specialties
10 75 00 Flagpoles
10 80 00 OTHER SPECIALTIES
10 81 00 Pest Control Devices
10 82 00 Grilles and Screens
10 83 00 Flags and Banners
10 86 00 Security Mirrors and Domes
10 88 00 Scales
10 90 00 Unassigned

DIVISION 11—EQUIPMENT

11 00 00 EQUIPMENT
11 01 00 Operation and Maintenance of Equipment
11 05 00 Common Work Results for Equipment
11 06 00 Schedules for Equipment
11 08 00 Commissioning of Equipment
11 10 00 VEHICLE AND PEDESTRIAN EQUIPMENT
11 11 00 Vehicle Service Equipment
11 12 00 Parking Control Equipment
11 13 00 Loading Dock Equipment
11 14 00 Pedestrian Control Equipment
11 15 00 SECURITY, DETENTION AND BANKING EQUIPMENT
11 16 00 Vault Equipment
11 17 00 Teller and Service Equipment
11 18 00 Security Equipment
11 19 00 Detention Equipment
11 20 00 COMMERCIAL EQUIPMENT
11 21 00 Mercantile and Service Equipment
11 22 00 Refrigerated Display Equipment
11 23 00 Commercial Laundry and Dry Cleaning Equipment
11 24 00 Maintenance Equipment
11 25 00 Hospitality Equipment
11 26 00 Unit Kitchens
11 27 00 Photographic Processing Equipment
11 28 00 Office Equipment
11 29 00 Postal, Packaging, and Shipping Equipment
11 30 00 RESIDENTIAL EQUIPMENT
11 31 00 Residential Appliances
11 33 00 Retractable Stairs
11 40 00 FOODSERVICE EQUIPMENT
11 41 00 Food Storage Equipment
11 42 00 Food Preparation Equipment
11 43 00 Food Delivery Carts and Conveyors
11 44 00 Food Cooking Equipment
11 46 00 Food Dispensing Equipment
11 47 00 Ice Machines
11 48 00 Cleaning and Disposal Equipment
11 50 00 EDUCATIONAL AND SCIENTIFIC EQUIPMENT
11 51 00 Library Equipment
11 52 00 Audio-Visual Equipment
11 53 00 Laboratory Equipment

11 55 00 Planetarium Equipment
11 56 00 Observatory Equipment
11 57 00 Vocational Shop Equipment
11 59 00 Exhibit Equipment
11 60 00 ENTERTAINMENT EQUIPMENT
11 61 00 Theater and Stage Equipment
11 62 00 Musical Equipment
11 65 00 ATHLETIC AND RECREATIONAL EQUIPMENT
11 66 00 Athletic Equipment
11 67 00 Recreational Equipment
11 68 00 Play Field Equipment and Structures
11 70 00 HEALTHCARE EQUIPMENT
11 71 00 Medical Sterilizing Equipment
11 72 00 Examination and Treatment Equipment
11 73 00 Patient Care Equipment
11 74 00 Dental Equipment
11 75 00 Optical Equipment
11 76 00 Operating Room Equipment
11 77 00 Radiology Equipment
11 78 00 Mortuary Equipment
11 79 00 Therapy Equipment
11 80 00 COLLECTION AND DISPOSAL EQUIPMENT
11 82 00 Solid Waste Handling Equipment
11 90 00 OTHER EQUIPMENT
11 91 00 Religious Equipment
11 92 00 Agricultural Equipment
11 93 00 Horticultural Equipment

DIVISION 12—FURNISHINGS

12 00 00 FURNISHINGS
12 01 00 Operation and Maintenance of Furnishings
12 05 00 Common Work Results for Furnishings
12 06 00 Schedules for Furnishings
12 08 00 Commissioning of Furnishings
12 10 00 ART
12 11 00 Murals
12 12 00 Wall Decorations
12 14 00 Sculptures
12 17 00 Art Glass
12 19 00 Religious Art
12 20 00 WINDOW TREATMENTS
12 21 00 Window Blinds
12 22 00 Curtains and Drapes
12 23 00 Interior Shutters
12 24 00 Window Shades
12 25 00 Window Treatment Operating Hardware
12 30 00 CASEWORK
12 31 00 Manufactured Metal Casework
12 32 00 Manufactured Wood Casework
12 34 00 Manufactured Plastic Casework
12 35 00 Specialty Casework
12 36 00 Countertops
12 40 00 FURNISHINGS AND ACCESSORIES
12 41 00 Office Accessories
12 42 00 Table Accessories
12 43 00 Portable Lamps

12 44 00 Bath Furnishings
12 45 00 Bedroom Furnishings
12 46 00 Furnishing Accessories
12 48 00 Rugs and Mats
12 50 00 FURNITURE
12 51 00 Office Furniture
12 52 00 Seating
12 53 00 Retail Furniture
12 54 00 Hospitality Furniture
12 55 00 Detention Furniture
12 56 00 Institutional Furniture
12 57 00 Industrial Furniture
12 58 00 Residential Furniture
12 59 00 Systems Furniture
12 60 00 MULTIPLE SEATING
12 61 00 Fixed Audience Seating
12 62 00 Portable Audience Seating
12 63 00 Stadium and Arena Seating
12 64 00 Booths and Tables
12 65 00 Multiple-Use Fixed Seating
12 66 00 Telescoping Stands
12 67 00 Pews and Benches
12 68 00 Seat and Table Assemblies
12 70 00 Unassigned
12 80 00 Unassigned
12 90 00 OTHER FURNISHINGS
12 92 00 Interior Planters and Artificial Plants
12 93 00 Site Furnishings

DIVISION 13—SPECIAL CONSTRUCTION

13 00 00 SPECIAL CONSTRUCTION
13 01 00 Operation and Maintenance of Special Construction
13 05 00 Common Work Results for Special Construction
13 06 00 Schedules for Special Construction
13 08 00 Commissioning of Special Construction
13 10 00 SPECIAL FACILITY COMPONENTS
13 11 00 Swimming Pools
13 12 00 Fountains
13 13 00 Aquariums
13 14 00 Amusement Park Structures and Equipment
13 17 00 Tubs and Pools
13 18 00 Ice Rinks
13 19 00 Kennels and Animal Shelters
13 20 00 SPECIAL PURPOSE ROOMS
13 21 00 Controlled Environment Rooms
13 22 00 Office Shelters and Booths
13 23 00 Planetariums
13 24 00 Special Activity Rooms
13 26 00 Fabricated Rooms
13 27 00 Vaults
13 28 00 Athletic and Recreational Special Construction
13 30 00 SPECIAL STRUCTURES
13 31 00 Fabric Structures
13 32 00 Space Frames
13 33 00 Geodesic Structures
13 34 00 Fabricated Engineered Structures
13 36 00 Towers

13 40 00 INTEGRATED CONSTRUCTION
13 42 00   Building Modules
13 44 00   Modular Mezzanines
13 48 00   Sound, Vibration, and Seismic Control
13 49 00   Radiation Protection
13 50 00 SPECIAL INSTRUMENTATION
13 51 00   Stress Instrumentation
13 52 00   Seismic Instrumentation
13 53 00   Meteorological Instrumentation
13 60 00   Unassigned
13 70 00   Unassigned
13 80 00   Unassigned
13 90 00   Unassigned

DIVISION 14—CONVEYING EQUIPMENT

14 00 00 CONVEYING EQUIPMENT
14 01 00   Operation and Maintenance of Conveying Equipment
14 05 00   Common Work Results for Conveying Equipment
14 06 00   Schedules for Conveying Equipment
14 08 00   Commissioning of Conveying Equipment
14 10 00 DUMBWAITERS
14 11 00   Manual Dumbwaiters
14 12 00   Electric Dumbwaiters
14 14 00   Hydraulic Dumbwaiters
14 20 00   ELEVATORS
14 21 00   Electric Traction Elevators
14 24 00   Hydraulic Elevators
14 26 00   Limited-Use/Limited-Application Elevators
14 27 00   Custom Elevator Cabs
14 28 00   Elevator Equipment and Controls
14 30 00 ESCALATORS AND MOVING WALKS
14 31 00   Escalators
14 32 00   Moving Walks
14 33 00   Moving Ramps
14 40 00 LIFTS
14 41 00   People Lifts
14 42 00   Wheelchair Lifts
14 43 00   Platform Lifts
14 44 00   Sidewalk Lifts
14 45 00   Vehicle Lifts
14 50 00   Unassigned
14 60 00   Unassigned
14 70 00 TURNTABLES
14 71 00   Industrial Turntables
14 72 00   Hospitality Turntables
14 73 00   Exhibit Turntables
14 74 00   Entertainment Turntables
14 80 00 SCAFFOLDING
14 81 00   Suspended Scaffolding
14 82 00   Rope Climbers
14 83 00   Elevating Platforms
14 84 00   Powered Scaffolding
14 90 00 OTHER CONVEYING EQUIPMENT
14 91 00   Facility Chutes
14 92 00   Pneumatic Tube Systems

DIVISION 21—FIRE SUPPRESSION

21 00 00 FIRE SUPPRESSION
21 01 00   Operation and Maintenance of Fire Suppression
21 05 00   Common Work Results for Fire Suppression
21 06 00   Schedules for Fire Suppression
21 07 00   Fire Suppression Systems Insulation
21 08 00   Commissioning of Fire Suppression
21 09 00   Instrumentation and Control for Fire-Suppression Systems
21 10 00 WATER-BASED FIRE-SUPPRESSION SYSTEMS
21 11 00   Facility Fire-Suppression Water-Service Piping
21 12 00   Fire-Suppression Standpipes
21 13 00   Fire-Suppression Sprinkler Systems
21 20 00 FIRE-EXTINGUISHING SYSTEMS
21 21 00   Carbon-Dioxide Fire-Extinguishing Systems
21 22 00   Clean-Agent Fire-Extinguishing Systems
21 23 00   Wet-Chemical Fire-Extinguishing Systems
21 24 00   Dry-Chemical Fire-Extinguishing Systems
21 30 00 FIRE PUMPS
21 31 00   Centrifugal Fire Pumps
21 32 00   Vertical-Turbine Fire Pumps
21 33 00   Positive-Displacement Fire Pumps
21 40 00 FIRE-SUPPRESSION WATER STORAGE
21 41 00   Storage Tanks for Fire-Suppression Water
21 50 00   Unassigned
21 60 00   Unassigned
21 70 00   Unassigned
21 80 00   Unassigned
21 90 00   Unassigned

DIVISION 22—PLUMBING

22 00 00 PLUMBING
22 01 00   Operation and Maintenance of Plumbing
22 05 00   Common Work Results for Plumbing
22 06 00   Schedules for Plumbing
22 07 00   Plumbing Insulation
22 08 00   Commissioning of Plumbing
22 09 00   Instrumentation and Control for Plumbing
22 10 00 PLUMBING PIPING AND PUMPS
22 11 00   Facility Water Distribution
22 12 00   Facility Potable-Water Storage Tanks
22 13 00   Facility Sanitary Sewerage
22 14 00   Facility Storm Drainage
22 15 00   General Service Compressed-Air Systems
22 20 00   Unassigned
22 30 00 PLUMBING EQUIPMENT
22 31 00   Domestic Water Softeners
22 32 00   Domestic Water Filtration Equipment
22 33 00   Electric Domestic Water Heaters
22 34 00   Fuel-Fired Domestic Water Heaters
22 35 00   Domestic Water Heat Exchangers
22 40 00 PLUMBING FIXTURES
22 41 00   Residential Plumbing Fixtures
22 42 00   Commercial Plumbing Fixtures
22 43 00   Healthcare Plumbing Fixtures

22 45 00   Emergency Plumbing Fixtures
22 46 00   Security Plumbing Fixtures
22 47 00   Drinking Fountains and Water Coolers
22 50 00 POOL AND FOUNTAIN PLUMBING SYSTEMS
22 51 00   Swimming Pool Plumbing Systems
22 52 00   Fountain Plumbing Systems
22 60 00 GAS AND VACUUM SYSTEMS FOR LABORATORY AND HEALTHCARE FACILITIES
22 61 00   Compressed-Air Systems for Laboratory and Healthcare Facilities
22 62 00   Vacuum Systems for Laboratory and Healthcare Facilities
22 63 00   Gas Systems for Laboratory and Healthcare Facilities
22 66 00   Chemical-Waste Systems for Laboratory and Healthcare Facilities
22 67 00   Processed Water Systems for Laboratory and Healthcare Facilities
22 70 00   Unassigned
22 80 00   Unassigned
22 90 00   Unassigned

DIVISION 23—HEATING, VENTILATING, AND AIR-CONDITIONING (HVAC)

23 00 00 HEATING, VENTILATING, AND AIR-CONDITIONING (HVAC)
23 01 00   Operation and Maintenance of HVAC Systems
23 05 00   Common Work Results for HVAC
23 06 00   Schedules for HVAC
23 07 00   HVAC Insulation
23 08 00   Commissioning of HVAC
23 09 00   Instrumentation and Control for HVAC
23 10 00 FACILITY FUEL SYSTEMS
23 11 00   Facility Fuel Piping
23 12 00   Facility Fuel Pumps
23 13 00   Facility Fuel-Storage Tanks
23 20 00 HVAC PIPING AND PUMPS
23 21 00   Hydronic Piping and Pumps
23 22 00   Steam and Condensate Piping and Pumps
23 23 00   Refrigerant Piping
23 24 00   Internal-Combustion Engine Piping
23 25 00   HVAC Water Treatment
23 30 00   HVAC AIR DISTRIBUTION
23 31 00   HVAC Ducts and Casings
23 32 00   Air Plenums and Chases
23 33 00   Air Duct Accessories
23 34 00   HVAC Fans
23 35 00   Special Exhaust Systems
23 36 00   Air Terminal Units
23 37 00   Air Outlets and Inlets
23 38 00   Ventilation Hoods
23 40 00 HVAC AIR CLEANING DEVICES
23 41 00   Particulate Air Filtration
23 42 00   Gas-Phase Air Filtration
23 43 00   Electronic Air Cleaners
23 50 00 CENTRAL HEATING EQUIPMENT
23 51 00   Breechings, Chimneys, and Stacks
23 52 00   Heating Boilers
23 53 00   Heating Boiler Feedwater Equipment
23 54 00   Furnaces
23 55 00   Fuel-Fired Heaters

23 56 00 Solar Energy Heating Equipment
23 57 00 Heat Exchangers for HVAC
23 60 00 CENTRAL COOLING EQUIPMENT
23 61 00 Refrigerant Compressors
23 62 00 Packaged Compressor and Condenser Units
23 63 00 Refrigerant Condensers
23 64 00 Packaged Water Chillers
23 65 00 Cooling Towers
23 70 00 CENTRAL HVAC EQUIPMENT
23 71 00 Thermal Storage
23 72 00 Air-to-Air Energy Recovery Equipment
23 73 00 Indoor Central-Station Air-Handling Units
23 74 00 Packaged Outdoor HVAC Equipment
23 75 00 Custom-Packaged Outdoor HVAC Equipment
23 76 00 Evaporative Air-Cooling Equipment
23 80 00 DECENTRALIZED HVAC EQUIPMENT
23 81 00 Decentralized Unitary HVAC Equipment
23 82 00 Convection Heating and Cooling Units
23 83 00 Radiant Heating Units
23 84 00 Humidity Control Equipment
23 90 00 Unassigned

DIVISION 25—INTEGRATED AUTOMATION

25 00 00 INTEGRATED AUTOMATION
25 01 00 Operation and Maintenance of Integrated Automation
25 05 00 Common Work Results for Integrated Automation
25 06 00 Schedules for Integrated Automation
25 08 00 Commissioning of Integrated Automation
25 10 00 INTEGRATED AUTOMATION NETWORK EQUIPMENT
25 11 00 Integrated Automation Network Devices
25 12 00 Integrated Automation Network Gateways
25 13 00 Integrated Automation Control and Monitoring Network
25 14 00 Integrated Automation Local Control Units
25 15 00 Integrated Automation Software
25 20 00 Unassigned
25 30 00 INTEGRATED AUTOMATION INSTRUMENTATION AND TERMINAL DEVICES
25 31 00 Integrated Automation Instrumentation and Terminal Devices for Facility Equipment
25 32 00 Integrated Automation Instrumentation and Terminal Devices for Conveying Equipment
25 33 00 Integrated Automation Instrumentation and Terminal Devices for Fire-Suppression Systems
25 34 00 Integrated Automation Instrumentation and Terminal Devices for Plumbing
25 35 00 Integrated Automation Instrumentation and Terminal Devices for HVAC
25 36 00 Integrated Automation Instrumentation and Terminal Devices for Electrical Systems
25 37 00 Integrated Automation Instrumentation and Terminal Devices for Communications Systems
25 38 00 Integrated Automation Instrumentation and Terminal Devices for Electronic Safety and Security Systems
25 40 00 Unassigned
25 50 00 INTEGRATED AUTOMATION FACILITY CONTROLS
25 51 00 Integrated Automation Control of Facility Equipment
25 52 00 Integrated Automation Control of Conveying Equipment

25 53 00   Integrated Automation Control of Fire-Suppression Systems
25 54 00   Integrated Automation Control of Plumbing
25 55 00   Integrated Automation Control of HVAC
25 56 00   Integrated Automation Control of Electrical Systems
25 57 00   Integrated Automation Control of Communications Systems
25 58 00   Integrated Automation Control of Electronic Safety and Security Systems
25 60 00   Unassigned
25 70 00   Unassigned
25 80 00   Unassigned
25 90 00 INTEGRATED AUTOMATION CONTROL SEQUENCES
25 91 00   Integrated Automation Control Sequences for Facility Equipment
25 92 00   Integrated Automation Control Sequences for Conveying Equipment
25 93 00   Integrated Automation Control Sequences for Fire-Suppression Systems
25 94 00   Integrated Automation Control Sequences for Plumbing
25 95 00   Integrated Automation Control Sequences for HVAC
25 96 00   Integrated Automation Control Sequences for Electrical Systems
25 97 00   Integrated Automation Control Sequences for Communications Systems
25 98 00   Integrated Automation Control Sequences for Electronic Safety and Security Systems

DIVISION 26—ELECTRICAL

26 00 00 ELECTRICAL
26 01 00   Operation and Maintenance of Electrical Systems
26 05 00   Common Work Results for Electrical
26 06 00   Schedules for Electrical
26 08 00   Commissioning of Electrical Systems
26 09 00   Instrumentation and Control for Electrical Systems
26 10 00 MEDIUM -VOLTAGE ELECTRICAL DISTRIBUTION
26 11 00   Substations
26 12 00   Medium-Voltage Transformers
26 13 00   Medium-Voltage Switchgear
26 18 00   Medium-Voltage Circuit Protection Devices
26 20 00 LOW—VOLTAGE ELECTRICAL DISTRIBUTION
26 21 00   Low-Voltage Overhead Electrical Power Systems
26 22 00   Low-Voltage Transformers
26 23 00   Low-Voltage Switchgear
26 24 00   Switchboards and Panelboards
26 25 00   Enclosed Bus Assemblies
26 26 00   Power Distribution Units
26 27 00   Low-Voltage Distribution Equipment
26 28 00   Low-Voltage Circuit Protective Devices
26 29 00   Low-Voltage Controllers
26 30 00 FACILITY ELECTRICAL POWER GENERATING AND STORING EQUIPMENT
26 31 00   Photovoltaic Collectors
26 32 00   Packaged Generator Assemblies
26 33 00   Battery Equipment
26 35 00   Power Filters and Conditioners
26 36 00   Transfer Switches
26 40 00 ELECTRICAL AND CATHODIC PROTECTION
26 41 00   Facility Lightning Protection
26 42 00   Cathodic Protection
26 43 00   Transient Voltage Suppression
26 50 00 LIGHTING
26 51 00   Interior Lighting
26 52 00   Emergency Lighting

26 53 00 Exit Signs
26 54 00 Classified Location Lighting
26 55 00 Special Purpose Lighting
26 56 00 Exterior Lighting
26 60 00 Unassigned
26 70 00 Unassigned
26 80 00 Unassigned
26 90 00 Unassigned

DIVISION 27—COMMUNICATIONS

27 00 00 COMMUNICATIONS
27 01 00 Operation and Maintenance of Communications Systems
27 05 00 Common Work Results for Communications
27 06 00 Schedules for Communications
27 08 00 Commissioning of Communications
27 10 00 STRUCTURED CABLING
27 11 00 Communications Equipment Room Fittings
27 13 00 Communications Backbone Cabling
27 15 00 Communications Horizontal Cabling
27 16 00 Communications Connecting Cords, Devices and Adapters
27 20 00 DATA COMMUNICATIONS
27 21 00 Data Communications Network Equipment
27 22 00 Data Communications Hardware
27 24 00 Data Communications Peripheral Data Equipment
27 25 00 Data Communications Software
27 26 00 Data Communications Programming and Integration Services
27 30 00 VOICE COMMUNICATIONS
27 31 00 Voice Communications Switching and Routing Equipment
27 32 00 Voice Communications Telephone Sets, Facsimiles
          and Modems
27 33 00 Voice Communications Messaging
27 34 00 Call Accounting
27 35 00 Call Management
27 40 00 AUDIO-VIDEO COMMUNICATIONS
27 41 00 Audio-Video Systems
27 42 00 Electronic Digital Systems
27 50 00 DISTRIBUTED COMMUNICATIONS AND MONITORING SYSTEMS
27 51 00 Distributed Audio-Video Communications Systems
27 52 00 Healthcare Communications and Monitoring Systems
27 53 00 Distributed Systems
27 60 00 Unassigned
27 70 00 Unassigned
27 80 00 Unassigned
27 90 00 Unassigned

DIVISION 28—ELECTRONIC SAFETY AND SECURITY

28 00 00 ELECTRONIC SAFETY AND SECURITY
28 01 00 Operation and Maintenance of Electronic Safety
          and Security
28 05 00 Common Work Results for Electronic Safety and Security
28 06 00 Schedules for Electronic Safety and Security
28 08 00 Commissioning of Electronic Safety and Security

28 10 00 ELECTRONIC ACCESS CONTROL AND INTRUSION DETECTION
28 13 00 Access Control
28 16 00 Intrusion Detection
28 20 00 ELECTRONIC SURVEILLANCE
28 23 00 Video Surveillance
28 26 00 Electronic Personal Protection Systems
28 30 00 ELECTRONIC DETECTION AND ALARM
28 31 00 Fire Detection and Alarm
28 32 00 Radiation Detection and Alarm
28 33 00 Fuel -Gas Detection and Alarm
28 34 00 Fuel-Oil Detection and Alarm
28 35 00 Refrigerant Detection and Alarm
28 40 00 ELECTRONIC MONITORING AND CONTROL
28 46 00 Electronic Detention Monitoring and Control Systems
28 50 00 Unassigned
28 60 00 Unassigned
28 70 00 Unassigned
28 80 00 Unassigned
28 90 00 Unassigned

DIVISION 31—EARTHWORK

31 00 00 EARTHWORK
31 01 00 Maintenance of Earthwork
31 05 00 Common Work Results for Earthwork
31 06 00 Schedules for Earthwork
31 08 00 Commissioning of Earthwork
31 09 00 Geotechnical Instrumentation and Monitoring of Earthwork
31 10 00 SITE CLEARING
31 11 00 Clearing and Grubbing
31 12 00 Selective Clearing
31 13 00 Selective Tree and Shrub Removal and Trimming
31 14 00 Earth Stripping and Stockpiling
31 20 00 EARTH MOVING
31 21 00 Off-Gassing Mitigation
31 22 00 Grading
31 23 00 Excavation and Fill
31 24 00 Embankments
31 25 00 Erosion and Sedimentation Controls
31 30 00 EARTHWORK METHODS
31 31 00 Soil Treatment
31 32 00 Soil Stabilization
31 33 00 Rock Stabilization
31 34 00 Soil Reinforcement
31 35 00 Slope Protection
31 36 00 Gabions
31 37 00 Riprap
31 40 00 SHORING AND UNDERPINNING
31 41 00 Shoring
31 43 00 Concrete Raising
31 45 00 Vibroflotation and Densification
31 46 00 Needle Beams
31 48 00 Underpinning
31 50 00 EXCAVATION SUPPORT AND PROTECTION
31 51 00 Anchor Tiebacks
31 52 00 Cofferdams

31 53 00 Cribbing and Walers
31 54 00 Ground Freezing
31 56 00 Slurry Walls
31 60 00 SPECIAL FOUNDATIONS AND LOAD-BEARING ELEMENTS
31 62 00 Driven Piles
31 63 00 Bored Piles
31 64 00 Caissons
31 66 00 Special Foundations
31 68 00 Foundation Anchors
31 70 00 TUNNELING AND MINING
31 71 00 Tunnel Excavation
31 72 00 Tunnel Support Systems
31 73 00 Tunnel Grouting
31 74 00 Tunnel Construction
31 75 00 Shaft Construction
31 77 00 Submersible Tube Tunnels
31 80 00 Unassigned
31 90 00 Unassigned

DIVISION 32—EXTERIOR IMPROVEMENTS

32 00 00 EXTERIOR IMPROVEMENTS
32 01 00 Operation and Maintenance of Exterior Improvements
32 05 00 Common Work Results for Exterior Improvements
32 06 00 Schedules for Exterior Improvements
32 08 00 Commissioning of Exterior Improvements
32 10 00 BASES, BALLASTS, AND PAVING
32 11 00 Base Courses
32 12 00 Flexible Paving
32 13 00 Rigid Paving
32 14 00 Unit Paving
32 15 00 Aggregate Surfacing
32 16 00 Curbs and Gutters
32 17 00 Paving Specialties
32 18 00 Athletic and Recreational Surfacing
32 20 00 Unassigned
32 30 00 SITE IMPROVEMENTS
32 31 00 Fences and Gates
32 32 00 Retaining Walls
32 34 00 Fabricated Bridges
32 35 00 Screening Devices
32 40 00 Unassigned
32 50 00 Unassigned
32 60 00 Unassigned
32 70 00 WETLANDS
32 71 00 Constructed Wetlands
32 72 00 Wetlands Restoration
32 80 00 IRRIGATION
32 82 00 Irrigation Pumps
32 84 00 Planting Irrigation
32 86 00 Agricultural Irrigation
32 90 00 PLANTING
32 91 00 Planting Preparation
32 92 00 Turf and Grasses
32 93 00 Plants
32 94 00 Planting Accessories
32 96 00 Transplanting

DIVISION 33—UTILITIES

33 00 00 UTILITIES
33 01 00 Operation and Maintenance of Utilities
33 05 00 Common Work Results for Utilities
33 06 00 Schedules for Utilities
33 08 00 Commissioning of Utilities
33 09 00 Instrumentation and Control for Utilities
33 10 00 WATER UTILITIES
33 11 00 Water Utility Distribution Piping
33 12 00 Water Utility Distribution Equipment
33 13 00 Disinfecting of Water Utility Distribution
33 16 00 Water Utility Storage Tanks
33 20 00 WELLS
33 21 00 Water Supply Wells
33 22 00 Test Wells
33 23 00 Extraction Wells
33 24 00 Monitoring Wells
33 25 00 Recharge Wells
33 26 00 Relief Wells
33 29 00 Well Abandonment
33 30 00 SANITARY SEWERAGE UTILITIES
33 31 00 Sanitary Utility Sewerage Piping
33 32 00 Wastewater Utility Pumping Stations
33 33 00 Low Pressure Utility Sewerage
33 34 00 Sanitary Utility Sewerage Force Mains
33 36 00 Utility Septic Tanks
33 39 00 Sanitary Utility Sewerage Structures
33 40 00 STORM DRAINAGE UTILITIES
33 41 00 Storm Utility Drainage Piping
33 42 00 Culverts
33 44 00 Storm Utility Water Drains
33 45 00 Storm Utility Drainage Pumps
33 46 00 Subdrainage
33 47 00 Ponds and Reservoirs
33 49 00 Storm Drainage Structures
33 50 00 FUEL DISTRIBUTION UTILITIES
33 51 00 Natural-Gas Distribution
33 52 00 Liquid Fuel Distribution
33 56 00 Fuel-Storage Tanks
33 60 00 HYDRONIC AND STEAM ENERGY UTILITIES
33 61 00 Hydronic Energy Distribution
33 63 00 Steam Energy Distribution
33 70 00 ELECTRICAL UTILITIES
33 71 00 Electrical Utility Transmission and Distribution
33 72 00 Utility Substations
33 73 00 Utility Transformers
33 75 00 High-Voltage Switchgear and Protection Devices
33 77 00 Medium-Voltage Utility Switchgear
and Protection Devices
33 79 00 Site Grounding
33 80 00 COMMUNICATIONS UTILITIES
33 81 00 Communications Structures
33 82 00 Communications Distribution
33 83 00 Wireless Communications Distribution
33 90 00 Unassigned

DIVISION 34—TRANSPORTATION

34 00 00 TRANSPORTATION
34 01 00   Operation and Maintenance of Transportation
34 05 00   Common Work Results for Transportation
34 06 00   Schedules for Transportation
34 08 00   Commissioning of Transportation
34 10 00 GUIDEWAYS/RAILWAYS
34 11 00   Rail Tracks
34 12 00   Monorails
34 13 00   Funiculars
34 14 00   Cable Transportation
34 20 00 TRACTION POWER
34 21 00   Traction Power Distribution
34 23 00   Overhead Traction Power
34 24 00   Third Rail Traction Power
34 30 00   Unassigned
34 40 00 TRANSPORTATION SIGNALING AND CONTROL EQUIPMENT
34 41 00   Roadway Signaling and Control Equipment
34 42 00   Railway Signaling and Control Equipment
34 43 00   Airfield Signaling and Control Equipment
34 48 00   Bridge Signaling and Control Equipment
34 50 00 TRANSPORTATION FARE COLLECTION EQUIPMENT
34 52 00   Vehicle Fare Collection
34 54 00   Passenger Fare Collection
34 60 00   Unassigned
34 70 00 TRANSPORTATION CONSTRUCTION AND EQUIPMENT
34 71 00   Roadway Construction
34 72 00   Railway Construction
34 73 00   Airfield Construction
34 75 00   Roadway Equipment
34 76 00   Railway Equipment
34 77 00   Transportation Equipment
34 80 00 BRIDGES
34 81 00   Bridge Machinery
34 82 00   Bridge Specialties
34 90 00   Unassigned

DIVISION 35—WATERWAY AND MARINE CONSTRUCTION

35 00 00 WATERWAY AND MARINE CONSTRUCTION
35 01 00   Operation and Maintenance of Waterway and Marine Construction
35 05 00   Common Work Results for Waterway and Marine Construction
35 06 00   Schedules for Waterway and Marine Construction
35 08 00   Commissioning of Waterway and Marine Construction
35 10 00 WATERWAY AND MARINE SIGNALING AND CONTROL EQUIPMENT
35 11 00   Signaling and Control Equipment for Waterways
35 12 00   Marine Signaling and Control Equipment
35 13 00   Signaling and Control Equipment for Dams
35 20 00 WATERWAY AND MARINE CONSTRUCTION AND EQUIPMENT
35 30 00   COASTAL CONSTRUCTION
35 31 00   Shoreline Protection
35 32 00   Artificial Reefs
35 40 00 WATERWAY CONSTRUCTION AND EQUIPMENT
35 41 00   Levees
35 42 00   Waterway Bank Protection

40 94 00 Digital Process Controllers
40 95 00 Process Control Hardware
40 96 00 Process Control Software
40 97 00 Process Control Auxiliary Devices

DIVISION 41—MATERIAL PROCESSING AND HANDLING EQUIPMENT

41 00 00 MATERIAL PROCESSING AND HANDLING EQUIPMENT
41 01 00 Operation and Maintenance of Material Processing and Handling Equipment
41 06 00 Schedules for Material Processing and Handling Equipment
41 08 00 Commissioning of Material Processing and Handling Equipment
41 10 00 BULK MATERIAL PROCESSING EQUIPMENT
41 11 00 Bulk Material Sizing Equipment
41 12 00 Bulk Material Conveying Equipment
41 13 00 Bulk Material Feeders
41 14 00 Batching Equipment
41 20 00 PIECE MATERIAL HANDLING EQUIPMENT
41 21 00 Conveyors
41 22 00 Cranes and Hoists
41 23 00 Lifting Devices
41 24 00 Specialty Material Handling Equipment
41 30 00 MANUFACTURING EQUIPMENT
41 31 00 Manufacturing Lines and Equipment
41 32 00 Forming Equipment
41 33 00 Machining Equipment
41 34 00 Finishing Equipment
41 35 00 Dies and Molds
41 36 00 Assembly and Testing Equipment
41 40 00 CONTAINER PROCESSING AND PACKAGING
41 41 00 Container Filling and Sealing
41 42 00 Container Packing Equipment
41 43 00 Shipping Packaging
41 50 00 MATERIAL STORAGE
41 51 00 Automatic Material Storage
41 52 00 Bulk Material Storage
41 53 00 Storage Equipment and Systems
41 60 00 MOBILE PLANT EQUIPMENT
41 61 00 Mobile Earth Moving Equipment
41 62 00 Trucks
41 63 00 General Vehicles
41 64 00 Rail Vehicles
41 65 00 Mobile Support Equipment
41 66 00 Miscellaneous Mobile Equipment
41 67 00 Plant Maintenance Equipment
41 70 00 Unassigned
41 80 00 Unassigned
41 90 00 Unassigned

DIVISION 42—PROCESS HEATING, COOLING, AND DRYING EQUIPMENT

42 00 00 PROCESS HEATING, COOLING, AND DRYING EQUIPMENT
42 01 00 Operation and Maintenance of Process Heating, Cooling, and Drying Equipment
42 06 00 Schedules for Process Heating, Cooling, and Drying Equipment
42 08 00 Commissioning of Process Heating, Cooling, and Drying Equipment

42 10 00 PROCESS HEATING EQUIPMENT
42 11 00 Process Boilers
42 12 00 Process Heaters
42 13 00 Industrial Heat Exchangers and Recuperators
42 14 00 Industrial Furnaces
42 15 00 Industrial Ovens
42 20 00 PROCESS COOLING EQUIPMENT
42 21 00 Process Cooling Towers
42 22 00 Process Chillers and Coolers
42 23 00 Process Condensers and Evaporators
42 30 00 PROCESS DRYING EQUIPMENT
42 31 00 Gas Dryers and Dehumidifiers
42 32 00 Material Dryers
42 40 00 Unassigned
42 50 00 Unassigned
42 60 00 Unassigned
42 70 00 Unassigned
42 80 00 Unassigned
42 90 00 Unassigned

DIVISION 43—PROCESS GAS AND LIQUID HANDLING, PURIFICATION, AND STORAGE EQUIPMENT

43 00 00 PROCESS GAS AND LIQUID HANDLING, PURIFICATION, AND STORAGE EQUIPMENT
43 01 00 Operation and Maintenance of Process Gas and Liquid Handling, Purification, and Storage Equipment
43 06 00 Schedules for Process Gas and Liquid Handling, Purification, and Storage Equipment
43 08 00 Commissioning of Process Gas and Liquid Handling, Purification, and Storage Equipment
43 10 00 GAS HANDLING EQUIPMENT
43 11 00 Gas Fans, Blowers and Pumps
43 12 00 Gas Compressors
43 13 00 Gas Process Equipment
43 20 00 LIQUID HANDLING EQUIPMENT
43 21 00 Liquid Pumps
43 22 00 Liquid Process Equipment
43 30 00 GAS AND LIQUID PURIFICATION EQUIPMENT
43 31 00 Gas and Liquid Purification Filtration Equipment
43 32 00 Gas and Liquid Purification Process Equipment
43 40 00 GAS AND LIQUID STORAGE
43 41 00 Gas and Liquid Storage Equipment
43 50 00 Unassigned
43 60 00 Unassigned
43 70 00 Unassigned
43 80 00 Unassigned
43 90 00 Unassigned

DIVISION 44—POLLUTION CONTROL EQUIPMENT

44 00 00 POLLUTION CONTROL EQUIPMENT
44 01 00 Operation and Maintenance of Pollution Control Equipment
44 06 00 Schedules for Pollution Control Equipment
44 08 00 Commissioning of Pollution Control Equipment

44 10 00 AIR POLLUTION CONTROL
44 11 00 Air Pollution Control Equipment
44 20 00 NOISE POLLUTION CONTROL
44 21 00 Noise Pollution Control Equipment
44 40 00 WATER TREATMENT EQUIPMENT
44 41 00 Packaged Water Treatment
44 42 00 General Water Treatment Equipment
44 43 00 Water Filtration Equipment
44 43 13 Water Filters
44 44 00 Water Treatment Chemical Systems Equipment
44 45 00 Water Treatment Biological Systems Equipment
44 46 00 Sludge Treatment and Handling Equipment for Water Treatment Systems
44 50 00 SOLID WASTE CONTROL
44 51 00 Solid Waste Control Equipment
44 60 00 Unassigned
44 70 00 Unassigned
44 80 00 Unassigned
44 90 00 Unassigned

DIVISION 45—INDUSTRY-SPECIFIC MANUFACTURING EQUIPMENT

45 00 00 INDUSTRY-SPECIFIC MANUFACTURING EQUIPMENT
45 08 00 Commissioning of Industry-Specific Manufacturing Equipment
45 11 00 Oil and Gas Extraction Equipment
45 13 00 Mining Machinery and Equipment
45 15 00 Food Manufacturing Equipment
45 17 00 Beverage and Tobacco Manufacturing Equipment
45 19 00 Textiles and Apparel Manufacturing Equipment
45 21 00 Leather and Allied Product Manufacturing Equipment
45 23 00 Wood Product Manufacturing Equipment
45 25 00 Paper Manufacturing Equipment
45 27 00 Printing and Related Manufacturing Equipment
45 29 00 Petroleum and Coal Products Manufacturing Equipment
45 31 00 Chemical Manufacturing Equipment
45 33 00 Plastics and Rubber Manufacturing Equipment
45 35 00 Nonmetallic Mineral Product Manufacturing Equipment
45 37 00 Primary Metal Manufacturing Equipment
45 39 00 Fabricated Metal Product Manufacturing Equipment
45 41 00 Machinery Manufacturing Equipment
45 43 00 Computer and Electronic Product Manufacturing Equipment
45 45 00 Electrical Equipment, Appliance, and Component Manufacturing Equipment
45 47 00 Transportation Manufacturing Equipment
45 49 00 Furniture and Related Product Manufacturing Equipment
45 51 00 Other Manufacturing Equipment
45 60 00 Unassigned
45 70 00 Unassigned
45 80 00 Unassigned
45 90 00 Unassigned

DIVISION 48—ELECTRICAL POWER GENERATION

48 00 00 ELECTRICAL POWER GENERATION
48 01 00 Operation and Maintenance for Electrical Power Generation
48 05 00 Common Work Results for Electrical Power Generation
48 06 00 Schedules for Electrical Power Generation

48 08 00   Commissioning of Electrical Power Generation
48 09 00   Instrumentation and Control for Electrical Power Generation
48 10 00 ELECTRICAL POWER GENERATION EQUIPMENT
48 11 00   Fossil Fuel Plant Electrical Power Generation Equipment
48 12 00   Nuclear Fuel Plant Electrical Power Generation Equipment
48 13 00   Hydroelectric Plant Electrical Power Generation Equipment
48 14 00   Solar Energy Electrical Power Generation Equipment
48 15 00   Wind Energy Electrical Power Generation Equipment
48 16 00   Geothermal Energy Electrical Power Generation Equipment
48 17 00   Electrochemical Energy Electrical Power Generation Equipment
48 18 00   Fuel Cell Electrical Power Generation Equipment
48 19 00   Electrical Power Control Equipment
48 20 00   Unassigned
48 30 00   Unassigned
48 40 00   Unassigned
48 50 00   Unassigned
48 60 00   Unassigned
48 70 00 ELECTRICAL POWER GENERATION TESTING
48 71 00   Electrical Power Generation Test Equipment
48 80 00   Unassigned
48 90 00   Unassigned

The Numbers and Titles used in this textbook are from *MasterFormat*™ 2004, published by the Construction Specifications Institute (CSI) and the Construction Specifications Canada (CSC), and are used with permission from CSI. For those interested in a more in-depth explanation of *MasterFormat*™ 2004 and its use in the construction industry, visit www.csinet.org/masterformat or contact:

The Construction Specifications Institute (CSI)
99 Canal Center Plaza, Suite 300
Alexandria, VA 22314
800-689-2900; 703-684-0300
http://www.csinet.org

APPENDIX

B

Common Plumbing Fixture and Piping Symbols

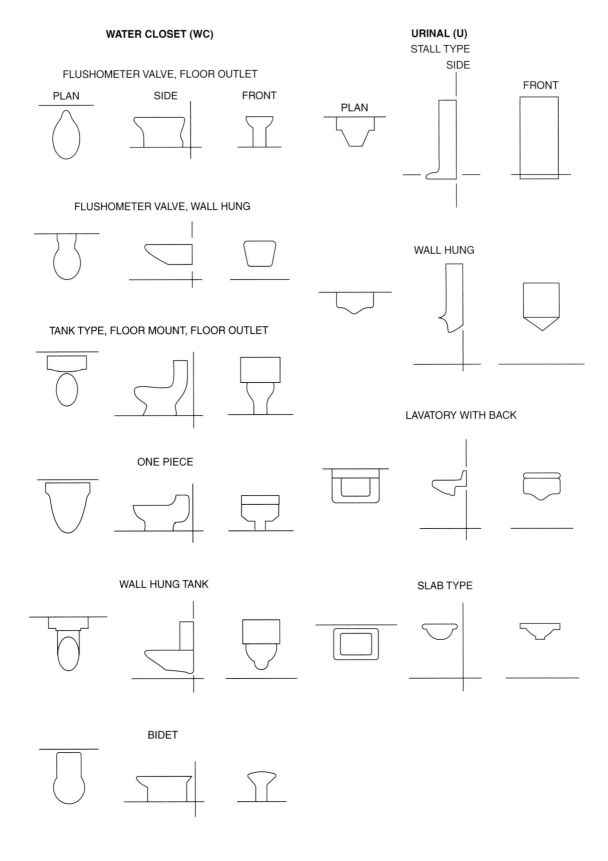

WATER CLOSET (WC)

FLUSHOMETER VALVE, FLOOR OUTLET

PLAN SIDE FRONT

FLUSHOMETER VALVE, WALL HUNG

TANK TYPE, FLOOR MOUNT, FLOOR OUTLET

ONE PIECE

WALL HUNG TANK

BIDET

URINAL (U)

STALL TYPE

SIDE

PLAN FRONT

WALL HUNG

LAVATORY WITH BACK

SLAB TYPE

LAVATORIES

CORNER

IN COUNTER

HANDICAPPED

SINKS, SINGLE COMPARTMENT

SINK WITH DRAINBOARD

DOUBLE COMPARTMENT TYPE

SINKS

DOUBLE COMPARTMENTS
WITH DRAINBOARDS

WITH LAUNDRY TRAY

SERVICE TYPE

WITH FOOD WASTE GRINDER

CIRCULAR WASH TYPE

SEMI-CIRCULAR WASH TYPE

SHOWER (SH) STALL TYPE

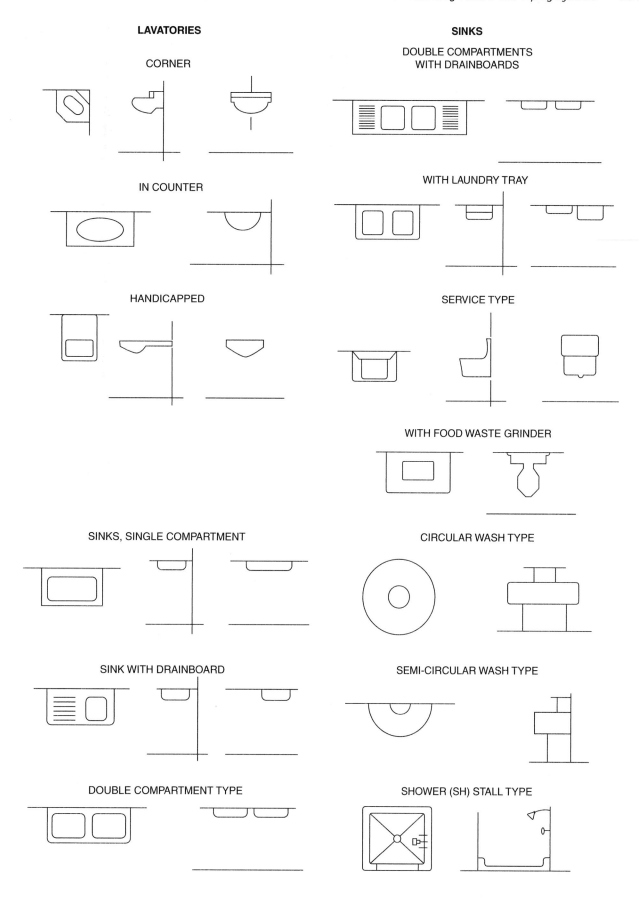

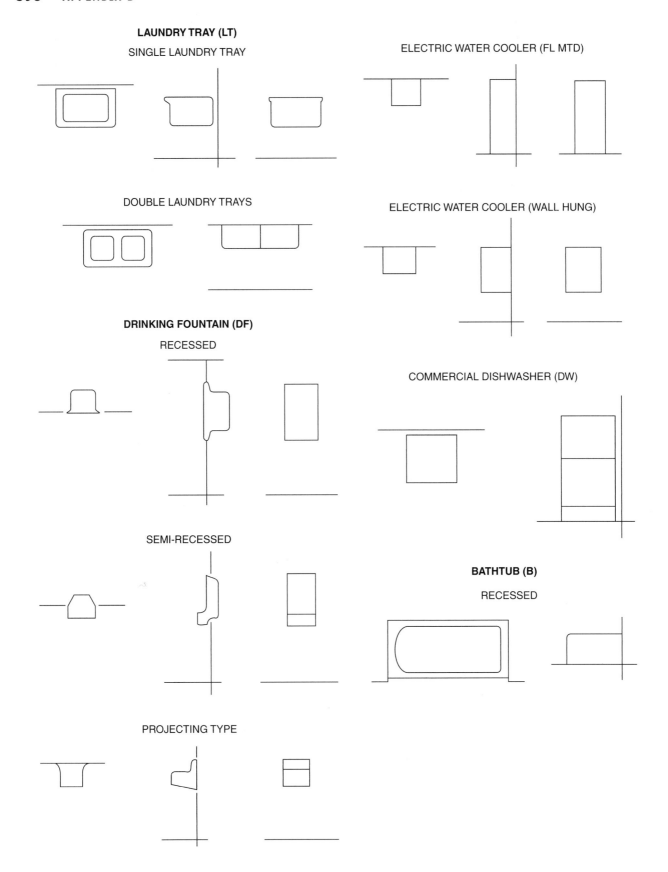

LAUNDRY TRAY (LT)

SINGLE LAUNDRY TRAY

ELECTRIC WATER COOLER (FL MTD)

DOUBLE LAUNDRY TRAYS

ELECTRIC WATER COOLER (WALL HUNG)

DRINKING FOUNTAIN (DF)

RECESSED

COMMERCIAL DISHWASHER (DW)

SEMI-RECESSED

BATHTUB (B)

RECESSED

PROJECTING TYPE

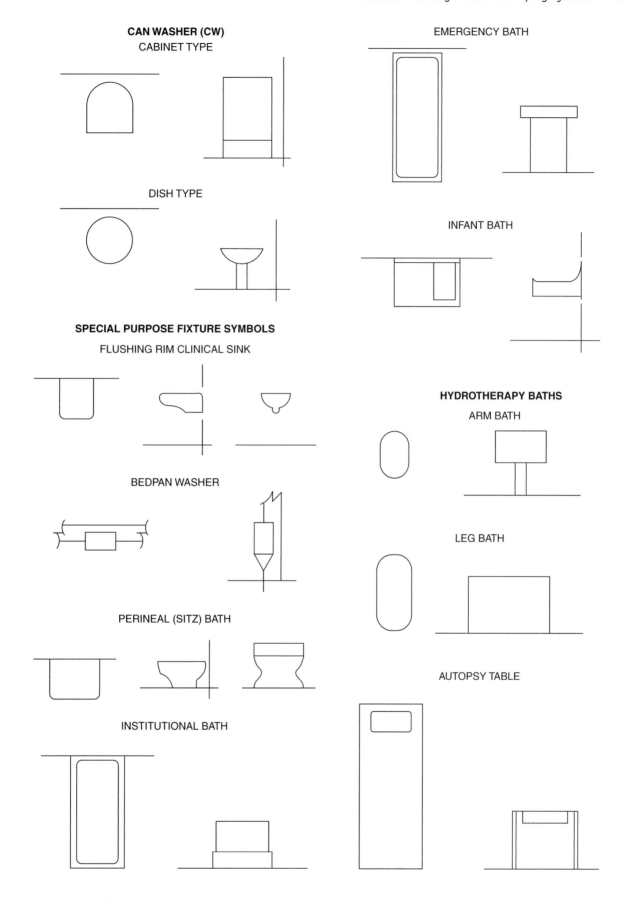

CAN WASHER (CW)
CABINET TYPE

DISH TYPE

SPECIAL PURPOSE FIXTURE SYMBOLS

FLUSHING RIM CLINICAL SINK

BEDPAN WASHER

PERINEAL (SITZ) BATH

INSTITUTIONAL BATH

EMERGENCY BATH

INFANT BATH

HYDROTHERAPY BATHS

ARM BATH

LEG BATH

AUTOPSY TABLE

PLUMBING

ACID WASTE	———————— ACID ————————
COLD WATER	——— – ———————— – ———
COMPRESSED AIR	———————— A ————————
FIRE LINE	——— F ———————— F ———
GAS	——— G ———————— G ———
HOT WATER	——— –– ———————— –– ———
HOT-WATER RETURN	––– – ––– – ––– – –––
SOIL, WASTE OR LEADER (ABOVE GRADE)	————————————
SOIL, WASTE OR LEADER (BELOW GRADE)	— — — — — — — — — — —
VACUUM	——— V ———————— V ———
VENT	– – – – – – – – – – – –

	FLANGED	SCREWED	BELL AND SPIGOT	WELDED	SOLDERED
TEE					
STRAIGHT SIZE					
OUTLET UP					
OUTLET DOWN					
REDUCING					
SIDE OUTLET (OUTLET DOWN)					
SIDE OUTLET (OUTLET UP)					
UNION					
ANGLE VALVE					
CHECK					
GATE (ELEVATION)					
GATE (PLAN)					
GLOBE (ELEVATION)					
GLOBE (PLAIN)					
CHECK VALVE ANGLE CHECK	←	SAME AS ANGLE VALVE (CHECK)			→
STRAIGHT WAY					
DIAPHRAGM VALVE					
GATE VALVE					
GLOBE VALVE					
SAFETY VALVE					

	FLANGED	SCREWED	BELL & SPIGOT	WELDED	SOLDERED
BUSHING		⊐	←	✳	⊙⊙
CAP		⊐	→		
CROSS — REDUCING					
CROSS — STRAIGHT SIZE					
CROSSOVER		⌢	⌢		
ELBOW — 45-DEGREE					
ELBOW — 90-DEGREE					
ELBOW — TURNED DOWN					
ELBOW — TURNED UP					
ELBOW — DOUBLE BRANCH					
ELBOW — LONG RADIUS					
ELBOW — REDUCING					
SIDE OUTLET (OUTLET DOWN)					
SIDE OUTLET (OUTLET UP)					
JOINT — CONNECTING PIPE					
JOINT — EXPANSION					
LATERAL					
REDUCING FLANGE					
REDUCER — CONCENTRIC					
REDUCER — ECCENTRIC					
SLEEVE					

APPENDIX

C

Specifications for the Suite 108B Bank Renovation at Commerce Court

HURTT-KENRICK AND ASSOCIATES, ARCHITECTS

DIVISION 0 GENERAL CONDITIONS OF CONTRACT
| 00710 | General Conditions—Preamble |
| 00851 | Drawing Index |

DIVISION 1 GENERAL REQUIREMENTS
01010	Summary of Work
01045	Cutting and Patching
01200	Project Meeting
01310	Construction Schedule
01340	Shop Drawings, Product Data and Samples
01370	Schedule of Values
01510	Temporary Utilities
01530	Barriers
01600	Material and Equipment
01630	Substitutions and Product Options
01700	Contract Closeout
01710	Cleaning
01720	Project Record Documents
01730	Operating and Maintenance
01740	Warranties and Bonds

DIVISION 2 DEMOLITION
| 02070 | Selective Demolition |

DIVISION 3 CONCRETE
| 03300 | Cast-in-Place Concrete |

DIVISION 4 MASONRY
| 04500 | Masonry Restoration |

DIVISION 5 METALS
| 05400 | Lightgauge Metal Framing |

DIVISION 6 WOOD AND PLASTICS
06100	Rough Carpentry
06200	Finish Carpentry
06220	Millwork
06410	Cabinet Work

DIVISION 7 THERMAL AND MOISTURE PROTECTION
| 07200 | Building Insulation |
| 07900 | Joint Sealants |

DIVISION 8 DOORS AND WINDOWS
08100	Metal Doors and Frames
08210	Wood Doors
08420	Aluminum Entrance Doors
08610	Wood Windows
08710	Finish Hardware
08800	Glass and Glazing

DIVISION 9 FINISHES
09260	Gypsum Wallboard System
09330	Quarry Tile
09600	Acoustical Treatment
09510	Acoustical Ceiling Tile
09660	Resilient Tile Flooring

09680	Carpeting
09900	Painting
09950	Wall Covering

DIVISION 12 FURNISHES

| 12677 | Matting |

DIVISION 15 MECHANICAL

15500	Fire Sprinkler System
15400	Plumbing
15600	Heating Ventilating and Air Conditioning

DIVISION 16 ELECTRICAL

16400	Electrical
16520	Electrical Fixtures
16750	Telephone System

General Conditions Preamble

1.01	Standard Form
A.	"The General Conditions of the Contract for Construction, AIA Document A-201, Thirteenth Edition, August 1976," Articles 1 thru 14 inclusive, are bound immediately following. They are included as part of the Contract. The Contractor and all subcontractors shall read and be governed by them and all of their addendums.
1.02	Availability
A.	Additional copies may be obtained from American Institute of Architects, 1735 New York Avenue NW, Washington, D.C. 20006, from Chapter offices in many cities, and some stationary stores throughout the country.
1.03	
A.	In event of conflicts between the Contract Documents, General Conditions and other parts of Specifications and Drawings, the Architect reserves the right to determine which governs, and in what order of precedence.

THE AMERICAN INSTITUTE OF ARCHITECTS

AIA Document A201

General Conditions of the Contract for Construction

This document has important legal consequences; consultation
with an attorney is encouraged with respect to its modification.

1976 EDITION
Table of Articles

This document has been approved and endorsed by the Associated General
Contractors of America

AIA DOCUMENT A201- GENERAL CONDITIONS OF THE CONTRACT FOR
CONSTRUCTION-THIRTEENTH EDITION-AUGUST 1976 AIA 1976-THE AMERICAN
INSTIUTE OF ARCHITECTS-1735 NEW YORK AVENUE. NW, WASHINGTON D.C 2006

INDEX

GENERAL CONDITIONS OF THE CONTRACT FOR CONSTRUCTION

Article 1

Contract Documents

1.1 Definitions

1.1.1 The Contract Documents

The Contract Documents consist of the Owner-Contractor Agreement, the Conditions of the Contract (General, Supplementary and other Conditions), the Drawings, the Specifications, and all Addenda issued prior to and all Modifications issued after execution of the Contract. A Modification is (1) a written amendment to the Centric, signed by both parties, (2) a Change Order, (3) a written interpretation issued by the Architect pursuant to Subparagraph 2.2'S, or (4) a written order lot a minor change in the Work issued by the Architect pursuant to Paragraph 12.4. The Contract Documents do not Include Bidding Documents such as the Advertisement or Invitation to bid, the instructions to Bidders, sample forms, the Contractor's Bid or portions of Addenda relating to any of these, or any other documents, unless specifically enumerated in the Owner-Contractor Agreement.

1.1.2 The Contract

The Contract Documents-form the Contract for Construction, this Contract represents the entire and integrated agreement between the parties hereto and supersedes all prior negotiations. representations, or agreements either written or oral. The Contract may be amended or modified only by a Modification as defined in Subparagraph 1,1_i, the Contract Documents shall not be construed to create joy contractual relationship of *any* kind between the Architect and the Contractor, but the Architect shall be enabled to performance of obligations intended for his benefit, and to enforcement thereof. Nothing contained a*s* the Contract Documents shall create any contractual relationship between the Owner or the Architect and any Subcontractor or Sub-subcontractor,

1.1.3 The Work

The Work comprises the completed construction required by the Contract Documents and includes all labor necessary to produce such construction, and all materials and equipment incorporated or to be incorporated in such construction.

1.1.4 The Project

The Project is the total construction of which the Work performed under the Contract Documents may be the whole or a part.

1.2 Execution, Correlation and Intent

1.2.1 The Contract Documents shall be signed in not less than triplicate by the Owner and Contractor. If either the Owner or the Contractor or both do not sign the Conditions of the Contract, Drawings Specifications, or any of the other Contract Documents, the Architect shall identify such Documents.

1.2.2 By executing the Contract, the Contractor represents that he has visited the tile, familiarized himself with the local conditions under which the Work is to be performed, and correlated his observations with the requirements of the Contract Documents.

1.2.3 The Intent of the Contract Documents is to include all items necessary for the proper execution and completion of the Work. The Contract Documents are complementary, and what is required by any one shall be as binding as if required by. Work not covered in the all Contract Documents will not be required unless it is consistent therewith and is reasonably inferable there from as being necessary to produce the intended results. Words and abbreviations which have well-known technical or trade meanings are used in the Contract Documents in accordance with such recognized meanings.

1.2.4 The organization of the Specifications into division, sections and articles, and the arrangement of Drawings shall not control the Contractor in dividing the Work among Subcontractors or in establishing the extent of Work to be performed by any trade.

1.3 Ownership and use of Documents

1.3.1 All Drawings, Specifications and copies thereof furnished by the Architect are and shall remain his property. They are to be used only with respect to this Project and are not to be used on any other project. With the exception of one contract set for each party to the Contract, such documents are to be returned or suitably accounted to the Architect on request at the completion of the Work. Submission or distribution to meet official regulatory requirements or for other purposes in connection with the Project is not to be construed as publication in derogation of the Architect's common law copyright or other reserved rights.

Article 2

Architect

2.1 Definition

2.1.1 The Architect is the person lawfully licensed to practice architecture, or all entity lawfully practice architecture identified as, such in the Owner-Contractor Agreement, and is referred to throughout the Contract Documents as if singular in number and masculine in gender. The term Architect means the Architect or his authorized representative.

2.2 Administration of the Contract

2.2.1 The Architect will provide administration of the Contract as hereinafter described.

2.2.2 The Architect 'viii he the Owner's representative during construction and until final payment is due. The Architect will advise and consult with the Owner. The Owner's instructions to the Contractor shall be forwarded through the Architect. The Architect will have authority to act on behalf of the Owner only to the extent provided in the Contract Documents, unless otherwise modified by written instrument in accordance with Sub paragraph 2.2.18.

2.2.3 The Architect will visit the site at intervals appropriate to the stage of construction to familiarize himself generally with the progress and quality of the Work and to determine in general if the Work is proceeding in accordance with the Contract Documents. However, the Architect will not be required to make exhaustive or continuous on-site inspections to check the quality or quantity of the Work. On the basis of his on-site observations as an architect, he will keep the Owner informed of the progress of the Work, and will endeavor to guard the Owner against defects and deficiencies in the Work of the Contractor.

2.2.4 The Architect will not be responsible for and will not have control or charge of construction means, methods, techniques, sequences or procedures, or for safety precautions and programs in connections with the Work, and he will not be responsible for the Contractors failure to carry out the Work in accordance with the Contract Documents. The Architect will not be responsible for or have control or charge over the acts or omissions of the Contractor, Subcontractors, or any of their agents or employees, or any other persons performing any of the Work.

2.2.5 The Architect shall at all times have access to the Work wherever it is in preparation and progress. The Contractor shall provide facilities for such access so the Architect may perform his functions under the Contract Documents.

2.2.6 Based on the Architect's observations and an evaluation of the Contractor's Application's for Payment, the Architect will determine the amounts owing to the Contractor and will issue Certificates for Payment in such amounts, as provided in Paragraph 9.4.

2.2.7 The Architect will be the interpreter of the requirements of the Contract Documents and the judge of the performance thereunder by both the Owner and Contractor.

2.2.8 The Architect will render interpretations necessary for the proper execution or progress of the Work, with reasonable promptness and in accordance with any time limit agreed upon. Either party to the Contract may make written request to the Architect for such interpretations.

2.2.9 Claims, disputes and other matters in question between the Contractor and the Owner relating to the execution or progress of the Work or the interpretation of the Contract Documents shall be referred initially to the Architect for decision which he will render in writing within a reasonable time.

2.2.10 All interpretations and decisions of the Architect shall be consistent with the intent of and reasonably inferable from the Contract Documents and will be in writing or in the form of drawings. In his capacity as interpreter and judge, he will endeavor to secure faithful performance by both the Owner and the Contractor, will not show partially to either, and will not be liable for the result of any interpretation or decision rendered in good faith in such capacity.

2.2.11 The Architects decisions in matters relating to artistic effect will be final if consistent with the intent of the Contract Documents.

2.2.12 Any claim, dispute or other matter in question between the Contractor and the Owner referred to the Architect, except those relating to artistic effect as provided in Subparagraph 2.2.11 and except those which have been waived by the making or acceptance of final payment is provided in Subparagraphs 9.9.4 and 9.9.5, shall be subject to arbitration upon the written demand of either party. However, no demand for arbitration of any such claim, dispute or other matter may be made until the earlier of (1) the date on which the Architect has rendered a written decision, or (2) the tenth day after the parties have presented their evidence to the Architect or have been given a reasonable opportunity to do so, if the Architect has not rendered his written decision by that date. When such a written decision of the Architect states (1) that the decision is final but subject to appeal, and (2) that any demand for arbitration of a claim, dispute or other matter covered by such decision must be made within thirty days after the date on which the party making the demand receives the written decision, failure to demand arbitration within said thirty days' period will resulting the Architect's decision becoming final and binding upon the Owner and the Contractor. If the Architect renders a decision after arbitration proceedings have been initiated, such decision may be entered as evidence but will not supersede any arbitration proceedings unless the decision is acceptable to all parties concerned.

2.2.13 The Architect will have authority to reject Work which does not conform to the Contract Documents. Whenever, in his opinion, he considers it necessary or advisable for the implementation of the intent of the Contract Documents, he will have authority to require special inspection or testing of the Work in accordance with Subparagraph 7.7.2 whether or not such Work be then fabricated, installed or completed. However, neither the Architect's authority to act under this Subparagraph 2.2.13, nor any decision made by him in good faith rather to exercise or not to exercise such authority, shall give rise to any duty responsible of the Architect to the Contractor, any other person performing any of the Work.

2.2.14 The Architect will review and approve or take other appropriate action upon Contractor's submittals such as Shop Drawings, Product Data and Samples, but only for conformance with the design concept of the Work and with the information given in the Contract Documents. Such action shall be taken with reasonable promptness so as to cause no delay. The Architect's approval of a specific item shall not indicate approval of an assembly of which the item is component.

2.2.15 The Architect will prepare Change Orders in accordance with Article 12, and will have authority to order minor changes in the Work as provided by Subparagraph 12.4.1.

2.2.16 The Architect will conduct inspections to determine the dates of Substantial Completion and final completion, will receive and forward to the Owner for the Owner's review written warranties and related documents required by the Contract and assembled by the Contractor, and will issue a final Certificate for Payment upon compliance with the requirements of Paragraph 9.9.

2.2.17 If the Owner and Architect agree, the Architect will provide one or more Project Representatives to assist the Architect in carrying out his responsibilities at the site. The duties, responsibilities and limitations of authority of any such Project Representatives shall be as set forth in an exhibit to be incorporated in the Contract Documents.

2.2.18 The duties, responsibilities and limitations of authority of the Architect as the Owner's representative during construction as set forth in the Contract Documents will not be modified or extended without written consent of the Owner, the Contractor and the Architect.

2.2.19 In case of the termination of the employment of the Architect, the Owner shall appoint an architect against whom the Contractor makes no reasonable objection whose status under the Contract Documents shall be that of the former architect. Any dispute in connection with such appointment shall be subject to arbitration.

Article 3

Owner

3.1 Definition

3.1.1 The Owner is the person or entity identified as such in the Owner-Contractor Agreement and is referred to throughout the Contract Documents as if singular in number and masculine in gender. The term Owner means the Owner or his authorized representative.

3.2 Information and Services Required of the Owner

3.2.1 The Owner shall, at the request of the Contractor, at the time of execution of the Owner-Contractor Agreement, furnish to the Contractor reasonable evidence that he has made financial arrangements to fulfill his obligations under the Contract. Unless such reasonable evidence is furnished, the Contractor is not required to execute the Owner-Contractor Agreement or to commence the Work.

3.2.2 The Owner shall furnish all surveys describing the physical characteristics, legal limitations and utility locations for the site of the Project, and a legal description of the site.

3.2.3 Except as provided in Subparagraph 4.7.1, the Owner shall secure and pay for necessary approvals, easements, assessments, and charges required for the construction use or occupancy of permanent structures or for permanent changes in existing facilities.

3.2.4 Information or services under the Owner's control shall be furnished by the Owner with reasonable promptness to avoid delay in the orderly progress of the Work.

3.2.5 Unless otherwise provided in the Contract Documents, the Contractor will be furnished, free of charge, all copies of Drawings and Specifications reasonably necessary for the execution of the Work.

3.2.6 The Owner shall forward all instructions to the Contractor through the Architect.

3.2.7 The foregoing are in addition to other duties and responsibilities of the Owner enumerated here in and especially those in respect to Work by Owner or by Separate Contractors, Payments, and Completion, and Insurance in Articles 6, 9, and 11 respectively.

3.3 Owner's Right to Stop the Work

3.3.1 If the Contractor fails to correct defective Work as required by Paragraph 13.2 or persistently fails to carry out the Work in accordance with the Contract Documents, by the Owner, by a written order signed personally or by an agent specifically so empowered by the Owner in writing, may order the Contractor to stop the Work, or any portion thereof, until the cause of for such order has been eliminated; however, this right of the Owner to stop the Work shall not give rise to any duty on the part of the Owner to exercise this right for the Contractor or any other person or entity, except to the extent required by Subparagraph 6.1.3.

3.4 Owner's Right to Carry Out the Work

3.4.1 If the Contractor defaults or neglects to carry out the Work in accordance with the Contract Documents and fails within seven days after receipt of written notice from the Owner to commence and continue correction of such default or neglect with diligence and promptness, the Owner may, after seven days following receipt by the Contractor of an additional written notice and without prejudice to any other remedy he may have, make good such deficiencies. In such case an appropriate Change Order shall be issued deducting from the payments then or thereafter due the Contractor the cost of correcting such deficiencies, including compensation for the Architect's additional services made necessary by such default, neglect or failure. Such action by the Owner and the amount charged to the Contractor are both subject to the prior approval of the Architect. If the payments then or thereafter due the Contractor are not sufficient to cover such amount, the Contractor shall pay the difference to the Owner.

Article 4

Contractor

4.1 Definition

4.1.1 The Contractor is the person or entity identified as such in the Owner-Contractor Agreement and is referred to throughout the Contract Documents as if singular in number and masculine in gender. The term Contractor means the Contractor or his authorized representative.

4.2 Review of Contract Documents

4.2.1 The Contractor shall carefully study and compare the Contract Documents and shall at once report to the Architect an error, inconsistency or omission he may discover. The Contractor shall not be liable to the Owner or the Architect for any damage resulting from any such errors, inconsistencies or omissions in the Contract Documents. The Contractor shall perform no portion of the Work at any time without Contract Documents or, here required, approved Shop Drawings, Product Data or Samples for such portion of the Work.

4.3 Supervision and Construction Procedures

4.3.1 The Contractor shall supervise and direct the Work, using his best skill and attention. He shall be solely responsible for all construction means, methods, techniques, sequences and procedures and for coordinating all portions of the Work under the Contract.

4.3.2 The Contractor shall be responsible to the Owner for the acts and omissions of his employees,

Subcontractors and their agents and employees, and other persons performing any of the Work under a contract with the Contractor.

4.3.3 The Contractor shall not be relieved from his obligations to perform the Work in accordance with the Contract Documents either by the activities or duties of the Architect in his administration of the Contract, or by inspections, tests or approvals required or performed under Paragraph 7.7 by persons other than the Contractor.

4.4 Labor and Materials

4.1 41 Unless otherwise provided in the Contract Documents, the Contractor shall provide and pay for all labor, materials, equipment, tools, construction equipment and machinery, water, heat, utilities, transportation, and other facilities and services necessary for the proper execution and completion of the Work, whether temporary or permanent and whether or not incorporated or to be incorporated in the Work.

4.4.2 The Contractor shall at all times enforce strict discipline and good order among his employees and shall not employ on the Work any unfit person or anyone not skilled in the task assigned to him.

4.5 Warranty

4.5.1 The Contractor warrants to the Owner and the Architect that all materials and equipment furnished under this Contract will be new unless otherwise specified, and that all Work will be of good quality, free from faults and defects and in conformance with the Contract Documents. All Work not conforming to these requirements, including substitutions nor properly approved and authorized, may be considered defective. If required by the Architect, the Contractor shall furnish satisfactory evidence as to the kind and quality of materials and equipment. This warranty is not limited by the provisions of Paragraph 13.2.

4.6 Taxes

4.6.1 The Contractor shall pay all sales, consumer, use and other similar taxes for the Work or portions thereof provided by the Contractor which are legally enacted at the time bids are received, whether or not yet effective.

4.7 Permits

4.7.1 Unless otherwise provided in the Contract Documents, the Contractor shall secure and pay for the building permit and for all other permits and governmental fees, licenses and inspections necessary for the proper execution and completion of the Work which are customarily secured after the execution of the Contract and which are legally required at the time the bids are received.

4.7.2 The Contractor shall give all notices and comply with all laws, ordinances, rules, regulations and lawful orders of any public authority bearing on the performance of the Work.

4.7.3 It is not the responsibility of the Contractor to make certain that the Contract Documents are in accordance with applicable laws, statues, building codes and regulations. If the Contractor observes that any of the Contract Documents are at variance there with in any respect, he shall promptly notify the Architect in writing, and any necessary changes shall be accomplished be appropriate Modification.

4.7.4 If the Contractor performs any Work knowing it to be contrary to such laws, ordinances, rules and regulations, and without such notice to the Architect, he shall assume full responsibility therefore and shall bear all costs attributable thereto.

4.8 Allowances

4.8.1 The Contractor shall include in the Contract Sum all allowances stated in the Contract Documents. Items covered by these allowances shall be supplied for such amounts and by such persons as the Owner may direct, but the Contractor will not be required to employ persons against whom he makes a reasonable objection.

4.8.2 Unless otherwise provided in the Contract Documents:

1. These allowances shall cover the cost to the Contractor, less any applicable trade discount, of the materials and equipment required by the allowance delivered at the site, and all applicable taxes;

2. The Contractor's costs for unloading and handling on the site, labor, installation costs, overhead, profit and other expenses contemplated for the original allowance shall be included in the Contract Sum and not in the allowance;

3. Whenever the cost is more than or less than the allowance, the Contract Sum shall be adjusted accordingly by Change Order, the amount of which will recognize changes, if any, in handling costs on the site, labor, installation costs, overhead, profit and other expenses.

4.9 Superintendent

4.9.1 The Contractor shall employ a competent superintendent and necessary assistants who shall be in attendance at the Project site during the progress of the Work. The Superintendent shall represent the Contractor and all communications given to the superintendent shall be as binding as if given to the Contractor. Important communications shall be so confirmed in writing. Other communications shall be so confirmed on written request in each case.

4.10 Progress Schedule

4.10.1 The Contractor, immediately after being awarded the Contract, shall prepare and submit for the Owner's and Architect's information an estimated progress schedule for the Work. The progress schedule shall be related to the entire Project to the extent required by the Contract Documents, and shall provide for expeditious and practicable execution of the Work.

4.11 Documents and Samples at the Site

4.11.1 The Contractor shall maintain at the site for the Owner one record copy of all Drawings, Specifications, Addenda, Change Orders and other Modifications, in good order and marked currently to record all changes made during construction and approved Shop Drawings, Product Data and Samples. These shall be available to the Architect and shall be delivered to him for the Owner upon completion of the Work.

4.12 Shop Drawings, Product Data and Samples

4.12.1 Shop Drawings are drawings, diagrams, schedules and other data specifically prepared for the Work by the Contractor or any Subcontractor, manufacturer, supplier or distributor to illustrate some portion of the Work.

4.12.2 Product Data are illustrations, standard schedules, performances charts, instructions, brochures, diagrams and other information furnished by the Contractor to illustrate a material, product or system for some portion of the Work.

4.12.3 Samples are physical examples which illustrate materials, equipment or workmanship and establish standards by which the Work will be judged.

4.12.4 The Contractor shall review, approve and submit, with reasonable promptness and in such sequence as to cause delay in the Work or in the work of the Owner or any separate contractor, all Shop Drawings, Product Data and Samples required by the Contract Documents.

4.12.5 By approving and submitting Shop Drawings, Product Data and Samples, the Contractor represents that he has determined and verified all materials, field measurements, and field construction criteria related thereto, or will do so, and that he has checked and coordinated the information contained within such submittals with the requirements of the Work and of the Contract Documents.

4.12.6 The Contractor shall no be relieved of responsibility for any deviation from the requirements of the Contract Documents by the Architect's approval of Shop Drawings, Product Data or Samples under Subparagraph 2.2.14 unless the Contractor has specifically informed the Architect in writing of such deviation at the time of submission and the Architect has given written approval to the specific deviation. The Contractor shall not be relieved from responsibility for errors or omissions in the Shop Drawings, Product Data or Samples by the Architect's approval thereof.

4.12.7 The Contractor shall direct specific attention, in writing or on resubmitted Shop Drawings, Product Data or Samples, to revisions other than those requested by the Architect on previous submittals.

4.12.8 No portion of the Work requiring submission of a Shop Drawing, Product Data or Sample shall be commenced until the submittal has been approved by the Architect as provided in Subparagraph 2.2.14. All such portions of the Work shall be in accordance with proved submittals.

4.13 Use of Site

4.13.1 The Contractor shall continue operations at the site to areas permitted by law, ordinances, permits and the Contract Documents and shall not unreasonably encumber the site with any materials or equipment.

4.14 Cutting and Patching of Work

4.14.1 The Contractor shall be responsible for all cutting, fitting or patching that may be required to complete the Work or to make its several parts fit together properly.

4.14.2 The Contractor shall not damage or endanger any portion of the Work or the work of the Owner or any separate contractors by cutting, patching or otherwise altering any work, or by excavation. The Contractor shall not cut or otherwise alter the work of the Owner or any separate contractor except with the written consent of the Owner and of such separate contractor. The Contractor shall not unreasonably withhold from the Owner or any separate contractor his consent to cutting of otherwise altering the Work.

4.15 Cleaning Up

4.15.1 The Contractor at all times shall keep the premises free from accumulation of waste materials or rubbish caused by his operations. At the completion of the Work he shall remove all his waste materials and rubbish from and about the Project as well as his tools, construction equipment, machinery, and surplus materials.

4.15.2 If the Contractor fails to clean up at the completion of the Work, the Owner may do so as provided in Paragraph 3.4 and the cost thereof shall be charged to the Contractor.

4.16 Communications

4.16.1 The Contractor shall forward all communications to the Owner through the Architect.

4.17 Royalties and Patents

4.17.1 The Contractor shall pay all royalties and license fees. He shall defend all suits or claims for infringement of any patent rights and shall save the Owner harmless from loss on account thereof, except the Owner shall be responsible for any such loss when a particular design process or the product of a particular manufacturer or manufacturers is specified, but if the Contractor has reason to believe that the design, process or product specified is an infringement of a patent, he shall be responsible for such loss unless he promptly gives such information to the architect.

4.18 Indemnification

4.18.1 To the fullest extent permitted by law, the Contractor shall indemnify and hold harmless the Owner and the Architect and their agents and employees from and against all claims, damages, losses and expenses, including but not limited to attorneys' fees, arising out of or resulting from the performance of the Work, provided that any such claim, damage, loss or expense (1) is attributable to bodily injury, sickness, disease or death, or to injury to or destruction of tangible property (other than the Work itself) including the loss of use resulting there from, and (2) is caused in whole or in part by any negligent act pr omission of the Contractor, and Subcontractor, anyone directly or indirectly employed by any of them or anyone for whose acts any of them may be liable, regardless of whether or not it is caused in part by a party indemnified hereunder. Such obligation shall not be construed to negate, abridge, or otherwise reduce any other right or obligation of indemnity which would otherwise exist as to any party or person described in this Paragraph 4.18.

4.18.2 In any and all claims against the Owner or the Architect or any of their agents or employees by any employee of the Contractor, any Subcontractor, anyone directly or indirectly employed by any of them or anyone for whose acts any of them may be liable, the indemnification obligation under this Paragraph 4.18 shall not be limited in any way by any limitation on the amount or type of damages, compensation or benefits payable by or for the Contractor or any Subcontractor under Workers' or workmen's compensation acts, disability benefit acts or oilier employee benefit acts.

4.18.3 The obligations of the Contractor under this Paragraph 4.18 shall not extend to the liability of the Architect, his agents or employees, arising out of (1) the preparation or approval of maps, drawings, opinions, reports, surveys, change orders, designs or specifications, or (2) the giving of or the failure to give directions or instructions by the Architect, his agents or employees provided such giving or failure to give is the primary cause of the injury or damage.

Article 5

Subcontractors

5.1 Definition

5.1.1 A Subcontractor is a person or entity who has a direct contract with the Contractor to perform any of the Work at the site. The term Subcontractor is referred to throughout the Contract Documents as if singular in number and masculine in gender and means a Subcontractor or his authorized representative. The term Subcontractor does not include any separate contractor or his subcontractors.

5.1.2 A Sub-subcontractor is a person or entity who has a direct or indirect contract with a Subcontractor to perform any of the Work at the site. The term Sub-subcontractor is referred to throughout the Contract Documents as if singular in number and masculine in gender and means a Sub-subcontractor or an authorized representative thereof.

5.2 Award of Subcontracts and Other Contracts for Portions of the Work

5.2.1 Unless otherwise required by the Contract Documents or the Bidding Documents, The Contractor, as soon as practicable after the award of the Contract, shall furnish to the Owner and the Architect in writing the names of the persons or entities (including those who are to furnish materials or equipment fabricated to a special design) proposed for each of the principal portions of the Work. The Architect will promptly reply to the Contractor in writing stating whether or not the Owner or the Architect, after due investigation, has reasonable objection to any such proposed person or entity. Failure of the Owner or Architect to reply promptly shall constitute notice of no reasonable objection.

5.2.2 The Contractor shall not contract with any such proposed person or entity to whom the Owner or the Architect has made reasonable objection under the provided Subparagraph 5.2.1. The Contractor shall not be required to contract with anyone whom he has a reasonable objection.

5.2.3 If the Owner or the Architect has reasonable objection to any such proposed person or entity, the Contractor shall submit a substitute to whom the Owner or the Architect has no reasonable objection, and the Contract Sum shall be increased or decreased by the difference in cost occasioned by such substitution and an appropriate Change Order shall be issued; however, no increase in the Contract Sum shall be allowed for any such substitution unless the Contractor has acted promptly and responsively in submitting names as required by Subparagraph 5.2.1.

5.2.4 The Contractor shall make no substitution for any Subcontractor, person or entity previously selected if the Owner or Architect makes reasonable objection to such substitution.

5.3 Subcontractual Relations

5.3.1 By an appropriate agreement, written where legally required for validity, the Contractor shall require each Subcontractor, to the extent of the Work to be performed by the Subcontractor, to be bound to the Contractor by the terms of the Contract Documents, and to assume toward the Contractor all the obligations and responsibilities which the Contractor, by these Documents, assumes toward the Owner and the Architect. Said agreement shall preserve and protect the rights of the Owner and the Architect under the Contract Documents with respect to the Work to be performed by the Subcontractor, unless specifically provided otherwise in the Contractor-Subcontractor agreement, the benefit of all rights, remedies and redress against the Contractor that the Contractor, by these Documents, has against the Owner. Where appropriate, the Contractor shall require each Subcontractor to enter into similar agreements with his Sub-subcontractors. The Contractor shall make available to each proposed Subcontractor, prior to the execution of the Subcontractor will be bound by this Paragraph 5.3, and identify to the Subcontractor any terms and conditions of the proposed Subcontract which may be at variance with the Contract Documents. Each Subcontractor shall similarly make copies of such Documents available to his Sub-subcontractors.

Article 6

Work by Owner or by Separate Contractors

6.1 Owner's Right to Perform Work and to Award Separate Contracts

6.1.1 The Owner reserves the right to perform work related to the Project with his own forces, and to award separate contracts in connection with other portions of the project or other work on the site under these or similar Conditions of the Contract. If the Contractor claims that delay or additional cost is involved because of such action by the Owner, he shall make such claim as provided elsewhere in the Contract Documents.

6.1.2 When separate contracts are awarded for different portions of the Project or other work on the site, the term Contractor in the Contract Documents in each case shall mean the Contractor who executes each separate Owner-Contractor Agreement.

6.1.3 The Owner will provide for the coordination of the work of his own forced and of each separate contractor with the Work of the Contractor, who shall cooperate therewith as provided in Paragraph 6.2.

6.2 Mutual Responsibility

6.2.1 The Contractor shall afford the Owner and separate contractor reasonable opportunity for the introduction and storage of their materials and equipment and the execution of their work, and shall connect and coordinate his work with theirs as required by the Contract Documents.

6.2.2 If any part of the Contractor's Work depends for proper execution of results upon the work of the Owner or any separate contractor, the Contractor shall, prior to proceeding with the Work, promptly report to the Architect any apparent discrepancies or defects in such other work that render it unsuitable for such proper execution and results. Failure of the Contractor so to report shall constitute an acceptance of the Owner's or separate contractors' work as fit and proper to receive his Work, except as to defects which may subsequently become apparent in such work by others.

6.2.3 Any costs caused by defective or ill-timed work shall be borne by the party responsible therefore.

6.2.4 Should the Contractor wrongfully cause damage to the work or property of the Owner, or to other work on the site, the Contractor shall promptly remedy such damage as provided in Subparagraph 10.2.5.

6.2.5 Should the Contractor wrongfully cause damage to the work or property of any separate contractor, the Contractor shall upon due notice promptly attempt to settle with such other contractor by agreement, or otherwise to resolve the dispute. If such separate contractor sues or initiates and arbitration proceeding against the Owner on account of any damage alleged to have been caused by the Contractor, the Owner shall notify the Contractor who shall defend such proceedings at the Owner's expense, and if any judgment or award against the Owner arises there from the Contractor shall pay or satisfy it and shall reimburse the Owner for all attorneys' fees and court or arbitration costs which the Owner has incurred.

6.3 Owner's Right to Clean Up

6.3.1 If a dispute arises between the Contractor and separate contractors as to their responsibility for cleaning up as required by Paragraph 4.15, the Owner may clean up and charge the cost thereof to the contractors responsible there for as the Architect shall determine to be just.

Article 7

Miscellaneous Provisions

7.1 Governing Law

7.1.1 The Contract shall be governed by the law of the place where the project is located.

7.2 Successors and Assigns

7.2.1 The Owner and the Contractor each binds himself, his partners, successors, assigns and legal representatives to the other party hereto and to be the partners, successors, assigns and legal representatives of such other party with respect to all covenants, agreements and obligations contained in the Contract Documents. Neither party to the Contract shall assign the Contract or sublet it as a whole without the written consent of the other, nor shall the Contractor assign any moneys due or to become due to him hereunder, without the previous written consent of the Owner.

7.3 Written Notice

7.3.1 Written notice shall be deemed to have been duly served if delivered in person to the individual or member of the firm or entity or to an officer of the corporation for whom it was intended, or if delivered at or sent by registered or certified mail to the last business address known to him who gives the notice.

7.4 Claims for Damages

7.4.1 Should either party to the Contract suffer injury or damage to person or property because of any act or omission of the other party or of any of his employees, agents or others for whose acts he is legally liable, claim shall be made in writing to such other party within a reasonable time after the first observance of such injury or damage.

7.5 Performance Bond and Labor and Material Payment Bond

7.5.1 The Owner shall have the right to require the Contractor to furnish bonds covering the faithful performance of the Contract and the payment of all obligations arising thereunder if and as required in the Bidding Documents or in the Contract Documents.

7.6 Rights and Remedies

7.6.1 The duties and obligations imposed by the Contract Documents and the rights and remedies available there under shall be in addition to and not a limitation of any duties, obligations, rights and remedies otherwise imposed or available by law.

7.6.2 No action or failure to act by the Owner, Architect or Contractor shall constitute a waiver of any right or duty afforded any of them under the Contract, nor shall any such action or failure to act constitute an approval of or acquiescence in any breach thereunder, except as may be specifically agreed in writing.

7.7 Tests

7.7.1 If the Contract Documents, laws, ordinances, rules, regulations or orders of any portion of the Work to be inspected, tested or approved, the Contractor shall give the Architect timely notice of its readiness so the architect may observe such inspection, testing or approval. The Contractor shall bear all costs of such inspections, tests or approvals conducted by public authorities. Unless otherwise provided, the Owner shall bear all costs of other inspections, tests or approvals.

7.7.2 If the Architect determines that any Work requires special inspection, testing or approval which Subparagraph 7.7.1 does not include, he will, upon written authorization from the Owner, instruct the Contractor to order such special inspection, testing or approval, and the Contractor shall give notice as provided in Subparagraph 7.7.1. If such special inspection or testing reveals a failure of the Work to comply with the requirements of the Contract Documents, the Contractor shall bear all costs thereof, including compensation for the Architect's additional services made necessary by such failure; otherwise the Owner shall bear such costs, and an appropriate Change Order shall be issued.

7.7.3 Required certificates of inspection, testing or approval shall be secured by the Contractor and promptly delivered by him to the Architect.

7.7.4 If the Architect is to observe the inspections, tests or approvals required by the Contract Documents, he will do so promptly and, where practicable, at the source of supply.

7.8 Interest

7.8.1 payments due and unpaid under the Contract Documents shall bear interest from the date payment is due at such rates as the parties may agree upon in writing or, in the absence thereof, at the legal rate prevailing at the place of the Project.

7.9 Arbitration

7.9.1 All claims, disputes and other matters in question between the Contractor and the Owner arising out of, or relating to, the Contract Documents or the breach thereof, except as provided in Subparagraph 2.2.11 with respect to the Architect's decisions on matters relating to artistic effect, and except for claims which have been waived by the making or acceptance of final payment as provided by Subparagraphs 9.9.4 and 9.9.5, shall be decided by arbitration in accordance with the Construction Industry Arbitration Rules of the American Arbitration Association then obtaining unless the parties mutually agree otherwise. No arbitration arising out of or relating to the Contract Documents shall include, by consolidation, joinder or in any other manner, the Architect, his employees or consultants except by written consent containing a specific reference to the Owner-Contractor agreement and signed by the Architect, the Owner, the Contractor and any other persons substantially involved in a common question of fact or law, whose presence is required if complete relief is to be accorded in the arbitration. No person other than the Owner or Contractor shall be included as an original third party or additional third to an arbitration whose interest or responsibility is insubstantial. Any consent to arbitration involving a person or persons shall not constitute consent to arbitration of any dispute not described therein. The foregoing agreement to arbitrate and any other agreement to arbitrate and any other agreement to arbitrate with an additional person or persons duly consented to by the parties to the Owner-Contractor Agreement shall be specifically enforceable under the prevailing arbitration law. The award rendered by the arbitrators shall be final, and judgment may be entered upon it in accordance with applicable law in any court having jurisdiction thereof.

7.9.2 Notice of the demand for arbitration shall be filed in writing with the other party to the Owner-Contractor Agreement and with the American Arbitration Association, and a copy shall be filed with the Architect. The demand for arbitration shall be made within the time and in all other limits specified in Subparagraph 2.2.12 where applicable, and in all other cases within a reasonable time after the claim, dispute or other matter in question has arisen, and in no event shall it be made after the date when institution of legal or quibble proceedings based on such claim, dispute or other matter in question would be barred by the applicable statute of limitations.

7.9.3 Unless otherwise agreed in writing, the Contractor shall carry on the Work and maintain its progress during any arbitration proceedings, and the Owner shall continue to make payments to the Contractor in accordance with the Contract Documents.

Article 8

Time

8.1 Definitions

8.1.1 Unless otherwise provided, the Contract Time is the periods of time allotted in the Contract Documents for Substantial Completion of the Work as defined in Subparagraph 8.1.3, including authorized adjustments thereto.

8.1.2 The date of commencement of the Work is the date established in a notice to proceed. If there is no notice to proceed, it shall be the date of the Owner-Contractor Agreement or such other date as may be established therein.

8.1.3 The Date of Substantial Completion of the Work or designated portion thereof is the date certified by the Architect when construction is sufficiently complete, in accordance with the Contract Documents, so the Owner can occupy or utilize the Work or designated portion thereof for the use for which it is intended.

8.1.4 The term day as used in the Contract Documents shall mean calendar day unless otherwise specifically designated.

8.2 Progress and Completion

8.2.1 All time limits stated in the Contract Documents are of the essence of the Contract.

8.2.2 The Contractor shall begin the Work on the date of commencement as defined in Subparagraph 8.1.2. He shall carry the Work forward expeditiously with adequate forces and shall achieve Substantial Completion within the Contract Time.

8.3 Delays and Extensions of Time

8.3.1 If the Contractor is delayed at any time in the progress of the Work by any act or neglect of the Owner or the Architect, or by any employee of either, or by any separate contractor employed by the Owner, or by changes ordered in the Work, or by labor disputes, fire, unusual delay in transportation, adverse weather conditions not reasonably anticipatable, unavoidable casualties, or any causes beyond the Contractor's control, or by any other cause which the Architect determines may justify the delay, then the Contract Time shall be extended by Change Order for such reasonable time as the Architect may determine.

8.3.2 Any claim for extension of time shall be made in writing to the Architect no more than twenty days after the commencement of the delay; only one claim is necessary. The Contractor shall provide an estimate of the probable effect of such delay on the progress of the Work.

8.3.3 If no agreement is made stating the dates upon which interpretations as provided in Subparagraph 2.2.8 shall be furnished, then no claim for delay shall be allowed on account of failure to furnish such interpretations until fifteen days after written request is made for them, and not then unless such claim is reasonable.

8.3.4 This Paragraph 8.3 does not exclude the recovery of damages by either under other provisions of the Contract Documents.

Article 9

Payments and Completion

9.1 Contract Sum

9.1.1 The Contractor Sum is stated in the Owner-Contractor Agreement and including authorized adjustments thereto, is the total amount payable by the Owner to the Contractor for the performance of the Work under the Contract Documents.

9.2 Schedule of Values

9.2.1 Before the first Application for Payment, the Contractor shall submit to the Architect a schedule of values allocated to the various portions of the Work, prepared in such form and supported by such data to substantiate its accuracy as the Architect may require. This schedule, unless objected to by the Architect, shall be used only as a basis for the Contractor's Applications for Payment.

9.3 Applications for Payment

9.3.1 At least ten days before the date for each progress payment established in the Owner-Contractor Agreement, the Contractor shall submit to the Architect an itemized Application for Payment, notarized if required, supported by such data substantiating the Contractor's right to payment as the Owner or the Architect may require, and reflecting retainage, if any, as provided elsewhere in the Contract Documents.

9.3.2 Unless otherwise provided in the Contract Documents, payments will be made on account of materials or equipment not incorporated in the Work but delivered and suitably stored at the site and, if approved in advance by the Owner, payments may similarly be made for materials or equipment stored on or off the site shall be conditioned upon submission by the Contractor of bills of sale or such other procedures satisfactory to the Owner to establish the Owner's title to such materials or equipment or otherwise protect the Owner's interest, including applicable insurance and transportation to the site for those materials and equipment off the site.

9.3.3 The Contractor warrants that title to all Work materials and equipment covered by an Application for Payment will pass to the owner either by incorporation in the construction or upon the receipt of payment by the Contractor, whichever occurs first, free and clear of all liens, claims, security interests or encumbrances herein after referred to in this Article 9 as "liens"; and that no work, materials or equipment covered by an application for payment will have been acquired by the contractor, or by any other person performing work at the site or furnishing materials and equipment for the project, subject to an agreement under which an interest therein or an encumbrance thereon is retained by the seller or otherwise imposed by the contractor or such other person.

9.4 Certificates For Payment

9.4.1 The Architect will, within seven days after the receipt of the Contractor's Application for Payment, either issue a Certificate for Payment to the Owner, with a copy to the contractor, for such amount as the architect determines is properly due, or notify the contractor in writing his reasons for withholding a certificate as provided in Subparagraph 9.6.1.

9.4.2 The issuance of a Certificate for Payment will constitute a representation by the architect to the owner, based on his observations at the site as provided in Subparagraph 2.2.3 and the date comprising the Application for Payment, that the work has progressed to the point indicated; that, to the best of his knowledge, information and belief, the quality of the work is in accordance with the Contract Documents (subject to an evaluation of the work for conformance with the Contract Documents upon Substantial Completion to the results of any subsequent tests required by or performed under the Contract Documents, to minor deviations from the Contract Documents correctable prior to completion and to any specific qualifications stated in his Certificate); and that the contractor is entitled to payment in the amount certified. However, by issuing a Certificate for Payment, the architect shall not thereby be deemed to represent that he has mad exhaustive or continuous on-site inspections to check the quality or quantity of the work or that he has reviewed the construction means, methods, techniques, sequences or procedures, or that he has made any examination to ascertain how or for what purpose that contractor has used the moneys previously paid on account of the contract sum.

9.5 Progress Payments

9.5.1 After the architect has issued a Certificate for Payment, the owner shall make payment in the manner and within the time provided in the Contract Documents.

9.5.2 The contractor shall promptly pay each subcontractor, upon receipt of payment from the owner, out of the amount paid to the contractor on account of such subcontractor's work, the amount to which said subcontractor is entitled, reflecting the percentage actually retained, if any from the payments to the contractor on account of such subcontractor's work. The contractor shall, by an appropriate agreement with each subcontractor, require each subcontractor to make payments to his sub-subcontractors in similar matter.

9.5.3 The architect may, on request and at his discretion, furnish to any subcontractor, or if practicable, information regarding the percentage of completion or the amounts applied for by the contractor and the action taken thereon by the architect on account of work done by each subcontractor.

9.5.4 Neither the owner nor the architect shall have any obligation to pay or to see to the payment of any moneys to any subcontractor except as may otherwise be required by law.

9.5.5 No certificate for a progress payment, nor any progress payment, nor any partial or entire use or occupancy of the project by the owner, shall constitute an acceptance of any work not in accordance with the Contract Document.

9.6 Payments Withheld

9.6.1 The architect may decline to certify payment and may withhold his Certificate in whole or in part, to the extent necessary reasonably to protect the owner, if in his opinion he is unable to make representations to the owner as provided in Subparagraph 9.4.2. If the architect is unable to make representations to the owner as provided in Subparagraph 9.4.1. If the contractor and the architect cannot agree on a revised amount, the architect will promptly issue a Certificate for Payment for the amount for which he is able to make such representations to the owner. The architect may also decline to certify payment or because of subsequently discovered evidence or subsequent observations he may nullify the whole or any part of any Certificate for Payment previously issued, to such extent as may be necessary in his opinion to protect the owner from loss because

1. defective work not remedied,

2. third party claims filed or reasonable evidence indicating probable filing of such claims,

3. failure of the contractor to make payments properly to subcontractors or for labor, materials or equipment,

4. reasonable evidence that the work cannot be completed for the unpaid balance of the Contract Sum,

5. damage to the owner or another contractor,

6. reasonable evidence that the work will not be completed within the Contract Tme, or

7. persistent failure to carry out the work in accordance with the Contract Documents.

9.6.2 When the above grounds in Subparagraph 9.6.1 are removed, payment shall be made for amounts withheld because of them.

9.7 Failure of Payment

9.7.1 If the architect does not issue a Certificate for Payment, through no fault of the contractor, within seven days after receipt of the contractors Application for Payment, or if the owner does not pay the contractor within seven days after the date established in the Contract Documents any amount certified by the architect or awarded by arbitration, then the contractor may, upon seven additional days' written notice to the owner and the architect, stop the work until payment of the amount owed has been received. The Contract Sum shall be increased by the amount of the contractors reasonable costs of shutdown, delay and start-up, which shall be effected by appropriate Change Order in accordance with Paragraph 12.3.

9.8 Substantial Completion

9.8.1 When the contractor considers that the work, or a designated portion thereof which is acceptable to the owner, is substantially complete as defined in Subparagraph 8.1.3, the contractor shall prepare for submission to the architect a list of items to be completed or corrected. The failure to include any items on such list does not alter the responsibility of the Contractor to complete all work in accordance with the Contract Documents. When the architect on the basis of an inspection determines that the work or designated portion thereof is substantially complete, he will the prepare a Certificate of Substantial Completion which shall establish the date of substantial completion, shall state the responsibilities of the owner and the contractor for security, maintenance, heat, utilities, damage to the work, and insurance, and shall fix the time within which the contractor shall complete the items listed therein. Warranties required by the Contract Documents shall commence on the date of substantial completion. The Certificate of Substantial Completion shall be submitted to the owner and the contractor for their written acceptance of the responsibilities assigned to them in such Certificate.

9.8.2 Upon substantial completion of the work or designated portion thereof and upon application by the contractor and certification by the architect, the owner shall make payment, reflecting adjustment in retainage if any, for such work or portion thereof as provided in the Contract Documents.

9.9 Final Completion and Final Payment

9.9.1 Upon receipt of written notice that the work is ready for final inspection and acceptance and upon receipt of a final Application for Payment, the architect will promptly make sure inspection and, when he finds the work acceptable under the Contract Documents and the contract fully performed, he will promptly issue a final Certificate for Payment stating that to the best of his knowledge, information and belief, and on the basis of his observations and inspections, the work has been completed in accordance with the terms and conditions of the Contract Documents and that the entire balance found to be due to the contractor, and noted in said final Certificate, is due and payable. The architect's final Certificate for Payment will constitute a further representation that the conditions precedent to the contractor's being entitled to final payment as set forth in Subparagraph 9.9.2 have been fulfilled.

9.9.2 Neither the final payment nor the remaining retained percentage shall become due until the contractor submits to the architect (1) an affidavit that all payrolls, bills for materials and equipment, and other indebtedness connected with the work for which the owner or his property might in any way be responsible, have been paid or otherwise satisfied, (2) consent of surety, if any, to final payment and (3) if required by the owner, other data establishing payment or satisfaction of all such obligations, such as receipts, releases and waivers of liens arising out of the contract, to the extent and in such form as may be designated by the owner. If any subcontractor refuses to furnish a release or waiver required by the owner, the contractor may furnish a release or waiver required by the owner, the contractor may furnish a bond satisfactory to the owner to indemnify him against any such lien. If any such lien remains unsatisfied after all the payments are made, the contractor shall refund to the owner all moneys that the latter may be compelled to pay in discharging such lien, including all costs and reasonable attorneys' fees.

9.9.3 If, after Substantial Completion of the work, final completion thereof is materially delayed through no fault of the contractor or by the issuance of Change Orders affecting final completion, and the architect so confirms, the owner shall, upon application by the contractor and certification by the architect and without terminating the contract, make payment of the balance due for that portion of the work fully completed and accepted. If the remaining balance for work not fully completed or corrected is less than the retainage stipulated in the Contract Documents, and if the bonds have been furnished as provided in Paragraph 7.5, the written

consent of the surety to the payment of balance due for that portion of the work fully completed and accepted shall be submitted by the contractor to the architect prior to certification of such payment. Such payment shall be made under the terms and conditions governing final payment, except that it shall not constitute a waiver of claims.

9.9.4 The making of final payment shall constitute a waiver of all claims by the owner except those arising from:

    1. unsettled liens,

    2. faulty or defective work appearing substantial completion,

    3. failure of the work to comply with the requirements of the Contract Documents, or

    4. terms of any special warranties required by the Contract Documents.

9.9.5 The acceptance of final payment shall constitute a waiver of all claims by the contractor except those previously made in writing and identified by the contractor as unsettled at the time of the final Application for Payment.

Article 10

Protection of Persons and Property

10.1 Safety Precautions and Programs

10.1.1 The contractor shall take all reasonable precautions for the safety of and shall provide all reasonable protection to prevent damage, injury or loss to:

    1. all employees on the work and all other persons who may be affected thereby;

    2. all the work and all materials and equipment to be incorporated therein, whether in storage on or off the site, under the care, custody or control of the contractor or any of his subcontractors or sub-subcontractors; and

    3. other property at the site or adjacent thereto, including trees, shrubs, lawns, walks, pavements, roadways, structures, and utilities not designated for removal, relocation or replacement in the course of construction.

10.2.2 The contractor shall give all notices and comply with all applicable laws, ordinances, rules, regulations and lawful orders of any public authority bearing on the safety of persons or property of their protection from damage, injury or loss.

10.2.3 The contractor shall erect and maintain, as required by existing conditions and progress of the work, all reasonable safeguards for safety and protection, including posting danger signs and other warnings against hazards, promulgating safety regulations and notifying owners and users of adjacent utilities.

10.2.4 When the use or storage of explosives or other explosives or other hazardous materials or equipment is necessary for the execution of the work, the contractor shall exercise the utmost care and shall carry on such activities under the supervision of property qualified personnel.

10.2.5 The contractor shall promptly remedy all damage or loss (other than damage or loss insured under Paragraph 11.3) to any property referred to in Clauses 10.2.1.2 and 10.2.1.3 caused in whole or in part by the contractor, any subcontractor, any sub-subcontractor, or anyone directly or indirectly employed by any of them, or by anyone for whose acts any of them may be liable and for which the contractor is responsible under clauses 10.2.1.2 and 10.2.1.3, except damages or loss attributable to the acts or omissions of the owner or architect or anyone directly or indirectly employed by either of them, or by anyone for whose acts either of them may be liable, and not attributable to the fault or negligence of the contractor. The foregoing obligations of the contractor are in addition to his obligations of the contractor are in addition to his obligation under Paragraph 4.18.

10.2.6 The contractor shall designate a responsible member of his organization at the site whose duty shall be the prevention of accidents. This person shall be the contractor's superintendent unless otherwise designated by the contractor in writing to the owner and the architect.

10.2.7 The contractor shall not load or permit any part of the work to be loaded so as to endanger its safety.

10.3 Emergencies

10.3.1 In any emergency affecting the safety of persons or property, the contractor shall act, at his discretion, to prevent threatened damage, injury or loss. Any additional compensation or extension of time claimed by the contractor on account of emergency work shall be determined as provided in Article 12 for Changes in the Work.

Article 11

Insurance

11.1 Contractors Liability Insurance

11.1.1 The contractor shall purchase and maintain such insurance as will protect him from claims set forth below which may arise out of or result from the contractor's operations under the contract, whether such operations by himself or by any subcontractor or by anyone directly or indirectly employed by any of them, or by anyone for whose acts any of them may be liable:

1. claims under workers' or workmen's compensation. Disability benefit and other similar employee benefit acts;

2. claims for damages because of bodily injury, occupational sickness or disease, or death of his employees;

3. claims for damages because of bodily injury, sickness or disease, or death of any person other than his employees;

4. claims for damages insured by usual personal injury liability coverage which are sustained (1) by any person as a result of an offense directly or indirectly related to the employment of such person by the contractor, or (2) by any other person;

5. claims for damages, other than to the work itself, because of injury to or destruction of tangible property, including loss of use resulting there from; and

6. claims for damages because of bodily injury or death of any person or property damage arising out of the ownership, maintenance or use of any motor vehicle.

11.1.2 The insurance required by Subparagraph 11.1.1 shall be written for not less than any limits of liability specified in the Contract Documents, or required by law, whichever is greater.

11.1.3 The insurance required by Subparagraph 11.1.1 shall include contractual liability insurance applicable to the contractor's obligations under Paragraph 1.18.

11.1.4 Certificate of insurance acceptable to the owner shall be filed with the owner prior to commencement of the work, These certificates shall contain a provision that coverages afforded under the policies will not be cancelled until at least thirty days' prior written notice has been given to the owner.

11.2 Owners Liability Insurance

11.2.1 The owner shall be responsible for purchasing and maintaining his own liability insurance and, at his option, may purchase and maintain such insurance as will protect him against claims which may arise from operations under the contract.

11.3 Property Insurance

11.3.1 Unless otherwise provided, the owner shall purchase and maintain property insurance upon the entire work at the site to the full insurable value thereof. This insurance shall include the interests of the owner, the contractor, subcontractor, and sub-subcontractors in the work and shall insure against the perils of fire and extended coverage and shall include all risk insurance for physical loss or damage including, without duplication of coverage, theft, vandalism and malicious mischief. If the owner does not intend to purchase such insurance for the full insurable value of the entire work, he shall inform the contractor in writing prior to commencement of the work. The contractor may then effect insurance which will protect the interests of himself, his subcontractors and the sub-subcontractors in the work and by appropriate change order the cost thereof shall be charged to the owner. If the contractor is damaged by failure of the owner to purchase or maintain such insurance and so notify the contractor, then the owner shall bear all reasonable costs properly attributable thereto. If not covered under the all risk insurance or otherwise provided in the Contract Documents, the contractor shall effect and maintain similar property insurance on portions of the work are to be included in an Application for Payment under Subparagraph 9.3.2.

11.3.2 The owner shall purchase and maintain such boiler and machinery insurance as may be required by the Contract Documents or by law. This insurance shall include the interests of the owner, the contractor, subcontractors and sub subcontractors in the work.

11.3.3 Any loss insured under Subparagraph 11.3.1 is to be adjusted with the owner and made payable to the owner as trustee for the insureds, as their interest may appear, subject to the requirements of any applicable mortgage clause and of Subparagraph 11.3.8. The contractor shall pay each subcontractor a just share of any insurance moneys received by the contractor, and by appropriate agreement, written where legally required for validity, shall require each subcontractor to make payments to his sub-subcontractors in similar matter.

11.3.4 The owner shall file a copy of all policies with the contractor before an exposure to loss may occur.

11.3.5 If the contractor requests in writing that insurance for risks other than those described in Subparagraphs 11.3.1 and 11.3.2 or other special hazards be included in the property insurance policy, the owner shall, if possible, include such insurance, and the cost thereof shall be charged to the contractor by appropriate Change Order.

11.3.6 The owner and contractor waive all rights against (1) each other and the subcontractors, sub-subcontractors, agents and employees each of the other, and (2) the architect and separate contractors, if any, subcontractors, agents and employees, for damages caused by fire or other perils to the extent covered by insurance obtained pursuant to this Paragraph 11.3 or any other property insurance applicable to the Work, except such rights as they may have to the proceeds of such insurance held by the Owner as trustee. The foregoing waiver afforded the Architect, his a gents and employees shall not extend to the liability imposed by Subparagraph 4.18.3. The Owner or the Contractor, as appropriate, shall require of the Architect, separate contractors, Subcontractors and Sub-subcontractors by appropriate agreements written where legally required for validity, similar waivers each in favor of all other parties enumerated in this Subparagraph 11.3.6.

11.3.7 If required in writing by any party in Interest, the Owner as trustee shall, upon the occurrence of an insured loss, give bond to the proper performance of his duties. He shall deposit in a separate account any money so received and he shall distribute it in accordance with such agreement as the parties in interest may reach, or in accordance with an award by arbitration in which ease the procedure shall be as provided in Paragraph 7.9. If after such loss no other special agreement is made, replacement of damaged work shall be covered by an appropriate Change Order.

11.3.8 The Owner as trustee shall have power to adjust and settle any loss with the insurers unless one of the parties in interest shall object in writing within five days after the occurrence of loss to the Owner's exercise of this power, and if such objection be made, arbitrators shall be chosen as provided in Paragraph 7.9. The Owner as trustee shall, in that case, make settlement with the insurers in accordance with the directions of such arbitrators. If distribution of the insurance proceeds by arbitration is required, the arbitrators will direct such distribution.

11.3.9 If the Owner finds it necessary to occupy or use a portion or portions of the Work prior to Substantial Completion thereof, such occupancy or use shall not commence prior to a time mutually agreed to by the Owner and Contractor and to which the Insurance company or companies providing the property insurance have consented by endorsement to the policy or policies. This insurance shall not be cancelled or lapsed on account of such partial occupancy or use-Consent of the Contractor and of the insurance company or companies to such occupancy or use shall not be unreasonably withheld.

11.4 Loss of Use Insurance

11.4.1 The Owner, at his option, may purchase and, maintain such insurance as will insure him against loss of use of his property due to fire or other hazards, however caused. The Owner waives all rights of action against the Contractor for loss of use of his property, including consequential losses due to fire or other hazards however caused, to the extent covered by insurance under this Paragraph 11.4.

Article 12

Changes in the Work

12.1 Change Orders

12.1.1 A Change Order is a written order to the contractor signed by the Owner and the Architect, issued after the execution of the Contract, authorizing a change in the work or an adjustment in the Contract Sum or the Contract Time. The Contract Sum and the Contract Time may be changed only by a change order. A change order signed by the contractor indicates his agreement there with, including the adjustment in the Contract Sum or the Contract Time.

12.1.2 The owner, without invalidating the contract, may order changes in the work within the general scope of the contract consisting of additions, deletions or other revisions, the Contract Sum and the Contract Time being adjusted accordingly. All such changes in the Work shall be authorized by Change Order, and shall be performed under the applicable conditions of the Contract Documents.

12.1.3 The cost or credit to the Owner resulting from a change in the Work shall be determined in one or more of the following ways:

1. by mutual acceptance of a lump sum properly itemized and supported by sufficient substantiating data to permit evaluation;

2. by unit prices stated in the Contract Documents or subsequently agreed upon;

3. by cost to be determined in a manner agreed upon by the parties and a mutually acceptable fixed or percentage fee; or

4. by the method provided in subparagraph 12.1.4.

12.1.4 If none of the methods set forth in Clauses 12.1.3.1, 12.1.3.2 or 12.1.3.3 is agreed upon, the Contractor, provided he receives a written order signed by the Owner, shall promptly proceed with the Work involved. The cost of such Work shall then be determined by the Architect on the basis of the reasonable expenditures and savings of those performing the Work attributable to the change, including, in the case of an increase in the Contract Sum, a reasonable allowance for overhead and profit. In such case, and also under Clauses 12.1.3.3 and 12.1.3.4 above, the Contractor shall keep and present, in such form as the Architect may prescribe, an itemized accounting together with appropriate supporting data for inclusion in a Change Order. Unless otherwise provided in the Contract Documents, cost shall be limited to the following: cost of materials, including sales tax and cost of delivery; cost of labor, including social security, old age and unemployment insurance, and fringe benefits required by agreement or custom; workers' or workmen's compensation insurance; bond premiums; rental value of equipment and machinery; and the additional costs of supervision and field office personnel directly attributable to the change. Pending final determination of cost to the Owner, payments on account shall be made on the Architect's Certificate for Payment. The amount of credit to be allowed by the Contractor to the Owner for any deletion or change which results in a net decrease in the Contract Sum will be the amount of the actual net cost as confirmed by the Architect. When both additions and credits covering related Work or substitutions are involved in any one

change, the allowance for overhead and profit shall be figured on the basis of the net increase, if any, with respect to that change.

12.1.5 If unit prices are stated in the Contract Documents or subsequently agreed upon, and if the quantities originally contemplated are so changed in a proposed Change Order that application of the agreed unit prices to the quantities of Work proposed will cause substantial inequity to the Owner or the Contractor, the applicable unit prices shall be equitably adjusted.

12.2 Concealed Conditions

12.2.1 Should concealed conditions encountered in the performance of the Work below the surface of the ground or should concealed or unknown conditions in an existing structure be at variance with the conditions indicated by the Contract Documents, or should unknown physical conditions below the surface of the ground or should concealed or unknown conditions in an existing structure of an unusual nature, differing materially from those ordinarily encountered and generally recognized as inherent in work of the character provided for in this Contract, be encountered, the Contract Sum shall be equitably adjusted by Change Order upon claim by either party made within twenty days alter the first observance of the conditions.

12.3 Claims Fog Additional Cost

12.3.1 If the Contractor wishes to make a claim for an increase in the Contract Sum, he shall give the Architect written notice thereof within twenty days after the occurrence of the event giving rise to such claim. This notice shall be given by the Contractor before proceeding to execute the work, except in an emergency endangering tile or property in which case the Contractor shall proceed in accordance with Paragraph 10.3. No such claim shall be valid unless so made. If the Owner and the Contractor cannot agree on the amount of the adjustment in the Contract Sum, it shall be determined by the Architect. Any change in the Contract Sum resulting from such claim shall be authorized by Change Order.

12.3.2 If the Contractor claims that additional cost is involved because of, but not limited to, (1) any written interpretation pursuant to Subparagraph 2.2.0, (2) any order by the Owner to stop the Work pursuant to Paragraph 3.3 where the Contractor was not at fault, (3) any written order for a minor change in the Work issued pursuant to Paragraph 12.4, or (4) failure of payment by the Owner pursuant to Paragraph 9.7, the Contractor shall make such claim as provided in Subparagraph 12.3.1.

12.4 Minor Changes in the Work

12.4.1 The Architect will have authority to order minor changes in the Work not involving an adjustment in the

Contract Sum or an extension of the Contract Time and not inconsistent with the intent of the Contract Documents. Such changes shall be effected by written order, and shall be binding on the Owner and the Contractor. The Contractor shall carry out such written orders promptly.

Article 13

Uncovering and Correction of Work

13.1 Uncovering Of Work

13.1.1 If any portion of the Work should be covered contrary to the request of the Architect or to requirements specifically expressed in the Contract Documents, it must, if required in writing by the Architect, be uncovered for his observation and shall be replaced at the Contractor's expense.

13.1.2 If any other portion or the Work has been covered which the Architect has not specifically requested to observe prior to being covered, the Architect may request to see such Work and it shall be uncovered by the Contractor. If such Work be found in accordance with the Contract Documents, the cost of uncovering and replacement shall, by appropriate Change Order, be charged to the Owner. If such Work be found not in accordance with the Contract Documents, the Contractor shall pay such costs unless it be found that this condition was caused by the Owner or a separate contractor as provided in Article 6, in which event the Owner shall be responsible for the payment of such costs.

13.2 Correction of Work

13.2.1 The Contractor shall promptly correct all Work rejected, by the Architect as defective or as failing to conform to the Contract Documents whether observed before or after Substantial Completion end whether or not fabricated, installed or completed. The Contractor shall bear all costs of correcting such rejected Work, including compensation for the Architect's additional services made necessary thereby.

13.2.2 If, within one year after the Date of Substantial Completion of the Work or designated portion thereof or within one year after acceptance by the Owner designated equipment or within such longer period of time as may be prescribed by law or by the terms of any applicable special warranty required by the Contract Documents, any of the Work is found to be defective or not in accordance with the Contract Documents, the Contractor shall correct it promptly after receipt of a written notice from the Owner to do so unless the Owner has previously given the Contractor a written acceptance of such condition. This obligation shall survive termination of the Contract. The Owner shall give such notice promptly after discovery of the condition.

13.2.3 The Contractor shall remove from the site all portions of the Work which are defective or non-conforming and which have not been corrected under Subparagraphs 4.5.1, 13.2.1, and 13.2.2., unless removal is waived by the Owner.

13.2.4 If the Contractor fails to correct defective or non-conforming Work as provided in Subparagraphs 4.5.1, 13.2.1 and 13.2.2, the Owner may correct it in accordance with Paragraph 3.4. the Contractor promptly.

13.2.5 If the Contractor does not proceed with the correction of such defective or non-conforming Work within a reasonable time fixed by written notice from the Architect, the Owner may remove it and may store the materials or equipment at the expense of the Contractor. If the Contractor does not pay the cost of such removal and storage within ten days thereafter, the Owner may upon ten additional days' written notice sell such Work at auction or at private sale and shall account for the net proceeds thereof, after deducting all the costs that should have been borne by the Contractor, including compensation for the Architect's additional services made necessary thereby. If such proceeds of sale do not cover all costs which the Contractor should have borne, the difference shall be charged to the Contractor and an appropriate Change Order shall be issued. If the payments then or thereafter due the Contractor are not sufficient to cover such amount, the Contractor shall pay the difference to the Owner.

13.2.6 The Contractor shall bear the cost of making good all work of the Owner or separate contractors destroyed or damaged by such correction or removal.

13.2.7 Nothing contained in this Paragraph 13.2 shall be construed to establish a period of limitation with respect to any other obligation which the Contractor might have under the Contract Documents, including Paragraph 4.5 hereof. The establishment of the time period of one year after the Date of Substantial Completion or such longer period of time as may be prescribed by law or by the terms of any warranty required by the Contract Documents relates only to the specific obligation of the Contractor to correct the Work, and has no relationship to the time within which his obligation to comply with the Contract Documents may be sought to be enforced, nor in the time within which proceedings may be commenced, establish the Contractor's liability with respect to his obligations other than specifically to correct the work.

13.3 Acceptance of Defective or Non-Defective Work

13.3.1 If the Owner prefers to accept defective or non-conforming Work, he may do so instead of requiring its removal and correction, in which case a Change Order will be issued to reflect a reduction in the Contract Sum where appropriate and equitable. Such adjustment shall be effected whether or not final payment has been made.

Article 14

Termination of the Contract

14.1 Termination by the Contractor

14.1.1 If the Work is stoppedd for a period of thirty days under an order of any court or other public authority having jurisdiction, or as a result of an act of government, such as a declaration of a national emergency making materials unavailable, through no act or fault of the Contractor or a Subcontractor or their agents or employees or any other persons performing any of the Work under a contract with the contractor, or if the Work should be stopped for a period of thirty days by the Contractor because the Architect has not issued a Certificate for Payment as provided in Paragraph 9.7 or because the Owner has not made payment thereon as provided in Paragraph 9.7, then the Contractor may, upon seven additional days' written notice to the Owner and the Architect, terminate the Contract and recover from the Owner payment for all Work executed and for any proven loss sustained upon any materials, equipment, tools, construction equipment and machinery, including reasonable profit and damages.

14.2 Termination by Owner

14.2.1 If the Contractor is adjudged a bankrupt, or if he makes a general assignment for the benefit of his creditors, or if a receiver is appointed on account of his insolvency, or if he persistently or repeatedly, refuses or fails, except in cases for which extension of time is provided, to supply enough properly skilled workmen or proper materials, or if he fails to make prompt payment to Subcontractors or for materials or labor, or persistently disregards laws, ordinances, rules, regulations or orders of any public authority having jurisdiction, or otherwise is guilty of a substantial violation of a provision of the Contract Documents, then the Owner, upon certification by the Architect that sufficient cause exists to justify such action, may, without prejudice to any right or remedy and after giving the Contractor and his surety, if any, seven days' written notice, terminate the employment of the Contractor and take possession of the site and of all materials, equipment, tools, construction equipment and machinery thereon owned by the Contractor and may finish the Work by whatever method he may deem expedient. In such case the Contractor shall not be entitled to receive any further payment until the Work is finished.

14.2.2 If the unpaid balance of the Contract Sum exceeds the costs of finishing the Work, including compensation for the Architect's additional services made necessary thereby, such excess shall be paid to the Contractor. If such costs exceed the unpaid balance, the Contractor shall pay the difference to the owner. The amount to be paid to the Contractor or to the Owner, as the case may be, shall be certified by the Architect, upon application, in the manner provided in Paragraph 9.4, and this obligation for payment shall survive the termination of the Contract.

FARMER'S STATE BANK

T-1	Title Sheet
C-1	Site Plan
A-1	Area Location Plan
A-2	Demolition Plan
A-3	Floor Plan
A-4	Reflected Ceiling Plan
A-5	Building Section
	Interior Elevations
A-6	Wail Sections
A-7	Door and Perimeter Window Details
A-8	Millwork Details
A-9	Door and Finish Schedules
AM-1	Mechanical Plan
AP-1	Plumbing Plan
AE-1	Electrical Plan

THE SUITE 108B
FARMER'S STATE BANK
THE COMMERCE CENTER

SUMMARY OF WORK
01010 PAGE 1

1.01   WORK COVERED BY CONTRACT DOCUMENTS
A.   Work Included:
Perform all work required by the contract documents for the conversion of Suite 108B Renovation, Farmers State Bank, The Commerce Center including the demolition of select structures. Only items specifically indicated on Contract Documents as N.I.C. or "By Others" are not included.

B.   General Description of Work:
Work includes but is not limited to: Demolition of certain structures, interior and exterior finishes, concrete, wood and masonry construction; restoration of certain existing masonry and refurbishing of interior spaces including wood, light gauge metal framing and furring, gypsum drywall, plywood, millwork and certain new glazing and painting; new and complete plumbing system, mechanical, fire protection and electrical. Drawings and specifications are included for coordination of all work between prime contractors.

1.02   EXISTING UTILITIES

A.   Utilities of record are shown on the Drawings insofar as possible to do so. These, however, are shown for convenience only and the Owner assumes no responsibility for improper locations or failure to show utility locations on the Drawings. At contractor's expense, immediately repair utilities damaged during construction. Contractor to locate and define existing utilities for use in the project.

1.03   OBJECTIONS TO APPLICATION OF PRODUCTS

A.   Subcontractors submitting bids for this Project shall thoroughly familiarize themselves with specified products and installation procedures and submit to Architect any objections (in writing) no later than 10 days prior to Bid Date, Submittal of bid constitutes acceptance of products and procedures specified.

1.04   CONFLICTS AND OMISSIONS IN DRAWINGS AND SPECIFICATIONS

A.   Bring immediately to Architects attention; he will determine order of precedence and make final decisions.

   1.   Where conflicts and omissions have not been brought to Architects attention, it is understood that Contractor is responsible for any corrections necessary because of his judgment in the field.
   2.   See "Conflicts" 00710
   3.   Contractor to verify all dimensions.

SUMMARY OF WORK
01010 PAGE 2

1.05   MISCELLANEOUS

A.   Items not included, but are not limited to:
   1.   Not unreasonably encumbering site with materials or equipment.
   2.   Not loading structure with weight-endangering structure.

3.   Assuming full responsibility for protection and safe keeping of products stored on premises.

4.   Moving any stored products interfering with any other Contractors.

5.   Obtaining and paying for use of additional storage of work areas needed for operations.

6.   Installing any indicated Owner-finished items noted to be installed by Contractor.

SELECTIVE DEMOLITION
02070 PAGE 1

element 1: GENERAL

1.01 RELATED WORK

A. Work in other Sections: Cutting and Patching Section 01045; Barriers section 01530; Cleaning Section 01110.

B. Demolition Responsibility
The General Contractor is responsible for all demolition work and must retain and direct all trades, workmen and specialists necessary to accomplish the required work indicated on the drawings and as noted in this section, except:

 The mechanical and electrical contractors are required to disconnect, cap off or re-route all existing active utilities as necessary and as indicated on the drawings and noted in the specifications.

1.02 EXTENT OF WORK IN THIS SECTION

A. Remove designated building equipment and fixtures.

B. Remove designated partitions and components.

C. Cap and identify exposed utilities.

D. Scraping and cleaning of existing exposed structure.

E. Masonry wall penetrations for doors, windows, mechanical, etc.

1.03 CODES

A. Conform to codes and requirements of governing authority.

B. Obtain and pay for all permits for demolition; protection of the public and property; transportation and disposal of debris; and severance of utility Services.

1.04 SUBMITTALS

A. Permits and notices authorizing demolition.

B. Certificates of severance of utility services.

C. Permit for transport and disposal of debris.

D. Submit a Demolition Schedule (proposed sequence of demolition and removal operations to the Architect for review prior to start of the work.
 1. Include coordination for shut-off, capping and continuation of utility services as required.

1.05 JOB CONDITIONS

A. Building will be occupied but Tenant will be vacated prior to start or work.

B. Conditions of Building
 1. Condition existing at time of inspection for bidding purposes will be maintained by Owner in so far as practicable.
 2. The Owner assumes no responsibility for actual condition of the building.

C. Obstruction, Interference
 1. Conduct demolition operations and removal of debris to ensure minimum interference with corridors, roads, streets, walks and other adjacent occupied or used facilities.

    2.   Do not close or obstruct corridors, streets', walks or exits from other occupied or used facilities without permission from authorities having jurisdiction and the Owner.

1.05

D.   Protection
    1.   Ensure safe passage of persons around areas of demolition. Conduct operations to prevent injury to adjacent buildings, structures, other facilities and persons.
    2.   Provide, erect and maintain barricades, and guard rails as required by governing authority to protect occupants and workers.
    3.   Protect persons and property from damage and discomfort caused by dust through construction of dustproof partitions.
    4.   Erect weatherproof closures for exterior openings.
    5.   On completion, remove protective devices.

E.   Environmental Controls
    1.   Burning of materials on site is not permitted.
    2.   Any disposal chutes required for debris removal and disposal must meet or exceed all OSHA or other applicable codes and requirements.

F.   Damage
    1.   Provide interior and exterior shoring, bracing, and other support to prevent movement, settlement or collapse of structure and adjacent construction to remain.
    2.   Assume liability for such movement, settlement, damage or injury.
    3.   Cease operations and notify the Architect immediately, if safety of structure appears to be endangered. Do not resume operations until safety is restored.
    4.   Promptly repair damages caused to adjacent facilities by demolition operations at no cost to Owner.

PART 2 MATERIALS

2.01 SALVAGE

A.   The General Contractor has no obligation to retain salvage for the Owner except as specifically indicated on drawings or as noted below for reuse.

2.02 SALVAGED MATERIALS FOR RE-USE

A.   Remove work carefully preserving original materials in sound undamaged conditions. Store in protected area out of harms way. See Drawing Notes for specific items to be retained for use.

PART 3 EXECUTION

3.01 GENERAL

A.   Proceed with demolition in a systematic manner in accordance with the Demolition Schedule to accommodate new work.

B.   Remove all items of equipment and construction shown on the Drawings.

C.   Locate demolition equipment throughout structure and remove materials so as not to impose excessive loads to supporting walls, floors or framing.

3.02 UTILITIES AND SERVICES

A.   Place markers to indicate location of disconnected service. Identify service lines and capping locations on Project Record Documents.

3.03 PROCEDURE

A.   Carefully cut and remove portions of construction required to be removed, in a manner not to disturb adjacent areas of construction to remain.

3.04 REMOVAL

A.   Remove from site, contaminated, vermin infested or dangerous materials and dispose of by safe means not to endanger health of workers and public.

B.   Remove demolished materials, tools and equipment from site upon completion of work.

3.05 DAMAGE DURING CONSTRUCTION

A.   Replace entirely or satisfactorily repair as required by Architect, any portion of new or existing structure damaged during construction at no cost to owner.

3.06   SEE CUTTING AND PATCHING DIVISION 1, SECTION 01045; DIVISION 15

SECTION 03.04; DIVISION 16 SECTION 03.01—C.

3.07 CLEAN-UP

A.   Clean Adjacent improvements of dust, dirt, and debris caused by demolition operations, as directed by Architect, Building Manager, or governing authority. Return adjacent areas to condition existing prior to the start of work.

B.   See Cleaning Division 1, Section 01710.

PART 1 GENERAL

1.01 DESCRIPTION

A.   Work Included: The plumbing system for this work includes all cold water distribution, domestic water heating and distribution, vents and wastes, floor drainage, natural gas distribution, plumbing fixtures and trim, and all other plumbing items indicated on the drawings or described in these specifications. plus all other plumbing items needed for a complete and proper installation.

1.02 RELATED WORK DESCRIBED ELSEWHERE

A.   Fire Sprinkler System: Section 15500.

PART 2 PRODUCTS

2.01   MATERIALS

A.   Sewer Lines. Sewer lines shall be per Bulletin SE—13, "Planning Guide for Water Supply and Waste Disposal," and the latest edition of the Uniform Plumbing Code.

B.   Fixtures and Fittings. Fixtures, fitting and settings of fixtures to comply to the latest editions of the Uniform Building Code and Uniform Plumbing Code. Fixtures shall be approved by Owner and meet the quality standards as follows:
 1.   Water Closet: Eljer "Elvortex Water—Saver", floor-mounted water closet or equal. Eljer Model No- 111-1215.
 2.   Lavatories:
(a) Eljer "Erskine" cabinet mount lavatory or
equal. Elier Model.No. 213—0138.
(b) Elier "Bucknell" wall mount lavatory or equaL
Model No. 052-0164.
 3.   Eljer "Ultima Center-set Fitting with pop-up waste or equal. Model No- 552—1400.

C.   Water Heater. Shall have a temperature control thermostat to limit the outlet of hot water at 120 degrees Fahrenheit. and a standby loss not exceeding 4.0 watts per square foot of shell area, such as sears 10 gal, dual filement electric lowboy model or equal.

D.   Water Piping. Shall be type "K", "L", "M" copper or PVC piping for interior work above grade. Insulated connections are to be installpd between dissimilar metals. Piping and installation shall be as per the latest edition of the Uniform Plumbing Code.

E.   Sanitary Drainage and Vent Piping. All sanitary drainage and vent piping inside of building shall be either ABS or PVC plastic piping material.

F.   Shock Absorbers. Shock absorbers shall be provided at all fixtures and equipment in accordance with the latest editions of the Uniform Plumbing Code.

G.   Gas Piping. All gas piping shall be black iron, schedule 40, inside the building and treated for corrosion when used underground and through masonry walls. Plumbing Contractor shall run and connect with unions and valves all gas pipe to gas fired equipment furnished by him. Also, Plumbing Contractor shall run the gas pipe with a shut-off valve to within 24" of each gas fired unit furnished and connected by others.
 1.   Gas piping to be sized per N.F.P.A. 54, Table 2—D2 at 1 PSI and under with a pressure drop of 0.5" water column and 0.6 specific gravity. (Verify with the local gas company.)

H. All equipment shall be installed with a union connection and shut-off water valves.

PART 3 INSTALLATION

3.01   STANDARDS

A.   Comply with all State and Municipal codes. 15500 PAGE 1

FIRE SPRINKLER SYSTEM

PART 1 GENERAL
1.01 DESCRIPTION

A.   Work Included: Design, fabricate, install, and secure all necessary approvals of a complete fire protection automatic sprinkler system throughout the work acceptable to the project architect or his on-site representatives and in accordance with the standards set forth in this section of these specifications.

1.02   RELATED WORK SPECIFIED ELSEWHERE

A.   Electrical—section 16400

B.   Life Safety—Section 16400

1.03   QUALITY ASSURANCE & STANDARDS

A.   All material, equipment, methods of installation, and other like items: shall be in strict accordance with applicable requirements of:
  1.   NFPA (National Fire Protection Association) 13, 1983 Edition.
  2.   UBC (Uniform Building Code), 1980 Edition, as per State of Indiana Fire Marshall's Office.

B.   All design fabrication, installation and equipment must meet U.L. approved for Fire Safety.

C.   All pertinent requirements of the National Board of Fire Underwriters must be met.

D.   In the event of conflict, the most stringent codes and standard are to govern.

1.04   SHOP DRAWINGS

A.   Before proceeding with any of the fire protection work, prepare shop drawings completely illustrating the fire protection work to be installed under this contract. Hydraulic calculations shall be provided to accompany the shop drawings. Submit required documents to authorities having jurisdiction, for their review, comments and approval.
  1.   The Contractor shall show method of hanging sprinkler pipes and spac ing The sprinkler pipes shall be supported from structural beam or slabs.
  2.   If conditions require rearrangements, Contractor shall submit drawings of proposed changes to construction manager and obtain Architect/Engineer's review before starting changes.

B.   Record Drawings: Upon completion of installation of each system submit two (2) reproducible sepias of Record Drawings that reflect all changes and offsets made in field to Architect/Engineer.

1.05   DESIGN CRITERIA

A.   Automatic sprinkler protection system shall be designed on the following criteria, and hydraulic calculations shall reflect these criteria.
  1.   Office Areas; (Including areas of typically low combustibles):
  A densityof 0.1 gpm/sq.ft. over the most remote 3000.sq. ft. (wet system) plus a 250 gpm hose stream allowance, Sprinkler heads rated at 1650 F. should be used.

2. Utility Areas: (including air conditioning equipment areas and power and telephone switchgear rooms): A density of 0.2 gpm/sq. ft. over the most remote 3000 sq. ft. (or room area, whichever is less) plus a 500 gyp hose stream allowance. Sprinkler heads rated at –286 F. should be used except in areas where only one or two heads are needed. Sprinkler heads rated at l&0F. should be used in these small areas.

3. System demand should not exceed more than 20% over the average density and demand pressure should not exceed 100 psi at first floor level.

B. Standpipe riser shall provide 65 psi at the topmost outlet with 500 gpm flowing. Water supply requirements for standpipe system shall be in accordance with requirements of NFPA—l4.

1.06   MANUFACTURERS

A. The following is a list of manufacturers names of material and equipment that are acceptable.

ITEM OR MATERIAL OR EQUIPMENT	MANUFACTURER
Valves	Traverse City, Mueller
	Jenkins, Nibco
Sprinkler Heads	Reliable, Grunau, Viking
Flow Alarm & Valve	Simplex, Hotifier Co.
Signaling Units	Standard Electric Time.
Dry Pipe Valve	Viking, Reliable
Pre-Action Valve and	Viking, Reliable.
Accessories	Grinnel1
Maintenance Air Compressor	Viking, Reliable
Fire Extinguishers and	Potter—Roemer, Gardens
Cabinets	Seco
Fire Valve and Cabinets	Potter-Roemer, Garden's
	Seco; J & I Industries
Fire Pump/Jockey Pump	Peerless, Fairbanks,
	Morse, Aurora,
	Allis-Chalmers
Fire Pump Controller	Metron, Firetrol, Clark
Grooved End Pipe Fittings	Victaulic, Gustin-fiacan
	Stockham

PART 2 PRODUCTS

2.01 DESIGN

A. General: The design shall be complete in all regards, and shall include but not necessarily be limited to:

1. Connection to utility main, including all required valves, fittings and other items.
2. Overhead sprinkler system.
3. Column sprinkler system.

2.02 VALVES

A. Except as hereinafter specified, furnish and install valves where indicated on the drawings or as required in accordance with all applicable codes.

2.03 SPRINKLERS

A. Furnish and install all sprinkler systems as indicated on the drawings, or as required in accordance with all applicable codes.

2.04 FLOW ALARMS & VALVE SIGNALLING UNITS

A. Furnish and install flow alarms and valve signalling units for all risers and valves on the fire protection system.

B. Connection to fire alarm system will be provided as specified under other Sections.

C. Devices shall be as listed below:
 1. Waterflow indicators shall be installed on all standpipe and automatic sprinkler risers or branch mains, as indicated.

2.05 F.E.C. FIRE EXTINGUISHER CABINET

A. Fire extinguishers shall be Potter-Roemer's dry chemical, for Class A, B and C fires.
 1. Provide fire extinguishers, wall mounted in the electrical and mechanical rooms.

PART 3 EXECUTION

3.01 TEST CONNECTIONS & DRAIN PIPING

A. Provide test connections and drain piping from sprinkler system on each floor. Provide supervised shut-off valve and valved drain connection at the base of each standpipe.

3.02 SPRINKLER HEADS

A. Sprinkler heads shall be chrome and semi-recessed heads. Unless otherwise indicated, all sprinkler heads shall be arranged symmetrically within each room or space. All sprinkler heads to be installed in acoustical tile ceilings of any type shall be located at the center of 2×2 tiles as indicated on the architectural reflected ceiling plans.

3.03 SPRINKLER PIPING

A. At Contractor's option sprinkler mains and branch piping may be welded using welding fittings.

B. At Contractor's option fittings and sprinkler mains and branch lines may be grooved mechanical tees, fits or hookers.

3.04 TESTING

A. The system shall be tested hydrostatically at not less than 200 pounds per square inch pressure for two hours at 50 pounds per square inch in excess of maximum Static pressure. If weather does not permit, test dry system with 40 pounds air pressure at least for 24 hours as interim test. Submit a copy of test certificate and copy of pump certification after the installation is tested successfully and accepted by all authorities having jurisdiction.

HEATING, VENTILATING AND AIR CONDITIONING

PART 1 GENERAL

1.01 DESCRIPTION

A. Work Included: Heating, ventilating and air conditioning required for this work indicated on the drawings and includes, but is not necessarily limited to:

1. Central station air conditioning units
2. Filters
3. Exhaust fans
4. Duct furnaces
5. Controls
6. Ducts, dampers, grills, registers and diffusers
7. Insulation of ducts
8. All other items required for complete and operating heating, ventilating, and air conditioning systems for each individual tenant space and all common areas in this work.

PART 2 PRODUCTS

2.01 EQUIPMENT

A. General: All equipment shall be the capacity and types shown on the equipment schedule in the drawings, and shall be the listed manufacturer and model number or shall be equal approved in advance by the Owner or Architect.

2.02 CODES AND STANDARDS

A. Comply with all pertinent federal, state and municipal codes and requirements.

B. All material and installation shall comply with the latest edition of the Uniform Mechanical Code and the NFPA. All mechanical equipment to be installed per manufacturers specifications and recommendations. All equipment is to be left in perfect operation conditions.

C. Mechanical Contractor shall furnish all material and labor as required for a complete installation. All equipment and controls shall carry the approval of an approved testing agency.

D. Mechanical Contractor shall be responsible for all HVAC equipment, equipment controls and systems to provide the required heating, air conditioning, combustion air and combustion flues and vents.

E. Mechanical Contractor shall furnish the Engineer with shop drawings on all equipment installed.

F. Vibration and noise of equipment shall be isolated by rubber-in—shear equipment isolators and mounts, and flexible neoprene duct connectors where ducts connect equipment containing motors.

G. All magnetic starters and controls shall be furnished and installed by the Mechanical Contractor on all equipment provided by the Mechanical Contractor.

H. Mechanical Contractor to supply and install all thermostats as required. Wiring of thermostats to be by the Electrical Contractor. Thermostats to have a maximum setting of 75 degrees and a minimum setting of 55 degrees for heating and a maximum setting of 85 degrees and a minimum setting of 70 degrees for cooling.

I. Gas piping sized per N.F.P.A. 54, Table 2D2, I.P.S. and under with a pressure drop of 0.5 inch water column and 0.6 Specific Gravity. (Verify with the local Gas Company.)

J. All gas piping shall be black iron, Schedule 40 inside the building, and treated for corrosion when used underground and through masonry walls. Mechanical Contractor shall run and connect with unions and valves all gas pipe to gas fired equipment furnished by him. Also, Mechanical Contractor shall run the gas pipe with a shut off valve to within 24 of each gas fired unit furnished and connected by others.

K. Above ground, inside building ductwork and casing shall be fabricated from prime quality, galvanized steel sheet, except as herein specified. Insulated glass duct may be accepted with the consent of the Architect/Engineer. Flex duct may be used for branch duct work with the consent of the Architect/Engineer. Minimum wall thicknesses to be same as duct lining indicated elsewhere in these Mechanical Notes. Type, quality and quantity of supports and bracing shall be in accordance with SMACNA requirements.

L.   Ductwork shall be constructed In accordance with the requirements of the SMACHA Guide, "Schedule of Recommended Construction for Low Pressure Ductwork."

M.   Duct sizes shown on drawings are net-free clear area after lining, if any, and all lined ducts shall be increased in size to allow for lining.

N.   Fiberglass duct lining shall meet N.F.P.A. 90A, I.J.L. 723 and the latest edition of the Uniform Mechanical Code. Table 10-0 and shall be 1 1/2 pounds per cubic foot density.

0.   Internally line all ducts, housing, filter casings, plenums. etc., as follows:

Mixed Air - 1" Lining

Supply Air - 1" Lining

Return Air - 1/2" Lining

Exhaust and Transfer - 1/2" Lining

P.   Mechanical Contractor shall supply and install combustion air ducts in accordance with the latest edition of the Uniform Mechanical Code and the N.F.P.A. 54 when required.

Q.   Mechanical Contractor shall install N.B.F.U. approved fire dampers and access panels when and where required per the latest edition of the Uniform Mechanical Code.

R.   All exhaust flues shall be double wall "Type B" gas vents.

S.   Dampers are tabe in all branch lines with vanes mall square elbows of main duct.

T.   Diffusers and grills to be selected to handle the c.f.m. as indicated on the plans. Diffusers to be the self damping type.

U.   Mechanical Contractor shall coordinate ceiling and/or floor diffuser location with the Electrical and General Contractors.

V.   Mechanical Contractor shall balance system to within 5% at all outlets as indicated on plans.

W. Install sound attenuation dampers in all supply and return air ducts.

ELECTRICAL
PART 1 GENERAL
1.01 DESCRIPTION
Work included: Electrical work includes, but is not necessarily limited to all of the work shown on the drawings.
1. Change over from existing electrical system.

1.02 CODES AND STANDARDS
In addition to complying with all pertinent codes and regulations, comply with:
1. National Electric Code
2. Local Code
3. Local utility company regulations
4. NFPA 1981 guidelines
5. ANSI A17.l, NATIONAL ELECTRIC CODE

PART 2 PRODUCTS
2.01 Service entrances equipment and main distribution panel.
2.02 Comply with all pertinent codes and specifications on the drawings.
2.03 Grounding System: All equipment including switchboards, transformers, conduit system, motors, and other apparatus, shall be grounded by conduit or conductor to cold water main or independent grounding electrode, as indicated on the drawings.
2.04 Raceways & Fittings.
2.05 All conduit installed concealed in walls, above ceilings, or exposed in work areas, shall be rigid galvanized, or sheradized steel conduit, or electrical metallic tubing with compression or tap on-type fittings.
2.06 Conduit installed in the floor slab or underground shall be rigid galvanized. Conduit in direct contact with earth shall be cooled with an asphaltum Paint.
2.07 Electrical metallic tubing shall conform to all requirements of the National Electric Code.
2.08 Other materials;
All other material, not specifically described but required for a complete and proper installation of this section, shall be as selected by the contractor subject to the approval of the owner or architect. All work done and materials required must comply with all pertinent codes.

LIGHTING FIXTURES

PART 1 GENERAL
1.01 DESCRIPTION
A. Fixtures furnished under this division shall be complete with lamps and all necessary trim and mounting hardware, and installed where shown on the drawings.

PART 2 PRODUCTS
2.01 MATERIALS
A. All florescent ballasts shall be Class P. CBM-ETL H.P.F. approved and shall be of the automatic thermal resetting type "A" sound rated Ballasts to be energy-efficient type.
B. Ballasts furnished under this division shall be covered by a two year warranty against defects and warranty shall include payment for normal labor casts or replacement of inoperative in-warrant ballasts

2.02 ELECTRICAL FIXTURE SCHEDULE
A.   Fluorescent Light Fixture "A"
        Metalux —   Flange 2FS—2UIK A
                        Grid 26S—2UIM A
                        Lamps 2FU 40
                        Rapid Start 2' × 2'
B.   Fluorescent Light Fixture "B"
                        Metalux—Flange 2 FS—Z40IMA
                        Grid 2GS-Z40IMA
                        Lamps 2F40
                        Rapid Start 2' × 4'
C.   Exit Lights
                        Capri Lite EX20SBG
D.   Emergency Lighting
                        Cloride Nichel Cadmium SPU—10120/217
E.   Fluorescent Strip Fixture
                        Metalux—40W—HPF Rapid Start 48"
F.   Fluorescent Strip Fixture
                        Metalux—20W—LPF—Trigger Start 24"
                        c-i 20
G.   Fluorescent Strip Fixture
                        Metalux—30w-LPF-Rapid Start 36"
H.   Shallow Round Can lighting
                        Capri Lite R30 Open 75R30
                        P—White Finish
I.   Shallow Round Can Lighting
                        Capri Lite R38 75ER30
                        P—White Finish

PART 3 EXECUTION
3.01 INSTALLATION
A.   Fixtures shall be neatly and firmly mounted, using standard supports for outlets and fixtures.

APPENDIX

D

Answers to Review Questions

CHAPTER 1 ANSWERS

1. Sink assemblies, water closet assemblies, valve, and strainer assemblies.
2. A change order is a request for payment for work that was not originally outlined in the bid.
3. This shows the customer that you are organized and respect the value of his/her time.

CHAPTER 2 ANSWERS

1. They are more costly and require a deeper wall cavity for the tank.
2. 4"
3. 2.5 gpm.
4. A toilet that has a pump and grinder to discharge waste to a higher elevation.
5. It refers to the side of the tub that the drain and tub spout are on.

CHAPTER 3 ANSWERS

1. The flushometer valve has fewer moving parts, takes up less space, and has a shorter flushing cycle.
2. The introduction of air may spread airborne diseases.
3. It is a device that as water passes through it, creates suction on an attached hose.
4. The top rim of the bowl is 16"–18" above the floor.
5. Stainless steel.

CHAPTER 4 ANSWERS

1. Because it has to be put in place at the earliest stage of construction.
2. 21"
3. The combined weight of the bowl and tank is heavier.
4. True.
5. True.

CHAPTER 5 ANSWERS

1. The shut-off valve that controls water flow to the fixture.
2. A public restroom or any location prone to vandalism.
3. 2.5
4. Saves water, more sanitary (the user does not have to touch the fixture).
5. Whenever a handheld showerhead could be dropped below the flood level rim of the tub.

CHAPTER 6 ANSWERS

1. Oil-based putties.
2. Wood screws (#12 or #14 × 1-1/2" or 2").
3. 3/8" NPT.
4. Escutcheon.
5. Strap.

CHAPTER 7 ANSWERS

1. Computer Aided Design or Computer Aided Drafting.
2. 10 gallons.
3. A-5, elevation 13.
4. Architect.
5. Division 15, Mechanical.

CHAPTER 8 ANSWERS

1. The size and shape of the rooms, architectural features, location of appliances, etc.
2. Typical means that the feature shown is common among other areas of the building. Example: "typical" describing a sink would mean other sinks for the same task would be the same as the one shown.
3. 5'
4. Locations of existing gas, water, sewer lines, etc.
5. See Figure 8-6.

CHAPTER 9 ANSWERS

1. By the direction that the exposure is facing: North, South, East, West.
2. Platform, Balloon, Braced.
3. To know the type of wall construction that he/she will have to fasten hangers or make wall penetrations through.
4. Ductwork.
5. Humidifiers, built in coffee makers, etc.

CHAPTER 10 ANSWERS

1. From the local gas supplier.
2. British Thermal Unit: The amount of heat required to raise one pound of water one degree Fahrenheit.
3. Outside, underground.
4. Lighter than air.
5. The amount of gas that delivers 100,000 BTUs when burned.

CHAPTER 11 ANSWERS

1. 1"
2. 1-1/2"
3. 10'
4. 2.6'
5. 3/4" or 25 EHD.

CHAPTER 12 ANSWERS

1. All users.
2. FALSE.
3. TRUE. It can be, but according to code, it only has to be supported every 6 feet.
4. FALSE. It is preferred, but not required.
5. Because, while it is flexible, it is not rated as a flexible appliance connector.

CHAPTER 13 ANSWERS

1. Injection, Power, Luminous.
2. Primary.
3. Thermocouple.
4. Integrated Furnace Controller.
5. Rotary.

CHAPTER 14 ANSWERS

1. Lead.
2. Threaded joints, welds, joints of dissimilar materials.
3. Coatings can become damaged during installation.
4. Where an electric current is generated at the junction of two dissimilar metals.

CHAPTER 15 ANSWERS

1. Forced and induced.
2. The draft hood acts to reduce the variation in draft seen by the appliance outlet, and it dilutes flue gases.
3. Category I are non-positive vent static pressure with a vent gas temperature which avoids excessive condensate production in the vent. Category IV has a positive pressure vent system that allows the vent gas temperature to be low enough to have condensate in the vent.
4. The maximum appliance input rating of a Category I appliance equipped with a draft hood that could be attached to the vent. There is no minimum appliance input rating for the draft hood equipped appliances.
5. 9'

CHAPTER 16 ANSWERS

1. $(50,000/4000) = 12.5$ in^2
2. $(50,000/1000) = 50$ in^2
3. Heat, carbon dioxide, water.
4. It stops the cooling of the internal flue of an appliance because of natural draft.
5. All blowers will start and run continuously when power is first applied. In older model equipment, the electronic circuit board would be destroyed, so be careful in wiring.

CHAPTER 17 ANSWERS

1. American Gas Association.
2. If the flame of the match moves towards the draft hood, then dilution air is being added and the flue is probably venting ok.
3. The authority have jurisdiction.
4. 10 minutes.

CHAPTER 18 ANSWERS

No separate questions; refer back to examples throughout chapter.

CHAPTER 19 ANSWERS

1. 373.98'
2. 830.18'
3. B = 740.01'   C = 734.77'   D = 728.80'
4. 95°
5. It will typically allow for all of the lower elevation readings to remain positive. This allows the difference between any two elevation readings to represent the change in elevation between the two points.

CHAPTER 20 ANSWERS

1. Stable rock, type A, type B, and type C.
2. Argon, carbon dioxide, propane (any from the above chart).
3. Transit can measure vertical angles.
4. 25'
5. A competent person, meaning someone who has been trained on soils and other OSHA requirements.

CHAPTER 21 ANSWERS

No separate questions; refer back to examples throughout chapter.

CHAPTER 22 ANSWERS

No separate questions; refer back to examples throughout chapter.

CHAPTER 23 ANSWERS

No separate questions; refer back to examples throughout chapter.

CHAPTER 24 ANSWERS

1. A leader is located on the outside of the structure while a conductor is located inside.
2. 100 square feet.
3. When it is near an opening in the structure that would allow odors to be drawn in.
4. 18"
5. 6"
6. 15"
7. 12"

CHAPTER 25 ANSWERS

1. Heat is the amount of heat energy contained in a substance, both latent and sensible. Temperature is the intensity or level of heat.
2. Conduction, convection, and radiation.
3. The temperature at which the moisture in the air will begin to condense.

4. To measure wet bulb and dry bulb temperatures.
5. British Thermal Unit. [ml1]The amount of heat energy required to raise one pound of water one degree Fahrenheit.

CHAPTER 26 ANSWERS

1. Arsenic, antimony, cadmium, beryllium, copper, lead, nitrate.
2. A chlorinator and filter.
3. Hardness.
4. Cloudiness in the water.
5. Carbon dioxide.

CHAPTER 27 ANSWERS

1. Legionella bacteria can grow at temperatures below 122°F.
2. The dip tube directs incoming water to the bottom of the water heater and stirs up sediment. This allows hot water to leave through the top outlet and reduces the amount of sediment build up on the bottom of the tank.
3. The circulation loop provides a quicker supply of hot water at the point-of-use without the waste of running water.
4. Allowing both top and bottom heating elements to energize at the same time would increase the peak demand on the electrical system.
5. Stacking refers to the temperature variations in the water of a water heater. The hotter water will be nearest the top while cooler water will be nearest the bottom.

CHAPTER 28 ANSWERS

1. EMF (electro motive force).
2. .625 A.
3. 1.
4. Open capacitor, start relay, or start winding.
5. Sleeve.

CHAPTER 29 ANSWERS

1. White or gray.
2. 150 VA ÷ 24 V = 6.25 amps (Amperage = power ÷ voltage).
3. I = V ÷ R; therefore, the current (I) would decrease.
4. 0.010 amps.
5. Turn off the power.

CHAPTER 30 ANSWERS

1. The highest.
2. Bad. The only way to read a voltage across a fuse would be if the fuse were bad.
3. Turn off the power.
4. Sediment on the bottom of the tank.
5. The top.

CHAPTER 31 ANSWERS

1. It means that with no external sign, the contacts will be open and no current would flow through the circuit.
2. Step down means that the incoming voltage will be reduced.
3. 40 VA ÷ 24 V = 1.67 amps.
4. #14.
5. The loose connection makes it harder for electrons to flow; therefore, the resistance is increased. An increase in resistance will cause a drop in voltage across the connection.

Index